Prescribing by Numbers

Prescribing by Numbers

Drugs and the Definition of Disease

Jeremy A. Greene

The Johns Hopkins University Press
Baltimore

© 2007 The Johns Hopkins University Press
All rights reserved. Published 2007
Printed in the United States of America on acid-free paper

9 8 7 6 5 4 3 2 1

The Johns Hopkins University Press
2715 North Charles Street
Baltimore, Maryland 21218-4363
www.press.jhu.edu

Library of Congress Cataloging-in-Publication Data

Greene, Jeremy A., 1974–
    Prescribing by numbers : drugs and the definition of disease / Jeremy A.
Greene.
        p. ; cm.
    Includes bibliographical references and index.
    ISBN 0-8018-8477-2 (hardcover : alk. paper)
    1. Chemotherapy—Miscellanea.   2. Diseases—Miscellanea.
3. Drugs—Miscellanea.   I. Title
    [DNLM: 1. Drug Therapy—history.   2. History, 20th Century.
3. Chronic Disease—drug therapy.   4. Diagnosis.   5. Drug Industry—
ethics.   6. Drug Therapy—ethics.   7. History of Medicine.
8. Physician's Practice Patterns—history. WZ 64 G811p 2006]
    RM262.G74 2006
    615.5′8—dc22   2006011616

A catalog record for this book is available from the British Library.

Contents

Preface

The way to live the longest is to acquire a chronic disease and take good care of it.
                                              —SIR WILLIAM OSLER

This week, in clinics and hospitals across the country, thousands of perfectly healthy-feeling adults will receive a diagnosis for a disorder that they did not know they had. There are several such disorders: imperceptible to patients, they produce no fevers, no chills, no headaches, no stomachaches, no pains. Neither are they immediately perceptible to physicians or other health providers; there is no lesion to be seen with an ophthalmoscope or suspicious sound to be heard with a stethoscope, no tell-tale skin finding or sense to be made from piecing together disparate observations into a cohesive diagnosis. These are diseases that bear no immediate relation to symptoms but rather are connected to a statistical likelihood of developing symptoms in the future, pathologies—such as high blood pressure, mild diabetes, or elevated cholesterol—that are measurable only with the aid of an intervening diagnostic technology. Although patients who receive these diagnoses are typically encouraged to change their diets, get more exercise, and pursue other therapeutic lifestyle changes, for most people these diagnoses lead directly to the prescription of a drug they will take every day for an indefinite period, if not for the rest of their lives.

This book is concerned with the modern predicament of the subjectively healthy but highly medicated individual, a type that is becoming more and more common among the adult population of the United States. Americans on average filled ten prescriptions per person in 2003; those over age sixty-five filled an average of twenty-five prescriptions in that year. Dominant in this prescription practice are a set of drugs that modify conditions of risk and also happen to be the top-selling therapeutic agents in the pharmaceutical landscape. The widespread use of such agents supports an industry with worldwide sales rapidly approaching $500 billion and now represents the fastest-growing segment of health care expenditures. Because the preventive efficacy of these drugs has been determined only at the level of the population, individual patients who consume medications for asymptomatic conditions do so without knowing whether they will, in fact, ever receive any benefit from their pharmaceutical regimen. For the many thousands who experience side effects from these medications, the only certain result of their diagnostic and therapeutic experience is, ironically, a set of iatrogenic symptoms.

And yet the promotion of this pharmacopoeia of risk reduction is not merely a marketing ploy on the part of drug manufacturers or a bid by physicians for more office visits. Among those lobbying for the broader use of these drugs have been public health advocates, well-respected scientists, eminent clinicians, and many patient-activists and disease communities themselves. Although their actions have contributed to the endorsement of widespread use of prescription drugs, these actors have not all simply been "bought off" by the drug industry. An enormous wealth of data—hundreds of long-term, randomized, placebo-controlled clinical trials representing millions of patient-years—have indicated that for many populations of asymptomatic patients, steady consumption of risk-reducing drugs has generated visible benefits in the prevention of heart disease, stroke, blindness, and renal failure. In the past three decades, as broad guidelines have supported increasing use of such drugs on a preventive basis, the number of actual strokes and heart attacks in the United States has significantly declined.

The data behind the doctrine of pharmaceutical prevention are convincing, but the production and dissemination of that medical knowledge and its translation into medical practice is not insulated from the marketplace. In every step of the process we see an amalgamation of marketing and research: in the early stages of drug development, when a promising compound is conceived in terms of its potential market size; in the conduct of clinical trials, whose growing expense and largely private funding makes them increasingly accountable to shareholders as well as to scientists and regulators; and in the process of educating physicians and the public about the expanding use of these medications, which takes place largely through a promotional network of pharmaceutical representatives and direct-to-consumer advertising. In the course of several decades, disease has become simultaneously an epidemiological event and a marketing event.

This book follows three overlapping narratives of drugs and diseases to explore the central confluence of marketing and epidemiology that underlies the contemporary doctrine of pharmaceutical prevention. Each two-chapter part of the text revolves around a single drug and a single disease during a time period pivotal in their mutual definition. The aim of this case-study method is to offer enough detail and context to trace how both drug and disease came to alter each other in their therapeutic embrace. The three stories overlap, share attributes and actors, and weave together to describe a consistent set of structural developments and sea changes in therapeutic knowledge and practice over the past half century.

Part 1 hinges on the relationship between Diuril (chlorothiazide) and hypertension. By the end of the twentieth century, hypertension (high blood pressure) had become the paradigmatic disease of risk, overwhelmingly diagnosed and pharmacologically treated on an asymptomatic, preventive basis. But in the mid-1950s, when Diuril was being developed as a therapeutic compound, high blood pressure was considered a treatable diagnosis only in patients with felt symptoms. Chapter 1 traces the interplay of marketing and research in the development and launch of Diuril within the newly formed Merck Sharp & Dohme, the hybrid combination of the scientifically acclaimed Merck Research Laboratories and the well-honed marketing and sales institution of the Sharp & Dohme Pharmaceutical Company. This chapter is largely based on a close reading of internal company documents found at the Merck Archives and traces in detail the practices by which drug promotion and disease promotion became intertwined in the pharmaceutical corporation. Chapter 2 follows Diuril after its launch and describes the varied roles the drug itself played in the production of a widespread consensus on the treatment of asymptomatic hypertension.

Part 2 maps the role of Orinase (tolbutamide) and other new oral antidiabetic drugs in the 1960s and 1970s as the diagnosis of adult-onset diabetes transformed from a frankly symptomatic process into a practice of preventive screening. Chapter 3 examines the public documents of the Upjohn Company and the business and clinical literatures surrounding the launch of Orinase and its connection to expanding diagnosis of prediabetes and other asymptomatic forms of the disease. The community of American diabetologists, long divided over the proper relationship between the management of diabetes and the physiological control of blood sugar levels, viewed both the relative ease of an oral dosage and the asymptomatic category of prediabetes with ambivalence and some skepticism. A decade after the first oral antidiabetics were released, researchers conducting the University Group Diabetes Project (UGDP)—a large-scale, multi-arm clinical trial assessing the usefulness of oral antidiabetic agents—shocked the clinical world with a proclamation that treatment with Orinase did not decrease the risk of cardiovascular mortality in diabetic patients but instead dramatically increased that risk. As the Food and Drug Administration (FDA) attempted to alter Orinase's package labeling to limit the usage of oral drugs to overtly symptomatic patients, the subsequent publicity sparked a controversy over the role of clinical trials in the regulation of clinical practice that roiled for well over a decade. Chapter 4 looks at these disputes over Orinase's promotion, labeling, and cardiovascular effects in the aftermath of the UGDP study, with a focus on the role of the patient and the practicing physician caught between the arguments of Upjohn, the FDA, radical consumer groups, the American Medical Association, and other parties. Both private and public organizations, all claiming to represent the interests of the ultimate consumer, were now set against each other in the fray. As a commodity targeted toward a well-organized and well-identified patient population, Orinase highlights the contested role of the patient as consumer in the pharmaceutical negotiation of disease.

Part 3 is concerned with the fall and rise of high blood cholesterol (hypercholesterolemia) as a treatable clinical category. It narrates the contingent development of Mevacor (lovastatin) and the therapeutic consensus around the treatment of high cholesterol between the late 1970s and the 1990s. Chapter 5 documents the failure of earlier cholesterol-lowering formulations and dietary regimens to gain currency in clinical practice and the resultant collapse of consensus on the advisability of treating high blood cholesterol at all. This destabilization of high cholesterol as a diagnosis reached its nadir in the late 1970s, amid conflicting accounts of the utility of drug, diet, and regimen. When Merck began to mobilize its efforts to bring Mevacor to market in 1987, its marketing staff approached the task of rebuilding a therapeutic consensus through far-reaching public-private promotional campaigns in concert with the newly founded National Cholesterol Education Program of the National Institutes of Health, emphasizing the rational design of the drug and encouraging the general population to "know your number." The scope of this promotional campaign, and the subtlety of its integration with epidemiological and basic science research, represents a striking development in the style and content of pharmaceutical marketing in the years since the 1958 launch of Diuril. Chapter 6 follows the social lives of Mevacor and cholesterol in the decades since the launch of the statins, exploring the relationships between commercial clinical trials, expert guidelines, and promotional strategies in the expansion of the target populations for cholesterol-lowering therapy. As therapeutic reformers placed the randomized clinical trial at the center of public health strategies and clinical guidelines, the private funding of postmarketing trials became a vital link between economies of medical knowledge and economies of pharmaceutical development.

This complex nexus of drug and disease, risk and diagnosis, medicine and marketplace now lies at the foundation of mainstream American medical practice, but it is a structure that has only recently been set in place. Looking back a half century reveals an America with few pharmaceuticals of risk reduction, a nation that viewed chronic disease largely as a process of inevitable decay, and a pharmaceutical industry that concentrated much more on the development of new classes of drugs than on the expansion of its markets to encompass more and more subjectively healthy people. How did we arrive at a state where the line between the normal and the pathological became a numerical abstraction? How did these asymptomatic diseases come to be, and what new relationships between health and illness, doctor and patient, individual and population do they represent? What forces have allowed pharmaceuticals to become crucial to the definition of disease and the philosophy of health promotion? These historical questions are urgently relevant to our times.

Acknowledgments

This book, and along with it my own double life as a physician-historian, would not have been possible without the support of many individuals. There is not enough space here to properly acknowledge the many acts of generosity that have helped to eventually bring this work to publication. First I must thank my principal dissertation adviser, Allan M. Brandt, who introduced me to the powerful analytic possibility that history offers to the study of disease, medicine, and society; encouraged my dual life by his own example; and represented all that I could ask for in a mentor. I had the outrageous good fortune to include Charles E. Rosenberg and Arthur Kleinman as co-advisers for my dissertation, and this book owes a great deal to both of them: to Arthur for helping me to engage in the interdisciplinary play between history and ethnography that is involved when one writes about the very recent past, and to Charles for being an exceptionally insightful and close reader, whose iterative commentary on my evolving work has helped me to become a better writer as well as a better historian.

Various parts of this text have benefited from close readings and conversations with Bridie Andrews, Robbie Aronowitz, Conevery Bolton-Valencius,

Dan Carter, Jennifer Clark, Chris Crenner, Arthur Daemmrich, Joseph Dumit, Chris Feudtner, Jennifer Fishman, Nathan Greenslit, Katja Guenther, Orit Halpern, David Healy, Bill Helfand, Greg Higby, David S. Jones, Powell Kazanjian, Nick King, Andrew Lakoff, Ilana Lowy, Harry Marks, Mara Mills, Greg Mitman, Amber Mussa, Todd Olszeuski, Nicholas Rasmussen, Barbara Rosenkrantz, Hanna Shell, Grace Shen, John Swann, Jason Szabo, Carsten Timmermann, Nancy Tomes, Andrea Tone, Keith Wailoo, Elizabeth Watkins, and Nicholas Weiss. Particularly effusive thanks must go to Debby Levine, Abena Osseo-Asare, Scott Podolsky, Alisha Rankin, and Elly Truitt for their extensive and repetitive engagement with the manuscript as a whole. The work has also benefited from scholarly presentation and discussion in several venues, including the meetings of the American Association for the History of Medicine and the Society for the Social Study of Science, and colloquia at Kansas University Medical Center, McGill University, the History of Science Department at Harvard University, Dartmouth-Hitchcock Medical Center, and the New York Academy of Medicine. Any mistakes that remain in the present work are, of course, my own fault.

Many people made it easier for me to integrate the research for this book with my medical training. At Harvard's History of Science Department I am particularly indebted to Jude LaJoie. At the medical school I had the good fortune to gain the guidance and support of Claudia Galeas, Cavin Hennig, Orah Platt, and William Taylor of the Castle Society; Leon Eisenberg, Paul Farmer, Byron Good, Jim Kim, Helena Martins, and Christine Moreira of the Department of Social Medicine; Nancy Andrews, Linda Burnley, David Golan, Alan Michelson, and Christopher Walsh in the MD/PhD Program; and Gerard Coste, David Hirsch, and Barbara Ogur of the Cambridge Health Alliance. Financial support for earlier versions of this work was provided by the History of Science Department, the Merit and John Parker Scholarships from the Graduate Student Council, the Whiting Foundation for the Humanities, the Charles Warren Center for American History, the American Association for the History of Pharmacy, and a MSTP Grant from the National Institutes of Health. I would also like to thank the faculty of the Brigham and Women's Hospital for creating an environment where a medical intern could be actively encouraged to keep up as a historian, and the rest of the residents at the Brigham and Women's Hospital—particularly the others in my cohort of interns—for their support while the manuscript was making its way to press.

Public and private archivists gave invaluable assistance to the research for this project, and many individuals also granted me extensive interviews. I would like to thank in particular John Huck, Eugene Kuryloski, William Helfand, Edward Freis, Susan Freis, Edward Roccella, James Cleeman, and Mickey Smith, who provided oral histories and made available items from their own personal collections. Jeff Sturchio, Joe Ciccone, and Doreen Strang provided access to the Merck Archives. Other notable guides include Susan Speaker at the National Library of Medicine, John Swann and Susan White Junod at the Food and Drug Administration, Gregory Higby and Elaine Stroud at the American Institute of the History of Pharmacy, Chris Warren at the New York Academy of Medicine, Thomas Horrocks and Jack Eckert at the Countway Library of Medicine of Harvard Medical School, and the staffs of the Baker Business Library of Harvard Business School, the Kalamazoo Public Library, and Western Michigan University Archives.

I would like particularly to thank the anonymous reviewer for the Johns Hopkins University Press, who challenged me to make this work more than just a pared-down doctoral dissertation; Lois Crum, for her detailed copyediting; and my editor, Jacqueline Wehmueller, who took me under her wing while the dissertation was still unwritten, smoothly shepherded me through the process of publishing a first book, and personally edited several drafts of the manuscript. Finally, I would like to express thanks to my family for their patience and support, particularly to my grandmother Marilyn Freedman and her partner Gerry Buter, who assiduously copyedited the early drafts; to my father, Wayne, who has for his own inscrutable reasons always encouraged my writing career; to my mother, Doren, for her general confidence that even my more outlandish projects in life would typically work out well; to my brother David, my sisters Rachel, Amy, and Rebecca, and my stepmother Patti; and to my wife, Elizabeth, for her editorial suggestions, her consistent support, and her uncanny ability to keep me engaged with my own life.

Prescribing by Numbers

The Pharmacopoeia of Risk Reduction

> By means of his oracles, a Zande can discover the mystical forces which hang over a man and doom him in advance, and having discovered them he can counteract them or alter his plans to avoid the doom which awaits him in any particular venture. Hence a man's future health and happiness depend on future conditions that are already in existence and can be exposed by the oracles and altered.
>
> — E. E. EVANS-PRITCHARD, 1937

The audience assembled in the hotel ballroom October 29, 1957, came from surprisingly diverse backgrounds: some had started their careers as salesmen, others as financial analysts, basic research scientists, or practicing physicians. But the group gathered in the Boca Raton Club for the fifty-first annual meeting of the American Drug Manufacturer's Association was focused on a common set of interests: the financial and material future of the prescription drug industry. In the decade or so since the end of World War II, pharmaceutical manufacturers had seen the scope of their business expand dramatically, the structure of their firms grow outward, and the pace of their new-product development accelerate at an impressive rate. The industry was progressive and forward-leaning, and the audience assembled to hear the address given by Charles Mottley—Pfizer's chief of operations planning—was keen to hear what he had to say about the pharmaceutical future.

Mottley asked his audience to consider the paradoxical long-term influence that antibiotics—at that time the most lucrative sector within the pharmaceutical market—were beginning to exert upon broader health statistics. Antibiotics had effectively reduced the frequency and severity of infectious disease, but they had also effectively reduced their own potential market. "There seems to be an important lesson here for the drug industry," he continued. "As the industry does a good job of producing efficacious drugs and helps to win a given campaign . . . the net result is to limit the potential market." If the industry was to have a viable future, it would be necessary to grasp the nature of this irony and work to invert it. Drugs needed to *grow* their markets, not shrink them. Mottley told his audience that the expanding prevalence of chronic disease already evident by the late 1950s offered the perfect opportunity to redesign the drug-disease relationship. "Trends are developing in the cause of death statistics," he concluded, "which indicate that 'tomorrow' greater proportions of people are likely to suffer fatal accidents or be afflicted with diseases, such as cancer and cardiovascular involvements, for which there are, as yet, no really effective drugs."[1] As chronic diseases gained in importance from the 1950s onward, Mottley suggested, and as a chronic pharmacopoeia developed alongside them, new concepts of disease and treatment could be explored to maximize the long-term growth potential of pharmaceuticals. Conditions that patients would necessarily have for the rest of their lives—coupled with treatments that could be taken every day for an indefinite period—had the makings of a market that could result in sustainable growth.

The same year that Mottley addressed the crowd at Boca Raton, medical journals carried the early results of the Framingham Study, the first major effort of the recently established National Heart Institute and the American Heart Association in their joint endeavor to better understand the "modern epidemic" of coronary heart disease.[2] In the wake of receding morbidity and mortality of infectious diseases, heart disease had emerged as the foremost killer of modern times, and the search for a cure, or at least a cause, of this epidemic had gained widespread popular attention.[3] As the Framingham investigators followed the cardiovascular history of some six thousand residents of this small Massachusetts city, they began to single out the predictive "prepathological" categories that would eventually become known as coronary risk factors. Some of the categories were apparently immutable demographic characteristics of an individual: age, sex, and family history. Others, such as cigarette smoking, were potentially modifiable behaviors. The central risk factors, however, were physiological variants believed to be mechanistically connected to heart disease: hypertension (high blood pressure), hypercholesterolemia (high blood cholesterol), and, later, diabetes (uncontrolled elevation of blood sugar).[4] Implicit in the initial Framingham publications, then, was a tantalizing new possibility: control these deviations, and you can control chronic disease. As Jeremiah Stamler, a principal Framingham investigator, noted auspiciously in 1958: "It is highly feasible to assess risk of coronary heart disease in healthy persons—and to identify susceptibles . . . Elevated blood pressure and hypercholesterolemia can be lowered to or toward normal in many. Diabetes can be well controlled . . . Moreover, it is quite clear that the measures available for correcting abnormalities are simple, practicable, reasonable, and devoid of danger. It therefore seems entirely in order to propose that the medical profession apply the knowledge from recent studies to identify those susceptible to coronary heart disease and attempt to help them prophylactically."[5]

Mottley and the Framingham investigators were setting forth essentially the same program for the future of health care priorities. In the ensuing half century, their respective visions have become reality in the expanding diagnosis and pharmaceutical treatment of chronic diseases and their precursor states. Prescriptions for chronic disease categories now dominate the American pharmaceutical industry's domestic income, and the Framingham risk factors—particularly the three physiologically modifiable conditions of hypertension, diabetes, and hypercholesterolemia—have become common figures in contemporary clinical practice. Safe, effective, and specific therapeutic agents for each condition, unavailable in 1957, have since seen their markets multiply to represent three of the ten highest-grossing therapeutic categories in the world, collectively accounting for nearly $40 billion in sales in the year 2000 alone.[6] The midcentury proclamations of the marketer and the epidemiologist are fused together in the contemporary doctrine of pharmaceutical prevention.

The Therapeutic Transition

This book addresses the riddle that lies at the confluence of these two viewpoints. How is it that the priorities of public health—a field traditionally associated with the welfare state and private charity—have become so closely aligned with the marketing practices of the single most profitable industry in the American economy? What mechanisms have come to link pharmaceutical agents with the widespread detection and promotion of conditions of risk?

By most accounts, the emergence of a highly profitable set of therapeutics for previously untreated, asymptomatic, and flexibly defined disease entities occurred only as a result of the scientific achievements of clinical epidemiology. In this commonly received narrative, the epidemiological study of risk reduction preceded the development of risk-reducing therapeutics: disease entities were recognized first, and then drugs were developed to treat them. Medical historians have thus situated the Framingham Study as a pivotal moment in the articulation of risk in health and medicine, a moment when the laboratory and the clinic began to share their primacy in medical epistemology with biostatistics and the long-term epidemiological study.[7] The role played by pharmaceuticals in the renegotiation of disease has typically been left to speculation or discounted as an afterthought, the inevitable consequence of a natural "epidemiologic transition" from acute to chronic disease in economically developed nations.[8]

In the following chapters, I argue that that there was nothing inevitable about the development of a specific therapeutics of risk reduction and that the widespread adoption of these pharmaceutical agents speaks to a social history far more complex than a mere shift in the demography or in the epidemiological study of chronic disease. Rather, pharmaceuticals played a central and active role in the definition of these categories of illness. The adoption of mild hypertension as a disease was not automatic or self-evident: it hinged upon a set of promotional practices—somewhere between education and salesmanship—to give it credence in the eyes of medical practitioners and consumers alike. For diabetes and high cholesterol, asthma and dyspepsia, the same is true: our contemporary understanding of chronic disease is the product of epidemiological practices and marketing practices that have come to configure their common subject in increasingly similar terms.

Over the second half of the twentieth century, in concert with the emergence of specific, efficacious, and palatable oral medications, the domain of chronic diseases expanded from a core nucleus of long-suffering symptomatic patients to encompass broader and broader populations who bore no immediate symptoms. This book presents selected episodes in the emergence of the three principal treatable cardiovascular risk factors of our time—hypertension, diabetes, and elevated blood cholesterol—and the careers of three pharmaceutical products whose fates have become inseparable from the conditions they treated. All three conditions were ultimately transmuted by pharmaceutical agents: attenuated, expanded, and displaced from the realm of symptom, history, and treatment to one of screening, measurement, and prophylaxis.

Diuril, Orinase, and Mevacor are not historical actors in the traditional sense, but they were nonetheless crucial agents in the transformation of disease in the twentieth century. Each of these mass-produced tablets represents the intersection of several interested parties who have competing stakes and claims; it is a site where the divergent trajectories of researchers, clinicians, patients, regulatory bodies, manufacturers, and insurers necessarily connect. In the postwar, ostensibly postinfectious era, the historical punctuation formerly provided by epidemics was replaced by a new sort of historical punctuation provided by pharmaceutical launches and marketing developments. Pharmaceuticals can also serve as portals into a distinctly social history; they form collective "sampling devices" through which we can observe the social tectonics underlying contemporary politics of health and normality. The stories of these three agents, linking discrete clinical categories and successive historical moments, work together to offer a central insight into the expansion and contestation of chronic disease categories in the late twentieth century.[9]

The program of pharmaceutical prevention cannot be reduced simply to a clever marketing effort or a centrally planned medicalization that generated artificial disease categories in order to transform every healthy American into a multiple-drug consumer. That argument overestimates the power of the research-based pharmaceutical industry and minimizes its substantial investment in scientific inquiry; it also echoes the 1970s use of the term *medicalization* as a paranoid polemic describing an omnipotent medical profession constantly seeking to expand its province over the healthy.[10] Although it is apparent that the autonomous stature of the medical profession has declined over the past fifty years while the resources of the pharmaceutical industry have grown substantially, the current politics of health cannot be described as a simple transfer in power from physicians to PhRMA (the industry lobby, abbreviated from Pharmaceutical Research and Manufacturers of America). The expansion of hypertension, diabetes, and high cholesterol to include previously healthy populations was indeed a process of medicalization, but it was not a concerted or monolithic strategy emanating from the board room of a pharmaceutical company or the American Medical Association. It was instead part of an overdetermined process that illustrates the porous relationship between the science and the business of health care and the centrality of disease categories in contemporary conceptions of health.

Since the 1950s, a set of related changes occurred—in demography and epidemiology; in policy structures surrounding biomedical research, pharmaceutical regulation, and clinical practice; in the R&D and marketing practices of the pharmaceutical industry; and in disease-centered activism—and each change played a substantial role in generating support for the pharmaceutical prevention of asymptomatic disease states. The framework thus constructed naturally emphasized the importance of the asymptomatic patient—a radical restructuring of the normal and the pathological—and the philosophy of the risk factor in clinical practice and public life. This new philosophy of health and medicine has dramatically altered experiences of patienthood and has fundamentally reshaped both the practice of medicine and the ethical priorities of the doctor-patient relationship. More broadly, the pharmaceutical-centered program of risk reduction has propagated a new moral economy of health values and a set of surveillance structures by which not only patients but also clinicians, policymakers, and even pharmaceutical executives find themselves constrained in their abilities to make decisions about the proper means of promoting good health and quality of life.

At the beginning of the twentieth century, Max Weber famously depicted the construction of an "iron cage" of instrumental rationality, in which the intertwining forces of science, capitalism, and bureaucracy would gradually restrict the possibility of human agency and bleed the moral value from human life. The system of risk reduction constructed in the second half of the century in many ways conforms to Weber's vision: equal parts science, commercialism, and the extension of bureaucratic rationality, this system threatens to enclose humanity within a process of physiological monitoring and pharmaceutical consumption. However, whereas Weber's iron cage was built on the inflexible certainty of technological rationality, the structure we now inhabit is flexible, for its links are bound not in certainty but in uncertainty: in probability, statistics, and calculations of risks.[11] Within this contemporary understanding of health and medicine, the concept of disease itself enjoys far more freedom of motion than does either doctor or patient.

Disease without Symptoms

A central feature of current conceptions of health and illness is that a person is no longer required to notice symptoms or even manifest visible signs of pathology to receive a diagnosis. We live in an age of numerical diagnosis in which the popular imagination depicts disease as a thing reducible to a fundamental molecular logic, ideally detectable by a blood test. On one hand, this technological embrace has been liberating: if the process of disease identification and intervention is no longer limited by the perceptive abilities of the individual patient or physician, earlier and more effective forms of preventing cancer and heart disease become possible. On the other hand, this very act of liberation has also served to further detach the meaning of disease from the individual body. The locus of disease definition has shifted away from the intimate space of doctor and patient to be deliberated within wider and more abstract arenas of policy, guidelines, and markets, simultaneously distanced from the level of human experience by the very small (molecular diagnosis) and the very large (massive long-term population studies). This privileging of micro- and macro-scale knowledge over individual-level knowledge has created a large and growing rank of patients who are pursuing treatment courses for illnesses they have never felt and quite possibly never would have felt had they been left untreated.

We now treat as diseases loose categories that themselves have never been connected to symptoms, entities such as mild hypertension, elevated cholesterol, and mild diabetes. These are physiological markers with only probabilistic connections to other conditions that do bear symptoms, such as stroke, myocardial infarction, and frank diabetes.[12] The language of risk and pre-pathological conditions did not take concrete form until the second half of the twentieth century. The detection of high blood pressures in 1945, for example, did not necessarily lead to any course of treatment: a famous illustration comes from the medical history of Franklin Delano Roosevelt, whose four-year record of high blood pressures—at times exceeding 260/150 mm Hg—were recorded by his physician, Howard G. Bruenn, without recommendations for treatment.[13] Roosevelt's death on April 13, 1945, from a hypertensive crisis was not a case of medical mismanagement of a previously diagnosed condition. Rather, his diagnosis with clinically treatable hypertension began only that day, when he reported a "terrific" headache and collapsed, unconscious, with a blood pressure of 300/190 mm Hg. And for a long time after FDR's death, the detection of elevated blood pressure in the therapeutic encounter was not typically regarded as clinically significant in the absence of such symptoms as headache, nausea, palpitations, loss of vision, or loss of consciousness. In the 1940s, symptoms were also generally required for the diagnosis of diabetes and of xanthomatosis, the only form of high blood cholesterol listed definitively as a disease: a condition in which excess lipids precipitated out of the bloodstream to form small fatty tumors all over the body.

By the close of the twentieth century, however, the diagnosis of any of these conditions required only a numerical measurement above a statistically defined threshold. A blood pressure higher than 130/80 mm Hg was now hypertension. A blood LDL (low-density lipoprotein) cholesterol level greater than 160 mg/dL was pathologically elevated. A fasting blood sugar level over 126 mg/dL could determine diabetes. These numbers are now central to the practice of diagnosis, their precision and standardizability allowing for a definition of disease in which the physical perceptions of doctor and patient are irrelevant.

When a patient complains of chest pain, this *symptom* has immediate subjective significance in defining illness, a first-person voice. When a physician sees, through ophthalmoscopic examination of a patient's retina, the copper-wire appearance of sclerotic arteries, this pathognomonic (literally, "disease-naming") *sign* adds a second-person dimension, of one person directly addressing the objective pathology in another. A long tradition of medical texts has emphasized the hidden pathognomonic signs by which physicians can detect and definitively diagnose occult diseases in persons who feel healthy. Such diseases, though hidden from the patient, always manifested some clear sign of organic pathology visible to the physician. Diseases invisible to the physician and patient alike, discernible only to those versed in large-data-set statistical analysis, seem to be a phenomenon first encountered in the late twentieth century. Such *numerical* definitions of pathology offer a detached, third-person perspective, seemingly independent of both doctor and patient, connected instead to the anonymity of measuring devices and expert committees that define standards, thresholds, and guidelines.

Two centuries earlier, neither the average patient nor her physician would find much use for numbers in describing the nature of an affliction. By the late nineteenth century, however, two distinct uses of numbers had worked their way into everyday medical practice: large numbers of patient cases were summarized in statistical tabulations of diseases and therapeutic outcomes, and physiological and pathological "measurables" within the individual patient began to be referred to numerically.[14] The former would become tied to the rise of the epidemiologist and the biostatistician, the latter to the rise of the medical laboratory and the medical technician, but both of these moves prompted active concern on the part of medical practitioners dedicated to the here-and-now of clinical practice. Many early-twentieth-century physicians voiced striking ambivalence toward the growing reign of numbers in medicine, praising the precision of modern laboratory and clinical measurement techniques but adding the proviso, as Harvard's Richard C. Cabot suggested in 1907, that although laboratory medicine offered the promise of enhanced diagnostic powers, "all that tends to make us build up our diagnosis *at a distance from the pa-*

tient, and without the constant reminders of every side of his case given us by his actual presence before our eyes—all such tendencies, I say, are dangerous."[15]

With the rise of large-scale statistical analyses and microscopic precision assays, numerical thresholds now occupy a central place in medical practice, and it would appear that Cabot's hopes and fears have both been realized. Numerical diagnosis has both fundamentally reshaped the doctor-patient relationship and generated entirely new processes of patienthood—such as waiting for one's test results and "knowing one's numbers." And studies of the recently diagnosed suggest that many patients appear to develop psychosomatic symptoms after receiving lab results that place them in higher risk categories.[16] The process of disease-naming today often relies less on subjective evaluations or clinical skill than on a vast surveillance system that screens, codifies, compares the patient to population databanks and decision-analysis algorithms, and returns with a diagnosis and treatment guidelines.

The growing irrelevance of the symptom to the present medical system has offered particular opportunities and challenges for the pharmaceutical industry, and the rise of the clinical guideline—particularly salient in the cases of hypertension, diabetes, and high cholesterol—has been an instrumental aspect of this transformation. Whereas the definition of normal and pathological may have been criticized as an overly closed-door, doctors-only affair in the past, the more open negotiation of evidence-based clinical medicine now allows many others a seat at the table. Because it controls billions of dollars of research and development funds, the pharmaceutical industry now has a significant say in determining which data are on the table for these discussions—and which are not.[17] In addition to influencing the availability of data, the industry can fund key decision-makers on guideline committees to affect clinical practice even more broadly. When the 2004 Adult Treatment Panel-III guidelines of the National Cholesterol Education Program—supporting the extension of pharmaceutical prevention therapy to millions of previously untreated patients—were called into question because of financial relationships between panel members and the pharmaceutical industry, Barbara Alving, acting chair of the National Heart, Lung, and Blood Institute, explained that "the experts who are most knowledgeable in a subject area are also the same people whose advice is sought by industry, and most guideline panels include experts who interact with industry."[18]

The changes in the process of disease definition have created a system that is at once more egalitarian than the physician-controlled process of the early twentieth century and more exposed to the movements of the marketplace.[19] The greater transparency of the current system has revealed a new political economy of interest and influence in disease definition. But this marketplace of diseases did not simply emerge from the development of diagnostic technologies, marketing mechanisms, or the new creed of evidence-based medicine. It was also highly contingent upon a fundamental shift in both the demography and the therapeutic understanding of chronic disease that took place in the years after World War II: the coming of a logic of risk reduction.

Risk and the Reshaping of Chronic Disease

To understand how contemporary conceptions of chronic disease have become intimately bound to pharmaceutical knowledge, we need to see how chronic disease was understood before the rise of the risk factor. Demographers characterize the twentieth century as a period of massive epidemiological transition, in which the burden of disease confronting the American population shifted from acute, infectious causes of mortality to chronic, noninfectious diseases. But chronic diseases did not arise de novo during that period; they had always been with us. In addition to the major chronic infectious diseases (such as syphilis and tuberculosis), chronic noninfectious diseases including cancer, heart disease, arthritis, and diabetes had been known as scourges of humanity for centuries. Until the middle of the twentieth century, however, these chronic diseases were overwhelmingly understood in degenerative terms; that is to say, such conditions were usually described as part of the decay of the aging organism, which might play out variably in different individuals but was, overall, unalterable. Chronic disease was the means by which the mortal body, taking either a natural or an accelerated course, began its inevitable breakdown. Quacks and patent-medicine hucksters might promote age-defying creams and concoctions, but the best a responsible physician could do was to offer supportive care and recommendations for a healthy coping lifestyle.[20]

By the late 1920s and the 1930s, the decline in infectious disease mortality brought increased attention to the mounting death tolls from heart disease and cancer, which would inherit the titles of number one and number two killers of Americans. As the leading cause of mortality in the country, heart disease acquired a new prominence as a public health problem, although people initially viewed such an endemic "disease of civilization" as a natural and therefore unchangeable consequence of greater longevity. Harvard's G. C. Shattuck, for example, in his 1926 *Principles of Medical Treatment*, classified all heart disease not associated with infectious causes or birth defects as a "degenerative" form "most commonly found in old age" and characterized by a general picture of "senility and general arteriosclerosis."[21] The individual diagnosed with degenerative heart disease would receive a supportive regimen of rest, volume depletion, suitable diet, and regulation of mode of life to reduce shocks to the system. The principal medication available for such a patient was digitalis, a cardiac stimulant that Shattuck considered "more accurately spoken of as a heart tonic."[22] Digitalis was a prop for the failing heart, supportive therapy for a body in the process of inexorable decay.

Against this fatalistic backdrop, developments in the fields of epidemiology and public health worked to redefine chronic disease in general—and chronic cardiovascular disease in particular—into more specific and activist terms. Extensive prevention efforts mobilized in the 1920s and 1930s to control rheumatic heart disease—a set of chronic cardiovascular ailments caused by prolonged exposure to a streptococcal agent—helped to produce a substantial decline in the chronic sequelae of rheumatic heart disease over the course of the century. Epidemiological techniques initially developed to study acute infectious diseases began to translate into the study of chronic diseases as mechanistically modifiable and preventable species of disease. Occasionally, as with the case of Joseph Goldberger's research into pellagra in the 1920s, or as with silicosis research in the 1930s, techniques of infectious disease epidemiology successfully linked chronic diseases to germ equivalents: specific microscopic entities such as vitamin deficiencies or occupational toxins, which, like microbes, could be treated as minuscule agents of disease. These agents helped to describe mechanisms of chronic disease processes and thus constituted a concrete site for active research, treatment, and prevention.[23]

More often, however, no microscopic agent could be found, and the study of chronic disease became characterized by work of a more sweeping nature, such as the surveys begun by state and local public health officials in the 1920s and 1930s to estimate the prevalence and severity of various classes of chronic diseases in the adult population. The first National Health Survey, conducted by the U.S. Public Health Service in 1935–36, turned the nation's attention toward the growing significance of chronic disease as a threat to public health. By the late 1940s, chronic disease research had developed a significant foundation in the National Institutes of Health with the founding of the National Cancer Institute in 1937, the National Heart Institute in 1948, and the National Institute of Mental Health in 1949. In late 1946 representatives from the American Hospital Association, the American Medical Association, the American Public Health Association, and the American Public Welfare Association formed what would become the Joint Commission on Chronic Disease, a blue-ribbon panel charged with the task of issuing recommendations on the status of chronic disease in America and future prospects for treatment and prevention. The commission was widely influential over the 1940s and 1950s and published a four-volume encyclopedic summation of all of its findings in 1957; an entire volume, *Prevention of Chronic Illness,* set out goals of detection and treatment of early forms of chronic diseases.[24]

At the same time, a much different detection effort regarding the early forms of chronic disease was being developed by the American life insurance industry. As chronic conditions like heart disease gained importance as causes of mortality in the United States, they acquired particular importance in the studies of actuarial calculations on the proper management of life insurance premiums. The goals of the insurance industry were not to improve individual health but to exclude unhealthy (and therefore expensive) individuals from the insured population at the earliest stage possible. Nonetheless, the insurance industry would contribute key dimensions to the study and management of chronic disease, including the population-based mortality study and the annual physical examination. By the early twentieth century, companies like Metropolitan Life Insurance were able to use their own claims records as large data sets to measure physiological variables against mortality. In doing so, the insurance industry recast physiology in terms of *risk,* a term with specific financial connotations that the insurance industry had developed to analyze broad population-based policies in quantifiable terms. Although these data were not directly applicable to the treatment of an individual patient, such studies were increasingly cited in the public health and epidemiological literatures to highlight high blood pressure, obesity, and cigarette smoking as threats to the nation's well-being.[25] Furthermore, the actuarial practices of the insurance industry contributed greatly to medicine's increased use of periodic health examinations—or annual physical examinations—which would come to constitute a vital foundation to the structure of preventive medicine.[26]

Nowhere was this shift from degenerative to preventive understandings of chronic disease more evident than in the arena of heart disease, which had been elevated by the mid-twentieth century to a national obsession of American popular culture. As death rates from infectious diseases continued to decline after World War II, coronary heart disease was known to be responsible for one out of three deaths in the nation, prompting the influential senator Claude Pepper to refer to the disease as an enemy "worse than Hitler, so far as the lives of our people are concerned."[27] The sudden, seemingly unpredictable character of death by heart attack and its tendency to strike not at the margins of society but at the mainstream (showing, if anything, a predilection for an age-class-race demographic that included most policymakers and doctors) made it a subject of widespread public anxiety.[28] Indeed, even before the broad publicizing of President Eisenhower's 1955 heart attack and recovery—during which the cardiovascular system of the president became the subject of an extended national conversation—coronary heart disease had begun to be thought of as a particularly American disease, a modern epidemic for a modern nation.[29] As a central focus for the preventive medicine movement, combating heart disease through the screening and treatment of risk factors became one of the most dominant public health promotions—enacted through private primary care visits—that the country had yet attempted.

By the early postwar era, new pharmaceuticals had already begun to play a part in the redefinition of chronic disease. The ability of drugs to transform previously untreatable chronic disease processes into manageable conditions helped to generate a popular imagery and attitude of pharmaceutical triumphalism. This attitude itself encouraged the further conception of chronic diseases as preventable conditions. One physician who worked with the Commission on Chronic Disease noted in 1951: "In my student days medicine had very little to offer the patients with severe diabetes mellitus, pernicious anemia, congenital heart disease, Addison's disease, rheumatoid arthritis, cirrhosis of the liver, to name a few chronic illnesses. Today the situation is much different, as you well know. With this fine record over the past 40 years and the present pace of research, is it not possible that the medical student of 1975 or 2000 may add hypertension or arteriosclerosis or cancer or all three to the list of preventable or controllable chronic diseases?"[30] The examples he cited all represent relatively recent pairings of drug and chronic disease that, by 1951, had already begun to transform therapeutic nihilism into therapeutic optimism: the development of insulin for diabetes in the 1920s and the near-simultaneous emergence of Lilly's Liver Extract 343 as a cure for pernicious anemia, the manufacture of synthetic corticosteroids in the late 1940s as a treatment for rheumatoid arthritis and Addison's disease, and so forth. As the physician's statement implied, future use of pharmaceuticals to treat new chronic disease categories would be crucial in the process of replacing degenerative concepts of chronic disease with more activist visions of chronic disease detection, management, and prevention.

Drugs and Disease in Historical Context

The role played by pharmaceuticals in American life has received surprisingly little historical examination, and much of the existing literature tends toward thin narratives of success or scandal. Triumphalist accounts of drug discovery and disease conquest are readily found within the clinical literature, joined by a smaller number of slim, well-illustrated corporate histories, commissioned by individual pharmaceutical houses to chronicle their unfolding from nineteenth-century apothecary shops to twentieth-century standard-bearers of science and humanitarian concern.[31] One can also trace a simultaneous and parallel lineage of muckraking polemic that stretches from Samuel Hopkins Adams's 1903 *Collier's* series on the "Great American Fraud" of patent medicines to contemporary indictments of the multinational pharmaceutical industry as an evil empire manipulating medical knowledge. Although some of the more rigorous works in this field provide good documentary analysis, their commitment to advocacy and exposé often precludes more textured portrayals of pharmaceuticals in practice.[32] Many of these works represent ideological screeds that chastise the industry for profiteering or poisoning without attempting to reconcile the complex and dependent relationship between contemporary expectations of the health care system and the continued availability of pharmaceutical agents.[33]

Missing from all of these accounts is an awareness of the pharmaceutical as a complex social object, as something neither Promethean nor poisonous but somewhere in between, a reflection of the ambivalent connection between science, health, and capital in the contemporary period. Only a handful of historians have explored the close history of the drug-disease relationship, the moral entanglement of pharmaceuticals and the production of medical knowledge, and the social lives that pharmaceutical agents lead beyond their immediate therapeutic roles.[34] This book attempts neither apology nor attack; instead I will underscore the fundamental political, economic, ethical, and moral contradictions within American understandings of health that pharmaceuticals bring into sharp relief.

Treating Diuril, Orinase, and Mevacor as entities with their own life spans offers a methodology for exploring the processes that enabled these drugs and their related diagnoses to gain a foothold in American medical practice and public life. The role of the consumer and the commodity in the changing cultural landscape of twentieth-century American life, and specifically the practices of marketing, salesmanship, and advertising, are equally important to understanding the social history of pharmaceutical agents.[35] Recent anthropological efforts to found a discipline of "pharmaceutical anthropology," studying the social relations traceable within the exchange of objects and the social lives or cultural biography of consumer artifacts, have also yielded methodological and theoretical tools for writing the history of pharmaceuticals.[36]

The most visible examples of interplay between pharmaceutical promotion and disease categorization may belong to the field of psychiatry. Scholars "listening to" the social lives of psychopharmacological agents have begun to use well-bounded historical examples to illustrate this interconnection, with particular attention to the augmented marketing practices and to the diagnostic "bracket creep" that has expanded the definition of mental illness categories.[37] Curiously, however, this scholarship often refers to somatic disease as an unproblematic foil—asking, for example, whether psychoactive drugs cause people to titrate their mood as they titrate their blood pressure without asking how it is that blood pressure became a "thing" one could or would titrate in the first place. The interconnections between therapeutics and diagnostics examined by critics of psychopharmacology are not, in fact, unique to the domain of psychiatric diagnosis, but rather are always present in all areas of medicine, even those that seem most indisputably concrete in their mechanisms.[38] As I will demonstrate, they exist at the very heart of somatic medicine.

To some extent, drugs have been central to Western definitions of disease for centuries. By the early fourteenth century, *antidotaria* categorized ailments with reference to their corresponding therapeutics, and the later chemical therapeutics of Paracelsus often explicitly defined diseases and therapeutics in elemental relationship to one another.[39] At the dawn of the twentieth century, the search for what Paul Ehrlich called the "magic bullet"—an ideal drug that singles out only the disease-causing agent while leaving healthy tissue unaffected—immediately suggested the role of the pharmaceutical agent itself in differentiating between normal and pathological.[40] Many drugs have been used as tools to study and classify disease: the early antihypertensive agent reserpine later generated a research model for the study of depression, and oral antidiabetic agents helped catalyze a research program differentiating between insulin-dependent diabetes mellitus and non–insulin dependent diabetes mellitus. On a less formal basis, drugs are used as empirical diagnostic tools in clinical practice: if patients with stomach pain respond to a month-long course of an H2-blocker like Zantac, their symptoms are often dismissed as a mild gastritis or reflux disease; if they do not improve, a more significant diagnosis is likely and a further diagnostic workup indicated.

But in the past half century, it has become clear that drugs do not merely affect the categories by which we understand disease; they also play an active and material role in remolding the bodily consequence of disease into new forms. The widespread use of insulin in the years after its discovery helped to bring about, ironically, a series of diabetic conditions that had not existed before, if only because diabetics had not previously lived long enough to develop them.[41] Similarly, for HIV patients with regular access to medications, the onset of successful antiretroviral chemotherapeutics in the mid-1990s recast the diagnosis of HIV/AIDS from an inevitable death sentence into a manageable chronic disease. In addition to unveiling new dimensions of pathogenic natural histories, drugs can directly generate more dangerous illnesses: they can produce drug-resistant strains of infectious diseases and malignancies. Methicillin-resistant *Staphylococcus aureus* infection, a difficult-to-treat pathogen now commonly found in American hospitals, did not exist before the advent of the antibiotic methicillin, nor did multiple-drug-resistant tuberculosis pose a threat before effective antituberculosis medications were broadly employed. Indirectly or directly, drugs continue to affect the material epidemiology of disease in surprising and counterintuitive ways.

Beyond their diagnostic and epidemiological roles, however, pharmaceuticals bear a third essential relationship to diseases that has grown in relevance over the past fifty years: the relationship of commodity to market. The material success of the pharmaceutical industry and the expansion of both pharmaceutical research and pharmaceutical marketing have created an awareness of diseases as markets that can be redefined in terms of specific drug products. Direct-to-consumer pharmaceutical advertising now promotes drug-centered disease categories—such as erectile dysfunction, premenstrual dysphoric disorder, and low testosterone—to the general population. A drug's "therapeutic indication," formerly simply used by the FDA to regulate whether a company would be allowed to advertise the therapeutic claims of its drug, increasingly serves to define the limits of the drug's legitimate use in clinical practice. When pharmaceutical companies define their products' indicated diseases as markets, they bring a logic of brands and commodities into the definition of disease itself. Expansive guidelines transform risk factors into growth markets; key clinical trials are reported not only in medical journals but on the front pages and increasingly in the business pages of national newspapers.

The historical connections between diagnoses of risk and pharmaceuticals of risk reduction help explain how both categories have come to occupy central roles in the contemporary landscape of health and medicine. The intertwined histories of hypertension and Diuril, diabetes and Orinase, and hypercholesterolemia and Mevacor suggest that relationships between drug and disease are not merely a matter of pharmacological theory, clinical trials, epidemiological change, or marketing tactics but are instead overdetermined by some combination of these elements. As our own classifications of chronic disease continue to encompass increasing numbers of subjectively healthy individuals, and as the roles of numerical diagnosis, risk assessment, and strategies of pharmaceutical prevention continue to grow, it is of critical importance to understand the forces that have shaped and continue to guide this radical transformation in American health practices.

Part I / Diuril and Hypertension, 1957–1977

CHAPTER ONE

Releasing the Flood Waters

The Development and Promotion of Diuril

Among newsmen, there's a saying that there is only one thing colder than yesterday's mashed potatoes, and that's yesterday's news. But when the subject is "DIURIL," the newspaper never gets cold—because "DIURIL" makes news every day. The following reports provide additional evidence that "DIURIL" is still a "hot" news item, especially to the physician thinking about hypertension.   —*Diuril News Report*

Thousands of small white tablets of chlorothiazide—better known by the brand name Diuril—emerged from Merck Sharp & Dohme production plants in January of 1958 amid a dazzle of research symposia, journal advertising, and record prescription rates for a novel agent. Diuril represented the first palatable pill for hypertension, and although its story is less well known than the saga of antibiotics, the dramatic emergence of antipsychotic drugs, or the cultural hand-wringing surrounding the minor tranquilizers, the influence of this drug on clinical practice was equally profound.[1] As late as World War II, a patient with high blood pressure and no symptoms was not likely to receive any treatment. A few decades later, hypertension had become a radically different entity, detected routinely in primary care screening and increasingly treated with specific pharmaceutical agents long before any symptoms had developed. The story of this transformation—and the faint but groundbreaking shift it marks in the definition of disease in the postwar era—is highly contingent upon the emergence of new therapeutic agents like Diuril and the promotional apparatus set in place to market them to physicians and patients across the country. To understand the interplay between drug and disease in late-twentieth-century therapeutics—and the contested positions of doctors, patients, researchers, pharmaceutical executives, and others with a stake in the outcome—we must pay attention to the simultaneous interactions between the research, the clinical, and the market arenas in which pharmaceuticals operate.

Hypertension became a different disease after Diuril. It is equally true, however, that Diuril became a different drug after it encountered hypertension. If we look at a pharmaceutical as both a clinical entity and a branded consumer product, the relationship between drug and disease emerges not as a story of design or serendipity, control or production, but rather as a narrative of cumulative negotiation and reciprocal definition. The history of Diuril and hypertension presented in this chapter illustrates the mutually constitutive processes of clinical research, clinical practice, and medical marketing in the postwar American pharmaceutical industry and traces the evolution of a set of hybrid structures that became central institutions of pharmaceutical promotion in the second half of the twentieth century.

Antihypertensives amid the American "Drug Explosion"

The American pharmaceutical industry in the two decades following World War II witnessed an increase in the scope and pace of therapeutic innovation unanticipated by physicians, consumers, or even the industry itself. Wartime infrastructural links between academic medicine, industry, and governmental institutions—overseen by the Office of Scientific Research and Development during the wartime effort—were swiftly adapted to yield a robust "pipeline" of novel and efficacious medicines that proved highly profitable for growing American drug concerns.[2] Entirely new classes of therapeutic compounds were appearing almost every year, supported by a burgeoning research literature. By 1947 it was estimated that fifty cents of every dollar of pharmaceutical revenues came from products not available ten years earlier, and perhaps the most tangible example of this process was the growth of the antibiotic sector.[3] Within fifteen years of the mass production of penicillin in 1943, the antibiotic market contained nearly twenty novel drugs and represented $270 million in annual sales.[4] By 1958 the practicing clinician was faced with a promising but easily confusing set of novel antibiotics and branded combinations of agents to integrate into his or her practice.

The same process took place with antihypertensives, a category of therapeutics similarly nonexistent a few decades earlier. Until the late 1940s, treatment for hypertension was largely limited to sedatives, nitrates, or the complete surgical interruption of the sympathetic nerves running alongside the spinal cord. Over the ensuing decade, four entirely new classes of antihypertensive drugs—the ganglionic blockers, hydralazine, Rauwolfia derivatives, and Veratrum alkaloids—were combined and recombined in a staggering array of branded preparations.[5] Although severe side effects limited their widespread use, by 1958 these antihypertensives had more or less displaced surgical treatment for acute hypertensive crises.[6] The sheer variety of these new therapeutic products, however, left many physicians unsure of which therapeutic agent to use and when such therapy was truly warranted. In a survey of the field that year, a prominent cardiologist noted: "A review of the pharmacopoeia for a list of antihypertensive drugs six or seven years ago would have required little time. Other than potassium thiocyanate, the nitrates and Phenobarbital, few if any drugs were advocated for the treatment of hypertension. Today, however, there is a plethora of trade names for antihypertensive agents and combinations of them. I cannot begin to remember even a small portion of their names or to recognize their pills by sight." The sheer abundance of novel combination pharmaceuticals did not necessarily produce therapeutic optimism. "One may estimate the effectiveness of antihypertensive therapy by the number of compounds advocated for its use," the same author continued. "When and if at long last a really *specific* antihypertensive agent is discovered, the present formidable array of drugs will wither as an ice cube in boiling water."[7] Similar cries of dismay in the face of new therapeutic plenty became commonplace in the 1950s.

The expanded pharmacopoeia of the postwar era was accompanied by heightened rates of pharmaceutical consumption: between 1939 and 1959, domestic sales of pharmaceuticals increased from $300 million to $2.3 billion.[8] Ironically, this rapid expansion of the drug market—popularly termed the "drug explosion"—posed substantial challenges to the two institutions most poised to benefit from it: the medical profession and the pharmaceutical industry. In 1955 the American Medical Association (AMA) publicly cited bureaucratic overload due to the excess of unevaluated novel compounds and formally disbanded the Seal of Acceptance program that had served since 1908 to evaluate new drugs and educate physicians regarding their usage.[9] Meanwhile, in the midst of such extraordinary growth, pharmaceutical executives perceived a set of threats to their former promotional practices.[10]

Growth of the industry meant increased product competition and an increasing concern about the competitiveness of any particular product. In the swell of competing products, the old principles of "ethical promotion"—with emphasis on chemical name and dry, factual advertising in medical journals—were increasingly seen as insufficient ways to persuade physicians to use novel products. Marketing departments were bolstered and reorganized; more attention was focused on the pharmaceutical as a *brand*.[11] By the time of Diuril's emergence onto the therapeutic landscape in the late 1950s, pharmaceutical companies had developed for their products sophisticated and integrated promotion systems that called for a tight intertwining of research and marketing. The institutional history of Merck Sharp & Dohme at the time of Diuril's release illustrates particularly well the changing relationship between research and marketing that was coming to characterize the industry.

Diuril in the Life of the Firm

By the early 1950s, Merck's research laboratories were the pride of the American pharmaceutical industry; they contained an impressive array of Nobel laureates and played a substantial role in the production of such "miracle drugs" as penicillin, streptomycin, vitamin B12, and cortisone.[12] Financially, however, Merck had earned almost all of its revenues not from selling prepackaged pharmaceutical formulations but by supplying refined chemicals to pharmacists and the pharmaceutical industry. Merck was not alone in this trade: for the first half of the century, most of the pills, tablets, elixirs, or capsules that reached the ultimate consumer were physically assembled, or "compounded," in the pharmacy, even if their contents were provided by major pharmaceutical firms. But by the late 1940s, the role of chemical supply was declining. As fewer and fewer pharmacies compounded their own medications, the market presence of pharmaceutical "specialties"—prepackaged prescription medicines made by only one firm—increased dramatically.[13] The core product of the pharmaceutical industry was shifting from equivalent and chemically standardized medicinal substances toward a set of singularly branded and mass-marketed goods.

One of Merck's first forays into pharmaceutical specialties—and also, significantly, into brand-name consumer products—was the breakthrough steroid Cortone (cortisone). The marketing of this first synthetic corticosteroid was hailed as a pivotal moment in the history of chronic disease: popular and medical periodicals quickly pronounced Cortone a chemotherapeutic cure-all, and its anti-inflammatory properties seemed to find uses in every specialty of medical practice.[14] Relying on this widespread enthusiasm for cortisone as a "miracle drug," Merck invested little in promoting pharmaceutical sales. Consequently, when supplies of cortisone derivatives were made available by competitors, the company quickly lost its market position to Upjohn and Schering and struggled for the rest of the decade to make any significant profits from Cortone.[15] An internal postmortem analysis of the Cortone project suggested to Merck's board and its executives that a focus on pharmaceutical research at the expense of marketing would not carry well into the second half of the twentieth century. Something needed to change.

Rather than build up a marketing force de novo, Merck's directors set out to acquire one, ending their search on the other side of the Delaware River at the Philadelphia pharmaceutical house of Sharp & Dohme. Sharp & Dohme was not known for its research laboratories, and most of the products it sold by the early 1950s were available from other pharmaceutical houses in other formulations.[16] It was, however, known to have a dynamic and well-trained sales force that was highly effective at getting the Sharp & Dohme brands onto physicians' prescription pads. Executives at Sharp & Dohme also perceived much to gain from access to Merck's deep research pipeline, and they listened carefully to the proposal. The merger was announced in 1953, and analysts predicted that the combined firm would be an ideal marriage of research and marketing. The launch of Diuril was to consummate this union.[17]

Diuril, however, did not develop out of any targeted search for an antihypertensive therapy. The drug did not even have any connection to hypertension until after it had left the company's research laboratories.[18] Rather, Diuril's material genealogy can be traced back to wartime shortages of penicillin. Under the leadership of Karl Beyer, the Renal Program of Sharp & Dohme's newly minted research laboratories in West Point, Pennsylvania, was established in 1943 with the explicit task of devising an agent that could block the body's excretion of penicillin, in order to stretch the value of a single dose of the precious substance.[19] Although this project ultimately did produce an agent that effectively blocked penicillin excretion with minimal side effects, known as Benemid (probenicid), the product did not appear on pharmacy shelves until the mid-1950s. By that time, the wartime collaborative effort had already rendered penicillin widely available and relatively inexpensive.[20]

After this partial and costly victory, Beyer and his team quickly sought another field in which to demonstrate the utility of renal physiology for pharmaceutical development. A 1949 paper suggesting that sulfonamides caused diuretic side effects had sparked their interest: if that diuresis could be translated from side effect to therapeutic indication, the Renal Program's existence would be more than justified.[21] Diuretics were already known to capture a substantial market with many therapeutic indications—though hypertension was not among them—and all presently available diuretics were either heavily toxic or largely ineffective.[22] Beyer knew that a fluid-relieving drug that was both palatable and effective would fetch a large market, and by the time of the merger he had assembled a team of pharmacologists to work toward that goal.[23]

In spite of the failure of several precursors such as Dirnate and Daranide, the Renal Program announced in 1956 that in-house animal experimentation had demonstrated that a newly synthesized compound named Diuril was a safe and surprisingly efficacious diuretic: a compound with significant potential to be a prescription drug product. In October of that year, the drug was first demonstrated to a set of potential clinical researchers, and by the spring of 1957, Diuril had been distributed to a network of 250 clinicians in the United States and 54 physicians in eighteen other countries. Two million patient-days worth of chlorothiazide was moved through this network in 1957; by July 1957, as case-series data mounted into the hundreds, the team filed a new drug application with the Food and Drug Administration, and in early September Merck Sharp & Dohme received clearance to market Diuril as a novel diuretic agent.[24]

Attachment: Diuril Meets Hypertension

In the spring, however, two groups of clinical investigators had happened to give the drug to congestive heart failure patients who suffered from severe hypertension in addition to edema. A dramatic reduction in blood pressure was noted. These researchers, two in Boston and two in Washington, DC, followed up with a small case series of severe hypertensive patients and by the fall of 1957 had documented Diuril's antihypertensive effects in a broader population of severe hypertensives.[25] At that point, the Boston group went public, holding a press conference in Boston in early October that attracted widespread popular coverage, not only in the medical press but also in newspapers and news-magazines around the country.[26]

Paradoxically, this new indication introduced several complications into the .Merck Sharp & Dohme (MSD) marketing effort. As late as May 1957 a detailed marketing plan had been drawn up based solely on Diuril's value as a safe and effective fluid-remover; the document did not include information regarding blood pressure effects.[27] Diuretics had not previously been considered blood-pressure-lowering drugs, and having too many physiological activities—particularly something as clinically significant as a drop in blood pressure—could seriously undermine a drug's safety profile. The launch, originally slated for late October, was delayed indefinitely.

George Schott, of Merck's public relations office, distributed a memo to the marketing teams in the immediate aftermath of the Boston press release: "With interest in 'Diuril' picking up momentum, I feel it is high time to bring our side of the story into focus, so that we will be identified with these reports, as a company and as a trade mark specialty. *I fear that our failure to adjust our program to outside pressures which are beyond our control may lead to our loss of identity with the product.*" Schott worried that the drug-launch conference originally planned for November of that year might "be an anticlimax . . . although we plan to follow through with as much publicity as possible." "However," he continued, "interest is high now, and we should be flexible enough to take advantage of it. It may never return."[28] Karl Beyer, who later received a Lasker award for his role in Diuril's development, recalls how this collision with hypertension altered the drug's therapeutic trajectory: "It wasn't certain at all, prior to chlorothiazide . . . How in hell can a diuretic agent be anti-hypertensive from a marketing standpoint? How can we promote this thing, even if it works? And our top Nobel consultant in medicine thought that was a lot of nonsense too . . . But all you have to do was have those first two little papers by Freis and Wilson and by Hollander and Wilkins on hypertension come out, and we were in hypertension for whatever it cost. Everybody knew the size of the hypertension field."[29]

This is no story of serendipity. Though it was not planned as an antihypertensive, Diuril did not merely "happen upon" hypertension either; rather, it became an antihypertensive medication through a concerted confluence of clinical, research, and marketing practices. That a marketing attitude penetrates even to the bench researcher is particularly visible in Beyer's narration, which reflects a pragmatic understanding that already, by 1958, the research enterprise and the marketing enterprise were intimately interconnected in the process of new-product development.

As Beyer suggests, the hypertension field was indeed promising. A national health survey conducted in 1957–58 concluded that 5.3 million Americans were suffering from some form of hypertension.[30] Initial sales projections for Diuril as a diuretic had been set at $8 million; hypertension now added at least another projected $10 million.[31] A few days after the press conference, Schott sent out a second memo, in which he formulated a plan to refashion the Diuril campaign emphasizing the antihypertensive qualities of the drug as its main indication and opportunity for market expansion. Heart disease was to become the central claim of the drug: "If 'Diuril' does indeed live up to expectations and becomes a valuable adjunct in the treatment of congestive heart failure particularly *and in the treatment of hypertension,* we will have the basis for a sound, on-going public relations program . . . 'Diuril' will surely catapult us into the forefront of the battle against heart disease. This is a valuable platform. Heart disease is the number one killer in this country today. Coldly speaking, emphasis on 'Diuril' in the battle against heart disease (rather than generally as a diuretic) will enable us to take advantage of the existing interest in heart disease."[32] There was little popular romance around the kidney. The heart, however, was at the forefront of health concerns in the Eisenhower era, and the heart was where MSD now wanted to be. A new launch date was set for January of 1958, resources were made available to clinical researchers investigating Diuril for hypertension, and expanded focus was given to other heart-related indications for diuresis, particularly congestive heart failure.[33]

Schott's revised plan was circulated to marketing, research, and medical divisions. This program was meticulously orchestrated and set to simultaneously launch Diuril as a treatment for hypertension and MSD as a company visibly situated in the vanguard of heart disease and preventive medicine. To help ease the entry of an unfamiliar drug into the practice of a relatively conservative physician population, Diuril's marketers worked to develop strategies in multiple overlapping arenas of influence, staggered at different phases to form an integrated chronology of promotion with continuity between the prelaunch, launch, and postlaunch periods. MSD marketers then divided these tasks into domains and assigned them to various executives. Juxtaposed with the externally visible events of Diuril's launch, the content of internal documents serves to sketch out the tactical framework of pharmaceutical promotion taking shape in the 1950s.

Clinical Research as a Marketing Arena

As late as 1948, pharmaceutical marketing textbooks made no mention of the relationship between marketing and clinical research.[34] By the 1950s, however, the industry explicitly recognized that clinical research did not precede or determine pharmaceutical marketing but was instead an essential and inter-woven part of the marketing process. Paul de Haen, a marketing consultant to the industry and a frequent contributor to the journal *Drug and Cosmetic Industry*, illustrated the necessary parallels between research and marketing in the development of a new drug. "The successful development of a new pharmaceutical product," de Haen insisted, "depends not only on the coordination of research and marketing, but on the synchronization of these activities."[35] Over the course of the 1950s, as marketing departments grew and an organizational principal known as the marketing concept spread through the industry, pharmaceutical executives began to speak more confidently about the relationship between research and marketing. Surveying the spread of marketing in the field, one executive wrote in the late 1950s, "If Marketing should not dictate to or control the scientists in our laboratories, it is equally true that Research and Development must not assume control over marketing . . . [Product planning] should function in such a way as to permit the scientist and the commercial man to meet on common ground . . . The important consideration here is that under the marketing concept the laboratories will not put a finished product in a bottle, unveil it before a sales manager, and say, 'Here it is, take it out and sell it.'"[36]

The marketing concept did not transform the industry or even MSD in a single instant: Sharp & Dohme's marketing staff were only gradually allowed into meetings at Merck research facilities, and even then the first marketers admitted to MSD Research Laboratory conferences after the merger were expressly forbidden to talk.[37] But by the time of Diuril's launch, clinical research was clearly understood in explicit relation to marketing at Merck Sharp & Dohme. Fred Heath, the head of the medical division at MSD, articulated a vision of clinical research that was intended both to generate data for the more convincing promotion of Diuril and to serve as a promotional structure in itself. External researchers were all treated quite well by Merck—each was given a color TV as a bonus (no small prize in 1957)—but internally the marketers divided these clinician-researchers into a marketing structure with two concentric spheres. The outer ring involved lesser-known researchers of negligible influence, while the core of this structure was a group of eighty-four clinician-researchers selected as influentials within their local areas, carefully chosen to be "scattered in each of the Merck Sharp & Dohme marketing areas excepting Memphis and Minneapolis."[38] In the marketing literature of the time, influentials were those who might, in addition to performing research, act as models for their peers in their endorsement of a product and even circulate samples of the experimental drug within their academic and private practices. This geography of clinical research was then used to generate a precise map of Diuril's premarket acceptance.[39]

In addition to their roles in regional promotion, researchers were mobilized by the MSD marketing team to form a series of symposia that traveled around the country and received nationwide attention. This Diuril road show held its kickoff event at the New York Academy of Sciences on November 8, 1957, featuring a lineup of prominent cardiologists and nephrologists from influential academic medical institutions. Known as the "third-oldest scientific organization in America," the New York Academy of Sciences (NYAS) was seen as a prestigious neutral space, a conference setting with a more protected public relations image than Merck's Rahway, New Jersey, headquarters.[40] Following the NYAS symposium, a speaking tour of clinical researchers was mobilized with talks evenly spaced from January to November of 1958, covering twenty states chosen to represent all geographic regions of the country. Medical director Fred Heath attended every talk and sent detailed notes back to the Rahway offices after each meeting.[41]

Symposia helped to generate local publicity but were necessarily fixed in time and place; Diuril's marketers recognized the importance of journal publications as way to introduce larger numbers of physicians to novel medications. Sometimes the symposium itself could become a publication. For example, the proceedings of the NYAS conference were published two weeks later as a special volume of the *Annals of the New York Academy of Sciences.* Most articles in peer-reviewed journals, however, entailed a longer and more complicated process before publication. This, too, marketers sought to manage as well as they could, and Diuril's development timeline plotted out when and how news of their product would ideally hit the general and specialty journals.

Setting quotas and timetables for research publications was crucial to the prelaunch marketing project, and marketers kept careful track of publications featuring Diuril (see fig. 1.1). "The importance of publications in promotion of the drug was stressed," read an internal Diuril memo from early 1957. "It is estimated that some 6–12 more papers will be submitted this year. It is also estimated that we can reach a goal of some two dozen papers by the end of 1958."[42] This estimate proved to be conservative: by the end of 1958 at least 50 articles had appeared in the medical literature on the merits of Diuril, and by June of

Fig. 1.1. Physician exposure through clinical publications. This image, from a Merck Sharp & Dohme in-house magazine, demonstrates the role of journal publications in late-1950s pharmaceutical promotion. *Source:* "Help for the Heart," *Merck Review,* January–February 1958, 2–5. Courtesy of Merck Archives, Whitehouse Station, NJ.

1960 a literature of more than 150 peer-reviewed journal articles specifically detailed the attributes of Diuril in treating hypertension.[43]

Publications were especially important in establishing Diuril as an antihypertensive agent, because this was the newest and most vulnerable aspect of the drug's therapeutic profile. The early case studies by Edward Freis, Annmarie Wanko, and I. M. Wilson and by W. Hollander and Robert Wilkins connecting Diuril with hypertension needed to be substantiated to generate a solid base in the scientific literature. MSD's investment in hypertension research yielded results by early December of 1957, when Freis revealed a study of one hundred patients over an eight-month period, showing that Diuril in combination therapy reduced blood pressure "more effectively than any other drugs generally used."[44] At the time of launch, publications became even more important as a means of influencing potential prescribers. The crowning jewel of Diuril's journal campaign was the January 11 issue of the *Journal of the American Med-*

ical Association (*JAMA*), which was devoted almost exclusively to articles detailing the efficacy and safety of chlorothiazide. The timely publication of this issue, coincidental with the launch month of Diuril, may have been facilitated in part by the fact that *JAMA*'s editor-in-chief, Austin Smith, was also that year elected president of the Pharmaceutical Manufacturers Association.[45]

Public knowledge of interactions between pharmaceutical companies, clinical researchers, and the journals and institutions in which research was presented seems only occasionally to have raised eyebrows in the probusiness environment of the 1950s. One year after Diuril's paid launch conference at the New York Academy of Science, a playful editorial used Lewis Carroll to poke obliquely at the ethics of an antibiotic launch conference at the same locale:

> "Have some Science," said the March Hare in an encouraging tone.
>
> Alice looked all around the table but there was very little Science, only Salesmanship.
>
> "I don't see much Science," she remarked.
>
> "There isn't any," said the March Hare.
>
> "Then it wasn't very civil of you to offer it," said Alice angrily.
>
> "It wasn't very civil of you to sit down without being invited," replied the March Hare.
>
> "It said it was the New York Academy of Science table. I didn't know it was your table," said Alice. "It's laid for a great many more than three."[46]

The occasional literary jibe aside, it was common for the pharmaceutical industry–medical profession relationship, as embodied by Austin Smith, for example, to be represented as a natural alliance in the interest of medical progress. Public and professional concern about undue pharmaceutical influence over the conduct and presentation of scientific research appears to have been minimal in the years preceding the critical 1959 Senate hearings widely publicized by Estes Kefauver. In 1958 Austin Smith's dual position as editor-in-chief of one of the nation's most influential medical journals and as the top executive of the pharmaceutical industry's chief lobbying group was public knowledge, reported in the newspapers without any suggestion that such a position might represent a conflict of interest or might stain *JAMA*'s objectivity.

In addition to the literature peer-reviewed by general-practice and specialist physicians, marketers utilized "throwaway journals" as key product venues. Throwaway journals had increased greatly in number and scope in the 1940s and 1950s and now included two major variants: the medical newsmagazine and the house organ. Medical newsmagazines, such as Arthur Sackler's *Medical Tribune* and *Medical World News*, presented journalistic accounts of recent medical advances and subsisted on a base of subscribers and advertising income. The house organ, however, was entirely devoted to pharmaceutical promotion. Produced in-house and supported by a public relations budget, these journals mimicked peer-reviewed medical journals in style and presentation and were distributed free to physicians across the country. By 1945 most drug companies had house organs, though they varied in quality, size, and tone.[47] Like other aspects of pharmaceutical promotion, house organs walked a delicate line between education and salesmanship, but in general the editorial responsibility appears to have been taken quite seriously, and the tone of the articles was often therapeutically conservative.[48]

Merck Sharp & Dohme's house organ, the *MSD Seminar-Report* (formerly the *Merck Report*), was one of the oldest and most venerable of the genre: it had been published since the nineteenth century, and its editor, Charles Lyght, received numerous medical editing awards for his handling of the journal and *The Merck Manual*. Diuril and hypertension became focal topics for the *Seminar-Report* for much of late 1957 and early 1958. In addition to several glossy, full-color, multipage advertisements for Diuril, the *Seminar-Report* contained an array of articles on hypertension as a general topic and chlorothiazide in particular, including three cover articles in 1958 specifically related to hypertension.[49] These articles were typically authored by eminent academic clinicians and constituted a significant, if costly, means of raising physician awareness of both drug and disease.[50]

The Expanding Gaze of Market Research

To test this promotional framework, the market research wing of Merck Sharp & Dohme (known as the Economic Research Area), conducted a study during the fall and winter of 1957. In the last two weeks of December, a team headed by Edward J. Carroll conducted five hundred interviews with a panel of physicians carefully selected "in such a manner that we believe the answers reflect the knowledge and attitude of the one hundred twenty thousand physicians who will be writing prescriptions for this type of drug."[51] The sample was chosen to be proportionally representative of generalists and specialists and urban and rural practitioners, with a geographic spread representing the known distribution of physicians in the continental United States. Nearly half of the sample group had already heard of Diuril or chlorothiazide, even though the drug was not yet available on the market.

Carroll and his team were not alone in canvassing physicians for their response to prelaunch publicity; by the late 1950s, market research had become an important division within all major pharmaceutical firms.[52] Pharmaceutical market researchers considered their work to be a rigorous scientific practice; their methods were explicitly delineated and debated in a growing series of conferences, journals, and textbooks over the course of the 1950s.[53] Producing data that was standardized, reproducible, and objective was considered more important than obtaining the answers desired by the marketers themselves; as a result significant internal antagonism often arose between marketing and market research departments.

Whereas marketers of the 1940s had based their distribution and promotional strategies on wholesale figures and publicly available data sets, by the early 1950s new technology allowed them to visualize in finer detail the dynamics of physicians' prescribing habits.[54] The National Prescription Audit (NPA), a month-by-month survey of prescription rates culled from a representative sample of pharmacies, was marketed in 1953 as a rich data set to analyze how marketing strategies were playing out in real time. Meanwhile, the National Disease and Therapeutic Index (NDTI) came to the market in 1956 as a "Nielsen-family" version of clinical decision-making, enlisting a panel of three thousand physicians who used case record diaries to record "basic diagnostic and therapeutic information about all patient contacts made by them during assigned two-day periods of their practices." The NPA gave market researchers a way to track prescriptions by physician market, and the NDTI allowed them to carefully tie those market data to diagnoses and changes in therapeutic practice.[55]

Merck's market researchers purchased data from the NPA and the NDTI, but they supplemented these tools by in-house research. In the case of Diuril, Fred Heath's base of 250 clinical researchers formed a sophisticated test market with which the "naive" population of physicians randomly sampled by Carroll could be compared. In the naive population of physicians, the overall publicity for Diuril had been excellent. However, although almost all of the physicians who recognized the words *Diuril* or *chlorothiazide* knew that the compound was a diuretic, less than half of them had begun to think of Diuril as a treatment for hypertension. "While the utility of the drug as a general diuretic is widespread," Carroll noted, "its usage as an anti-hypertensive agent is not. This fact, we believe, is important in the planning of future promotion . . . It is felt that these specialized usages which 'DIURIL' possesses must be publicized in order to expand the existing market. If this is not done through clinical or other means, we will not be developing the full potential of the drug."[56]

Public Relations and Popular Media

Publicity was only partially controllable, but marketers worked to manage what variables they could. Press kits were prepared with publication-ready illustrations to explain hypertension and diuretics to a lay audience; symposia and press luncheons were planned in detail to maximize media attention.[57] Although the field of public relations had flourished in the first half of the century, many of these practices had only recently spread to the pharmaceutical industry.

In 1953 the American Pharmaceutical Manufacturer's Association (APMA) had commissioned a public relations "primer" that encouraged companies to build institutional relationships with science writers and other specialized journalists.[58] During Diuril's early development, MSD marketers set out to commission popular magazine articles and courted some of the best-known science writers of the day. In October of 1957, Diuril public relations manager George Schott ordered the creation of a Diuril "Research Report" for distribution to science writers for newspapers and newsmagazines. "This is frankly a fishing expedition," Schott explained, "designed to see if any writer may be gathering material for a piece on any of the areas for which 'Diuril' may be indicated. If we strike a responsive chord, we may find ourselves in a piece from which we would otherwise be excluded. Make whatever personal contacts may be desirable." In addition to the general-interest magazines, he argued, it was essential to "promote special stories for publications with limited interests," such as a story for *Lifetime Living* "stressing benefits to Senior Citizens" and a story for *Medical Economics* highlighting the role of the physician as clinical investigator in the development of Diuril.[59]

These efforts achieved considerable success. Paul de Kruif, one of the best-known science writers in the health field—who had written numerous popular articles in addition to his widely read *Microbe Hunters* and his collaboration with Sinclair Lewis on *Arrowsmith*—was engaged to promote chlorothiazide to the popular audience, as evidenced by this memo from Diuril product manager Gordon Klodt in early 1958: "Thank you for supplying me with the information regarding your recent meeting with Dr. de Kruif and Dr. Spies. It appears that an article on 'DIURIL' will be published in the *Readers Digest* as we desire. As you know, *Good Housekeeping* magazine features a section on medicine each issue. An article in this publication on 'DIURIL' could be very helpful in bringing our product to the attention of the average housewife. If we have contacts with the science writers who are responsible for this section in *Good Housekeeping* we possibly should consider contacting them in this regard."[60] *Good Housekeeping* never did publish an article on Diuril, but by April of 1958, *Reader's Digest* and the *Saturday Evening Post* had both published full-length articles, titled, respectively, "New Hope for Overloaded Hearts" and "The Pill with the Built-In Surprise." The former was authored by de Kruif, who rated Diuril as "the biggest medical breakthrough in recent years."[61] The built-in surprise in the title of the *Saturday Evening Post* article referred to Diuril's action as an antihypertensive; the article concluded with similar praise: "On the basis of the number of [prescriptions] being written, the drug, known commercially as Diuril, would be a contender for the title of Pill of the Year, if there were such a thing . . . if no long-range flaw turns up in its surprising and still unexplained action in reducing high blood pressure, many of the 20,000,000 Americans who suffer from that have hit a medical jack pot."[62]

Following publication of the *Saturday Evening Post* article, the magazine's publisher, Curtis Publications—who also produced the *Ladies' Home Journal* and several other large-circulation periodicals—sent an analysis of the article's reception to Merck executives. General readers had rated "The Pill with the Built-In Surprise" as the most interesting article in that issue of the *Saturday Evening Post*, although the article was ranked second among the male audience, behind "Del Crandall: Solid Man of the Braves." Curtis analysts estimated from their sample that more than 7 million people had learned of Diuril through this article.[63] *Reader's Digest* sent a similar memo to Merck in early 1959, suggesting that more than 1.5 million patients who used Diuril each month had likely learned of the drug through de Kruif's article.[64]

Technically, Merck's public relations department was not advertising Diuril, since advertising to the lay public was forbidden by convention of the American prescription drug industry. And yet these magazine articles, along with paid "institutional" advertisements that highlighted Diuril as a "development to showcase Merck research," constituted an acceptable and manipulable public relations operation that could be coordinated with Diuril's development schedule.[65] Lest the company be perceived as overstepping its bounds, however, Merck's public relations department was careful to monitor physicians' responses to media coverage on Diuril. The 1953 APMA primer had counseled that too much pharmaceutical publicity could arouse defensive responses from physicians, noting that "the physician resents pressure from the public for new drugs or methods of treatment."[66] The department quickly created a Diuril news circular, listing all drug-related publicity by radio, television, newspaper, and magazine, which it circulated every month to all salesmen and marketers. Called the *Diuril News Report,* the circular warned salesmen about particular articles that physicians might have recently read. One issue advised, "The prescribing physician sometimes objects to lay articles about drugs he uses in his practice . . . Review the article in the attached copy of CORONET so that you can intelligently discuss it, SHOULD a doctor bring up the subject."[67]

As Diuril's marketers worked with MSD public relations specialists, they sought to popularize and manage public expectations of the scientific and clinical breakthrough that chlorothiazide represented. But they performed these duties within a tightly constrained space, so they would not be perceived as overstepping their responsibilities as ethical drug manufacturers and would not jeopardize their relationship with the physician community. As we will see, Diuril's marketers had far more sophisticated means of getting their message out to physicians.

Ethical Marketing: The Detail Man and the Diuril Man

Since the founding of the American pharmaceutical industry in the mid-nineteenth century, marketing of "ethical" (i.e., prescription) pharmaceuticals had targeted the physician, and not the ultimate consumer.[68] At a time when physicians and pharmaceutical firms were joined in the project of establishing the legitimacy of their own services in contradistinction to quacks, alternative healers, and patent-medicine salesmen, both institutions agreed to codes of ethics forbidding the direct advertising of their services to the general public.[69] In addition to satisfying the regulatory demands of the Federal Trade Commission and, in the wake of the 1938 sulfanilamide scandal, the Food and Drug Administration, prescription drug marketing textbooks stressed the importance of satisfying the "ethical demands" of the AMA. "Unless acceptance is obtained from the medical and dental organizations and unless great care is taken to keep the sales approach definitely on the professional basis," a 1940s text on drug marketing observed, "ill will can be created that can well prove disastrous."[70] Initially, this professional basis referred mainly to factual, text-based journal advertising and direct mailings to physicians, but by the mid-twentieth century, visually complex advertising and the expanded use of the pharmaceutical sales representative became more important means of providing physicians with information about new drugs.[71]

The restriction of marketing to physicians was not a limitation to the industry. Instead, restricting the key decisions of pharmaceutical consumption to a small and well-bounded elite created clear advantages for marketers of prescription drugs. Richard Hull, director of marketing for Smith, Kline & French in the late 1950s, illustrated the benefits in a lecture given to the National Pharmaceutical Council. Placing a full-page advertisement in four medical journals in 1958, he argued, would cost less than three thousand dollars and provide a "very complete, duplicated coverage of their marketplace." Advertising the same material to the lay public in the four most prominent popular magazines—*Life, Time,* the *Saturday Evening Post,* and the *Reader's Digest*—would cost more than eighty thousand dollars and reach only a fraction of the consumer market. In addition to being a smaller target for promotion, physicians represented a visible and easily studied population. "Thanks to the fact that physicians must be licensed," Hull continued, "and because the American Medical Association maintains what has been called the best professional directory service in the world, we have accurate lists of physicians, together with good information on the nature of the individual doctor's practice and his specialty, if any."[72] Marketers used this biographical data to weed out older and retired physicians, focus marketing efforts on the 125,000 actively prescribing physicians, and further subdivide the profession to promote specific products.

Promotional materials for Diuril were designed for an audience of general practitioners as well as specialists in cardiology, internal medicine, and other relevant fields. Physicians licensed in these fields were sent a series of "Dear Doctor" letters and a pamphlet entitled *Information for the Physician on Diuril (Chlorothiazide), a New Diuretic and Antihypertensive Agent,* a technical publication that emphasized Diuril's safety and efficacy and provided a broad annotated bibliography of publications in peer-reviewed journals related to chlorothiazide.[73] Merck's own market research indicated that direct mail was an effective way to reach physicians; this finding would be confirmed by several other market research agencies.[74] And yet marketers recognized the limitations of direct mail as a modality for convincing conservative physicians to feel comfortable using a novel therapeutic agent. In the late 1950s, a growing

literature of pharmaceutical market research began to document the effectiveness of journal advertising and sales representatives in influencing the more
affective dimensions of prescribing. Using various sampling methods, these
studies—published in the trade literature or as monographs—measured the
influence of journal advertising at 9 to 25 percent of prescribing decision-
making and suggested that contact with salesmen provided 31 to 52 percent of
the information physicians used to make prescribing decisions.[75] As medical
journals became increasingly important to drug marketers, the journals themselves changed dramatically: the number of advertising pages per medical journal increased by 34 percent between 1953 and 1958 to make up nearly 40 percent of all printed pages in medical journals; by that time the pharmaceutical
industry accounted for more than $30 million of medical journal advertising
revenue.[76]

Whereas Merck had earlier spent a negligible amount on advertising, the Diuril campaign catapulted its budget for journal advertisements and direct mail
to $5 million a year.[77] Although Diuril was not advertised in the January 11,
1958, issue of the *Journal of the American Medical Association* that heralded the
drug's launch, the following week its advertising campaign began in earnest.
The resulting campaign was polished and penetrating, uniting all mailings, advertisements, and associated trinkets with the iconic figure of the "Diuril Man"
(see fig. 1.2). The Diuril Man was described to Merck employees in a feature article in their in-house monthly magazine, *Merck Review:* "Standing just a shade
under six inches on a tiny platform, the DIURIL man is a transparent figurine,
not quite filled with liquid, showing the heart, lungs, kidneys, ureters, and blad-
der. He has set the theme for the entire campaign, for he emphasizes wordlessly
how an edematous man can actually drown in his own excess body fluid. As a
result, hardly a promotional piece or an advertisement appears without this
picture."[78] As a visual aid in journal advertisements, the Diuril Man presented
an idealized image of physiology, health, and disease; as a six-inch-high desk
trophy, however, he took on added significance as an early example of the gifts
that detail men increasingly provided to physicians as part of their routine promotional activities. Like the Diuril Heart and the Diuril Kidney that would be
presented as gifts to physicians in future years, the Diuril Man could be argued
to serve as a prop for physicians in educating patients about their conditions:
the doctor could point out the organs involved in high blood pressure and the
site of action of chlorothiazide in the kidneys.[79] Perhaps the most ingenious
aspect of the Diuril statuette was its hollow base. When the figurine was in-

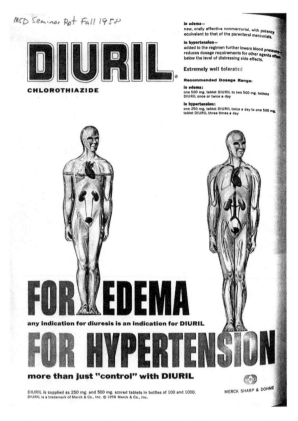

Fig. 1.2. The Diuril Man. *Source:* Diuril journal advertisement, 1958. Courtesy of Merck Archives, Whitehouse Station, NJ.

verted, fluid would fill the body, when he was righted, the excess fluid would slowly drain out of the body back into the base, cleverly simulating both the pathological mechanism of excess body fluid and its therapeutic removal via Diuril.

However much these promotional materials might be used to educate patients, though, their main purpose was to influence doctors to see hypertensive disease in general—and their patients in particular—in terms of Diuril. From Merck's perspective, half of the value of the Diuril Man was that he sat on the doctor's desk, continually radiating the name Diuril and the mechanism of fluid removal to all physicians, nurses, and patients who might pass by or pick him up and turn him over. Later ads for Diuril, which emphasized the accumulated "weight of evidence" (see fig. 1.3), clearly made a visual pun about the role of the Diuril Man as a paperweight commonly found in physician's offices.

Another illustration that appeared around the same time was a cartoon reprinted in the *MSD Sales Dispatch* (see fig. 1.4), in which a confused physician sees a Diuril Man (wearing pants) appear in his office instead of a human patient. Although it is humorous that the beleaguered doctor asks, "Haven't I seen you somewhere before?" the humor of the cartoon wryly underscores a sense that physicians were learning to see their patients in terms of the promotional imagery of the drug.

In addition to its function as an advertising surface, the Diuril Man served an equally significant symbolic function as a gift. By the late 1950s, a unidirectional gift economy had become an important part of the relationship between

Fig. 1.3. The Diuril Man. *Source:* Diuril journal advertisement, 1959, reproduced from the *Journal of the American Medical Association.*

Haven't I Seen You Somewhere Before?

Fig. 1.4. The Diuril Man as patient. *Source: MSD Sales Dispatch,* October 1958, 10. Courtesy of Merck Archives, Whitehouse Station, NJ.

pharmaceutical sales representatives and the physicians upon whom they called. This is not to say that physicians were bribed to prescribe drugs for the low price of a pen, a pad, a mug, or a desktop gimmick. Rather, marketers focused on the value of the gift in establishing a relationship between the physician and the salesman—and, by extension, the company and its products.[80] Merck salesmen themselves began to identify with the Diuril Man; for example, in retired MSD salesman Larry Clarke's self-portrait, his image as a Merck salesman is communicated by the Diuril Man that he holds on his lap and the Diuril Heart and the Diuril Kidney in the background. One cover of *MSD Sales Dispatch,* the company's in-house publication for sales staff, featured an enlarged Diuril Man holding a detail man's sample bag (see fig. 1.5). Thanks to a well-pitched combination of product innovation and promotional presentation, wrapped tightly in an integrated campaign of multimedia promotion, the drug appeared to sell itself.

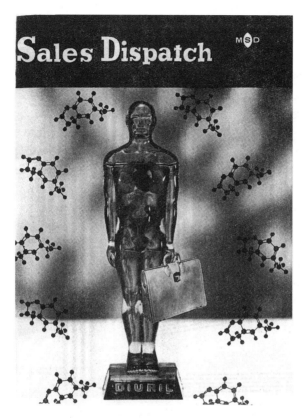

Fig. 1.5. The Diuril Man as detail man. *Source: MSD Sales Dispatch,* April 1958, front cover. Courtesy of Merck Archives, Whitehouse Station, NJ.

1958 as End and Beginning: Pharmaceutical Marketing before and after Kefauver

For the most part, physicians responded favorably to the advertising and publicity surrounding Diuril. In its first year of release, 13 million prescriptions for Diuril were written, and in 1958 MSD recorded $20 million in Diuril sales, making it the most financially successful drug launch the firm had yet seen. Whereas the entire market for diuretics had amounted to $7 million in 1957, a stunning $25 million worth of prescription diuretics were sold in 1958. By 1959 MSD's prescription volume had more than doubled and the company moved from thirteenth to fifth in size among American drug firms; in that year every other prescription written for a Merck drug was written for Diuril.[81] By 1961 Diuril was one of twelve drugs with annual sales over $10 million, described by industry analysts as the "ethical drug hit-parade." Diuril had not just done well; it had, in the words of one analyst, "create[d] its own market."[82] As we will see in the next chapter, hypertension was crucial to that lucrative and newly fabricated market, for Diuril's launch took place in the midst of a fundamental debate on the expanding diagnosis and treatment of hypertension.[83]

Diuril not only represented the first successfully developed and marketed product for the newly merged Merck Sharp & Dohme; it represented at the same time a final instantiation of the practice of pharmaceutical promotion in the atmosphere of relative public goodwill that preceded the maelstrom of the Kefauver investigation. In early 1959, as MSD's accountants were tallying the record-breaking first-year sales of their flagship drug, Senator Estes Kefauver addressed the nation with a list of searing indictments against the pharmaceutical industry, including charges of price-fixing and promotional misconduct at the highest levels. Over the next two years, executives from all major pharmaceutical companies were brought before his committee, and their putative ethical and professional shortcomings were broadcast on televised nightly news across the nation. For an industry that had, on the whole, enjoyed uncommonly good relations with the general public, regulatory bodies, and the medical and allied health professions since the end of World War II, the withering critique and exacting public scrutiny focused upon it was a rude surprise.[84]

Kefauver's Senate hearings anticipated the growing influence of consumerism in late 1950s and 1960s political culture, a significant departure from the more probusiness mood of the early Eisenhower years.[85] Although the hearings would investigate all aspects of the industry, including research, manufacturing, safety, and efficacy, the senator's initial motivating question dealt with price: how had prescription drugs become so profitable—and so unreachably expensive—in the past decade? Why, after a period of dazzling accomplishments in the development of therapeutics, did the majority of brand-name drugs remain unaffordable to a large part of the population? Kefauver openly suggested that the answer lay in a set of overly protective patenting and regulatory practices that had encouraged a few large companies to monopolize the industry, collude to maintain high prices, and spend a substantial and growing amount of their income on promotional and branding tactics.[86]

Over the course of the hearings, the promotional methods of the pharmaceutical industry became a publicly debated phenomenon in a manner unseen since Samuel Hopkins Adams's muckraking critiques of patent-medicine advertising, which had culminated in the passage of the 1906 Pure Food and Drug Act.[87] The actors whose contributions to pharmaceutical promotion we have just seen in the mechanics of Diuril's launch—the salesman, the marketer, the market researcher, the director of public relations, the product manager—were now introduced to the general public as seeds of commercial corruption within everyday medical practice. Ironically, although the reputation of its cast of supporting characters suffered greatly in the process, Diuril itself made it through the hearings unscathed. The drug would eventually become a rhetorical rallying point for the besieged industry as it regrouped to defend itself against the waves of negative publicity that followed Kefauver's accusations. Caught off guard by the senator's masterful use of the relatively new media of nightly television news, the industry scrambled for a spokesperson to proclaim the public good of the pharmaceutical industry and a platform to stand on. A central argument that the industry would eventually deploy was found in an earlier speech by John T. Connor, the president of Merck Sharp & Dohme.

On November 6, 1958, at a conference on the perceived "Soviet economic offensive" held by the American Management Association, Connor had described what he called "an early skirmish" in the biopolitics of the cold war. The setting of the skirmish was the Indian pharmaceutical market, a cold war hot spot highly contested between the internationally expansive American pharmaceutical industry and the Soviets, who had proposed to financially help establish a state-run Indian pharmaceutical infrastructure. The protagonists of Connor's tale were two officials of Merck Sharp & Dohme International who gradually, and successfully, led the Indian government away from its initial inclination to accept financing from the Soviet Union. According to Connor, Diuril was central to these negotiations, as Merck officials insisted that the novel drug and similar pharmaceutical breakthroughs would become available to the Indian people only if the Indian government chose to invest in a relationship with a profit-oriented, research-based pharmaceutical industry rather than with a socialist state program. This episode was described by Connor as an early battle against "the coming Russian drive on the other front—the war against disease." Connor portrayed the socialist health system as a vast army and warned that "when this well-staffed army sallies forth from its borders—as it will—carrying the nostrums of Communism in its medical kit, it will have a proposal to make that could be quite appealing. Reorganize your state along our lines, the proposal would go, and you, too, can do what we did—make the fastest progress in health achieved by any large nation in modern times."[88]

As the Kefauver investigations gained intensity, Connor publicly contrasted Sputnik and Diuril, arguing that a parallel "pharmaceutical race" was being conducted in the early cold war and that its significance was equal to that of the space race that had been frenetically reported in the popular press since the launch of the satellite earlier that year. "Many of our citizens are balking right now at the price we would be forced to pay to catch up with the Soviet Union's rate of discovery in the race for space," Connor noted, and "the reason why the Soviet Union has allowed itself to be blanked out completely in the international competition for new drugs is because it did not want to foot the bill. The United States has now drawn so far ahead in pharmaceutical research that it will take the Soviets more than ten years and more than a billion dollars to close the gap."[89] Countering Kefauver's accusations that the prices of prescription drugs were grossly elevated relative to the costs of raw materials, Connor insisted that drug prices were actually a *bargain* for the American consumer, citing the two decades of research and development that were required to make Diuril available to Western markets.[90] As Diuril became a potent symbol of American victories in the pharmaceutical cold war, Connor's talks were reproduced in industry trade publications and reprinted as pamphlets; he received letters of support and requests for speaking engagements.[91]

His central argument would be appropriated as the archetypal defense of the industry's pricing and promotional policies and as the Pharmaceutical Manufacturers of America's official stance on regulation and investment in innovation. PMA president Austin Smith began in 1960 to introduce Connor's cold war imagery into his stump speeches on behalf of the drug industry, here paraphrased by the *Oil, Paint, and Drug Reporter*: "The Russian welfare state is a 'farewell' as far as pharmaceuticals are concerned; the Soviets, since their revolution forty-two years ago, have not made a single major drug discovery. On the other hand, the United States, under free and aggressive competition, has placed more drugs in the hand of the average man than all the government-controlled systems of all other countries combined. Such, in essence, is the defense of the American drug industry as formulated by Dr. Austin Smith, president of the Pharmaceutical Manufacturers Association."[92]

The argument was echoed by Vannevar Bush—head of the Office of Scientific Research and Development during the World War II penicillin effort and subsequently chairman of Merck's Board of Directors—when he was questioned by the Kefauver panel as to why postwar pharmaceuticals should be sheltered by patent protection when penicillin had received no such protection. Bush explicitly invoked Diuril in his response, noting that if a patent for penicillin been provided, as it was for Diuril, the American people would have had penicillin ten years earlier. Diuril was a shining example of the value of free competition in producing innovation, a vital weapon in the cold war.[93]

Partly as a result of Connor's speech and the alignment of cold war politics it invoked, Kefauver's proposed regulatory reforms were deadlocked in Congress by late 1960.[94] In 1962, with public concern revived by the thalidomide scare, Congress passed a measure that had very little to do with Kefauver's original aims of price reduction and promotion-taming, focusing instead on increasing the regulatory barriers involved in getting a product to market. Paradoxically, this move seems to have subsequently helped the established pharmaceutical industry consolidate its interests and more effectively police its boundaries against newcomers. Moreover, the extended approval process mandated by the 1962 Kefauver-Harris Amendment has become instrumental in legitimating subsequent industry arguments for the necessity of higher drug prices.[95]

Conclusion: Detaching Diuril from Hypertension

In spite of Diuril's positive image, the overall effect of the Kefauver investigations tarnished the industry and made companies much more sensitive about image management, particularly regarding marketing tactics. Industry discussions of pharmaceutical marketing research and strategies largely went underground. After the Kefauver hearings used articles from Upjohn Company's own in-house sales magazine as evidence against the company, Upjohn ceased production of the journal, in part to prevent further potentially embarrassing texts available to congressional subpoena. Before, there had been a mostly respectful silence in the popular press regarding the financial interest of the prescription drug industry, but the Kefauver hearings prompted investigative journalists such as Morton Mintz to develop a new pharmaceutical muckraking literature with popular book-length accounts such as *Prescription for Profit* and *Medical Nightmare* and critical newspaper and magazine articles. The clinical literature had previously not commented about the propriety of promotional relationships between drug companies and physicians, but after the Kefauver publicity, a growing critical literature emerged in that venue as well.[96] It was no longer acceptable for public figures to hold prominent membership in both the AMA and the PMA, such as Austin Smith had held in 1958.[97] The probusiness environment of the Eisenhower administration had given way to the more proconsumer climate of the Kennedy years.

Although the hearings brought increased national scrutiny of the industry's marketing practices—criticism that would be amplified in future years with further Senate investigation in the late 1960s and Senator Edward Kennedy's hearings on "doctor-bribing" in the late 1970s—the basic scaffolding of pharmaceutical promotion utilized in the Diuril campaign has been preserved. With a few minor modifications such as stricter regulation of gifts, the refinement of marketing data provided by external market research firms such as IMS Health, and changing media of pharmaceutical promotion, the core set of structures exhibited by the Diuril campaign remains today as a template for the promotion of new medical products.

The proof of this continuity lies, ironically, in the subsequent submergence of Diuril and the other thiazide diuretics in the decades following their initial brand-name glory, due to the emergence of newer antihypertensive agents that utilized precisely the same promotional structure. As we saw earlier, Merck had initially promoted Diuril as the first *specific* antihypertensive drug, a magic bullet of vascular tone, and a popular magazine article proclaimed soon after its launch, "[A] fascinating fact about Diuril is that it also performs a mysterious selective action. It will lower blood pressure only when pressure is abnormally high; *it will not lower normal blood pressure.* How Diuril does this is not known. When it is explained, doctors will perhaps be close to the secret of the kidney–high-blood pressure relationship."[98] The very existence of this physiological action held out hope that somewhere in the mysteries of chlorothiazide's pharmacological fate lay the specific mechanism that produced hypertension. The year of Diuril's launch, the *Proceedings of the Council of High Blood Research* devoted an entire journal to "Drug Action," focusing largely on chlorothiazide.

But Diuril would not hold the limelight for long. As hope thinned that Diuril's specificity of action would uncover a unitary mechanism of hypertension, newer therapeutic agents offering more promising molecular explanations and longer patent periods began to dominate the scientific literatures and promotional efforts. Already by 1960 the introduction of two centrally acting catecholamine blockers, Ciba's Ismelin (guanethidine) and Merck's Aldomet (alpha-methyl-dopa), had begun to direct attention away from the kidney and toward the sympathetic nervous system as a site of intervention for hypertension. Other agents soon followed. By the end of the 1960s, as Diuril advertisements depicted chlorothiazide as an old and trusted companion to the practicing physician, these newer generations of antihypertensive drugs competed to claim their own molecular insights into the fundamental mechanism of hypertension. Later decades would bring subsequent avatars of rational drug design: beta-blockers, calcium channel blockers, angiotensin-converting enzyme inhibitors, and angiotensin receptor blockers—each accompanied by its own outpouring of promotional materials.

As these newer drugs displaced Diuril and the other thiazides as first-line antihypertensive agents, their manufacturers made use of precisely the same promotional tools that had offered Diuril's message to the prescribing physician: symposia, journal articles, sales representatives, journal advertising, and public relations. Moreover, in the 1980s competing pharmaceutical firms began to fund and promote studies suggesting that chlorothiazide and other generically available thiazide diuretics bore additional risks to patients—such as elevated blood cholesterol, cardiac arrhythmias, and diabetes mellitus—which the newer, more expensive antihypertensive medications did not.[99] Although most of these negative claims were clinically insignificant and did not bear up under long-term clinical evaluation, they were consistently used in the promotional materials of newer antihypertensive agents, to the detriment of Diuril and the other thiazides.[100]

Ironically, the recent publication of the largest comparative clinical trial to date, the National Institutes of Health's Antihypertensive and Lipid-Lowering Therapy to Reduce Heart Attacks Trial (ALLHAT), publicly vindicates thiazide diuretics as the most efficacious and appropriate first-line antihypertensive therapy for most patients, though their actual prescription rates now lag far behind those of newer, more heavily advertised drugs.[101] The examination of Diuril's career, then, leaves us with some unsatisfying ambivalence toward the drug promotional process: while we should applaud the fact that Diuril was launched into the world so effectively, it is clear that the same efficient machine of promotion was instrumental in the subsequent decline and neglect of the thiazide diuretics once they ceased to be a financial priority for the industry. We must grasp this irony to understand the dual nature of drug promotion as a process rooted in both education and salesmanship, a process that has since become an essential aspect of the circulation of knowledge and changing practice in American medicine.

As we have seen, during the course of Diuril's development, an unforeseen interaction with hypertension launched Merck Sharp & Dohme into the business of producing agents for the prevention of heart disease, transforming an agent initially best suited for hospital use into an agent broadly useful to an unexpected outpatient population. This transformation was neither immediate nor complete; it was, however, contingent on the interaction between drug and disease in the multiple spaces of clinic, laboratory, and marketplace. At the same time, the story of Diuril serves as a uniquely situated historical example of how, in the years preceding the Kefauver investigation, pharmaceutical marketing had already become fundamentally involved in the development of therapeutic agents and the conceptualization of disease. While Diuril's definition was changing from diuretic to antihypertensive, a parallel but related shift was occurring in the definition of hypertension as a disease.

CHAPTER TWO

Shrinking the Symptom, Growing the Disease

Hypertension after Diuril

> An important remaining question is, at what level of blood pressure should one begin treatment? If the life insurance statistics are heeded, perhaps treatment should be instituted at diastolic levels presently regarded as normal. —EDWARD FREIS, 1967

Physicians around the country were invited, in 1968, to mark Diuril's tenth anniversary with a brief birthday celebration for the drug. A new desktop model of the Diuril Man was released in conjunction with a blitz of journal advertisements proclaiming, in glittering metallic ink, the historic role of chlorothiazide in combating heart disease (see fig. 2.1).[1] Local receptions with Diuril birthday cakes were held, and early in the year every registered physician in the United States received a slim volume entitled *A Decade with Diuril,* which proudly outlined the expanding usage of the drug in its first ten years.[2] "Whatever the next ten years may bring," one of the volume's contributors proclaimed, "chlorothiazide deserves a place of honor in the annals of antihypertensive therapy as the first of a series of effective oral diuretics that have immensely facilitated the medical treatment of hypertension."[3]

Meanwhile, back at the Merck Sharp & Dohme Research Laboratories (MSDRL), Max Tishler was working hard to crown Diuril's birthday with more glittering finery: the coveted Lasker Prize, regarded as the American equivalent of the Nobel in clinical medicine. The Lasker Prize originated from the efforts of Mary Lasker, a prominent Washington saloniste who, aided by considerable

Fig. 2.1. Diuril's tenth anniversary. *Source:* Diuril journal advertisement, 1968. Courtesy of Merck Archives, Whitehouse Station, NJ.

social skills and a fortune amassed by her husband's advertising career, became an expert in mobilizing politicians, researchers, and lobbyists to expand the federal government's efforts to fight heart disease and other chronic conditions. In a series of letters to the prize committee and to Mary Lasker herself, Tishler detailed the accomplishments of MSDRL researchers in developing Diuril, adding, coyly, that receipt of such honors would be a perfect tenth birthday present for the Lasker Committee to bestow upon the drug and its developers. The committee did not agree. In a series of letters of decreasing politeness, they reminded Tishler that however momentous Diuril's historic role had been thus far, the Lasker Prize Committee was not inclined to make its decisions based on the convenience of Merck's marketing schedule. When the MSD researchers who developed Diuril were eventually crowned with the Lasker award in 1973, the honors were several years too late to aid in the promotion of the now-generic drug.[4]

It is nonetheless fitting that the Lasker Committee waited until 1973 to recognize the historic significance of Diuril, because in the intervening years the

Veterans' Administration (VA) Cooperative Study Group on Antihypertensive Agents conducted the first long-term randomized trial documenting the benefits of antihypertensive therapy in moderate hypertension—with Diuril as a key drug in the study regimen. Edward D. Freis, a VA cardiologist, a Diuril clinical researcher, and the director of the VA study, also won a Lasker award in the process. Freis's prize citation noted, "His recent contribution has been the definitive study and demonstration of the fact that even moderate hypertension is dangerous, and should, and can be treated successfully . . . It is an exemplary demonstration of the potential of preventive medicine for saving and prolonging the lives of tens of thousands of Americans."[5]

Diuril's launch had taken place in the midst of a fundamental debate on the diagnosis and treatment of hypertension.[6] The emergence of specific therapeutics with demonstrated ability to lower pressure—as well as a significant set of adverse effects—demanded a pragmatic consensus about which patients had a true *disease* that merited treatment and which had merely a blood pressure measurement that was above average. As the question of who to treat began to trump the question of what was normal, Diuril became materially involved in altering the definition of hypertension in America, helping to transform a degenerative and symptomatic condition into a symptomless and treatable category of risk.

This transformation did not occur overnight. Even by the time of Diuril's tenth birthday, several prominent cardiologists continued to insist that the broad medical treatment of symptomless hypertensive patients was itself unethical, "a huge uncontrolled clinical-pharmacological experiment . . . masquerading as a clinically acceptable therapy."[7] At stake in the debates of the late 1950s and 1960s over mild-to-moderate hypertension was the emergence of a new paradigm of pharmaceutical prevention for chronic disease, the identification of symptomless precursor states that then became viewed as diseases in their own right. In the life span of Diuril, the disease of hypertension was disengaged from its symptoms and redefined in terms of numerical thresholds, expert committees, and clinical guidelines.

As anthropologist Claude Lévi-Strauss has famously noted, objects often take on importance in the life of society that extends beyond their immediate utility precisely because they are "good to *think* with." Diuril not only altered the options available for the treatment of hypertension but also changed irreversibly the tools available to think "hypertension" with. By making antihypertensive therapy a sweeter pill to swallow, Diuril lowered the threshold for the prescription and consumption of antihypertensive medications, enlarged the population of potential hypertensive patients in both clinical trials and clinical practice, and contributed to the consolidation of a single threshold for the definition of hypertension. Its discrete oral form of administration lent itself easily to the outpatient setting and to the developing methodology of large-scale, multisite randomized clinical trials for chronic conditions. Data produced by these trials engendered a positive feedback cycle that allowed more physicians to diagnose hypertension with confidence and enroll more patients in therapeutic programs and further clinical trials.[8] Hypertension after the publication of the VA study would become a category incommensurate with the hypertension that came before.

Prior therapeutic limitations had already limited prevailing conceptions of pathology and normality. Diuril presented a pragmatic opportunity to transform this cycle from a negative, mutually nihilistic relationship to a positive, mutually reinforcing and potentiating one. The career of Diuril connects the role of clinical experience, clinical trials, and pharmaceutical promotion in expanding the widespread acceptance of asymptomatic hypertension as a treatable disease and illustrates the growing importance of postmarketing research in shaping drug and disease in the period after World War II.

From Sign to Disease

By the time of Diuril's launch, high blood pressure had become more than a sign of disease. It was increasingly regarded as a disease in its own right— known as "primary" or "essential" hypertension—though its definition and practice guidelines varied widely. Even the consensus that high blood pressure was a pathological state—and not, as formerly thought, a purely adaptive mechanism to aid a weakened heart in squeezing blood through hardened tissues—had only recently been agreed upon. As late as 1931, Paul Dudley White, representing the mainstream of American cardiology, had written that "hypertension may be an important compensatory mechanism which should not be tampered with, even were it certain that we could control it."[9] In the same year, a prominent British cardiologist also announced that "the greatest danger to a man with high blood pressure lies in its discovery, because then some fool is certain to try and reduce it."[10]

A concern with the prognosis and treatment of "hard pulses" had been a part of Western medical practice since Celsus—and can be been traced back within the Chinese medical literature to manuscripts dating around 2600 BC. The quantitative measurement of blood pressure, however, has a more recent origin, in Stephen Hales's eighteenth-century experiments on animals.[11] Clinical measurement of blood pressures was not widespread until the twentieth century; throughout the nineteenth century, blood pressure measurement was a largely experimental process carried out with invasive needle-in-the-artery devices such as the mercury hydrometer and the direct manometer. Only after Nikolai Korotkoff's 1905 popularization of the auscultatory method—a minimally invasive technique that used the recently developed Riva-Rocci inflatable cuff instead of an intra-arterial needle to measure pressure—did blood pressure became widely and pragmatically available as a clinically measurable entity. Shortly after the development of this portable, low-impact technique, large population studies of blood pressure began to accumulate documenting normal tables of blood pressure and the extremes of high and low.[12]

Nowhere was this practice more efficiently pursued than in the American life insurance industry, to which blood pressure—along with other measurable and graded populational variables such as height, weight, and age—offered a chance to quantify and consequently rationalize the actuarial risk present in apparently healthy populations. The life insurance industry had only recently begun to require all applicants to have a medical examination (a movement closely linked to the origins of the annual physical exam in primary care), and insurance examinations rapidly became a vehicle through which the sphygmomanometer found its way into physicians' offices nationwide. In 1905 John Walton Fisher, the medical director of the Northwestern Mutual Life Insurance Company, began to require blood pressure measurements in all examinations of applicants for insurance; similar policies soon became standard across the industry. By the end of the decade, Fisher announced that actuarial data on the high mortality of hypertensives had become convincing enough that Northwestern would no longer insure individuals with a systolic blood pressure of 170 mm Hg or higher. By the early 1920s, Fisher's work had established asymptomatic hypertensives as a "high-risk" group for all life insurance agencies; in the following thirty years, his initial research would be confirmed and highlighted by a series of larger and larger actuarial studies.[13]

It is important to recognize here that the risk Fisher wrote of was financial and not clinical. Although many physicians knew of the population-based correlation of blood pressure with mortality, this quantification of risk was understood to be significant only in establishing population-wide insurance premiums and was usually considered irrelevant when it came to the individual patient. At the heart of the disjuncture was the implied distance in intention and ethical responsibility between the physician and the actuary: the actuary's responsibility lay not in determining the causal basis of disease in any individual but rather in finding useful markers for reducing financial risk over an entire population of policyholders. Measurements of blood pressure had a definite prognostic function on the level of the population, but such prognosis was far less certain on the level of the individual, and it was seen as poor medicine to directly apply actuarial conclusions to one's clinical practice. "Medico-actuarial insurance studies have contributed much to various phases of hypertension," one review insisted as late as 1942, "but it has impressed us forcibly that insurance medicine and the actuaries have considered hypertension not as a disease or symptom of a disease with a natural and largely predictable course, but arbitrarily as merely a question of numerical units of blood pressure."[14]

In addition to the disjuncture between population-based studies and diagnosing disease in an individual, considerable controversy existed surrounding the definition of disease in terms of number without visible sign or symptom of pathology. In a 1916 prize-winning essay on blood pressure, one physician stated that hypertension is "not an illness, but merely evidence of it; not pathological in meaning, but rather a physiological, mechanical adjustment to an unknown diseased condition; not a true, but a sphygmomanometric disease."[15] The term "sphygmomanometric disease" implied that false confidence in the precision of clinical instrumentation could produce a cult of false diseases created by medical technologies themselves, much like the late-nineteenth-century diagnosis of fever, which developed in proportion to the clinical use of the precision thermometer and the spreading use of antipyretic drugs. One critic of numerical hypertension noted in 1926: "As in other fields of medicine drugs should be the last to receive consideration in the treatment of hypertension. To lower blood pressure by such means is like lowering fever by means of antipyretics. In both cases one only treats a symptom the cause of which one does not understand and the elimination of which by such means usually does more harm than good. In both cases one's efforts can only be very temporary and one's results are often negligible if not entirely futile."[16] Like the sphygmomanometer, the precision clinical thermometer had made possible a measurement-oriented definition of fever as a disease in its own right that had dominated therapeutic practice in the late nineteenth century before being displaced by specific bacteriological etiologies by the turn of the century.[17]

Ironically, the same microbe-driven nosology of specificity that demoted fever from a disease in itself to merely a pathological sign would prove crucial in the elevation of high blood pressure from pathological sign to self-evident disease. Using the tools with which bacteriologists had recently collected case histories of symptomatic patients around individual microbes to establish specific infectious disease etiologies, physiologically oriented physicians used clinical measurement of blood pressure to organize case histories of patients with symptomatic high blood pressure. Once they defined discrete hypertensive syndromes, they attempted to trace backward from the numbers to find the pathological manifestations—and, they hoped, the etiological mechanisms—associated with them. Theodore Janeway, a New York clinician who shared a large medical practice with his father, began to collect blood pressure measurements for all of his patients to describe what symptoms were common in high-blood-pressure states. In 1913 he published an influential description of the "hypertensive syndrome" based on symptomatic manifestations that he attempted to correlate with overall mortality. This line of research was continued by others seeking to delineate the natural history of symptomatic high blood pressure, and in the 1930s hypertension was described by several authors as a disease associated with a set of symptoms including headache, nervousness, fatigability, irritability, and dizziness.[18]

These symptomatic manifestations of high blood pressure were nonspecific, however, and many researchers complained that the early symptoms of arterial hypertension could not be distinguished from the insignificant aches and pains of the general population. Clinicians and investigators strived to define more objective clinical signs of hypertension, a project that met with its greatest success in the area of ophthalmic lesions. A collaborative effort between Norman Keith, an internist, and Henry Wagener, an ophthalmologist, delineated a set of identifiable lesions visible on the retinas of hypertensive patients using an ophthalmoscope; these could be ranked into four grades that correlated with differences in mortality rates. Using an ophthalmoscope to objectively study pathological changes was considered by many physicians to be a more convincing study of hypertension-as-disease than the data provided by the actuarial statistics of the insurance industry.[19] For decades after their original 1939 publication, these "Wagener's grades" were considered the firmest clinically measurable sign of hypertensive disease.

To be fully accepted as a disease category, however, hypertension still needed a credible etiological mechanism. As clinical and laboratory investigations into the nature of "essential" hypertension increased in the 1930s and 1940s, the diversity of etiological hypotheses soon spiraled out of control, stretching beyond cardiovascular and renal hypotheses to include endocrine, neurological, psychiatric, genetic, and social etiologies. A unitary cause was crucial to establish "essential hypertension" as a unitary disease, and yet the category continued to prove resistant to any single mechanistic explanation. Just as psychiatric researchers, searching for a mechanistic cause for psychiatric disorders, tended to whittle away the territory of their own subject with each discovery, so did researchers searching for a primary cause of hypertension tend, upon each discovery, to relegate their findings to secondary hypertension. Those who argued that the seat of hypertension lay in the kidney would, upon successfully demonstrating renovascular disease as a cause of hypertension, effectively remove renal disease as a possible cause of essential hypertension.[20] Those in favor of an adrenal-hormonal cause of high blood pressure, by linking hypertension to hyperadrenalism via pheochromocytoma or Conn's syndrome, similarly acted to add such categories to the list of secondary hypertensions.[21] The etiology of essential hypertension remained an elusive goal, ever receding from those who pursued it.[22]

The ideal of a unitary mechanism of essential hypertension found perhaps its ultimate avatar in the writings of Richard Platt, who maintained in a 1947 article entitled "Heredity in Hypertension" that humanity could be divided into two genetic pools, one normotensive and one hypertensive, whose body types deployed distinct physiological mechanisms for maintaining vascular tone. Platt's research on the distribution of blood pressure in the population produced a bimodal curve that supported his two-population hypothesis; a mechanism to explain this difference, he insisted, could not be far behind. In a well-publicized interchange with his foremost critic, George Pickering, Platt continued to argue for a unitary etiology of essential hypertension as late as 1959: "Pickering rightly says that arterial pressure is the resultant of a large number of variables . . . This does not argue at all against the possibility that only one of these variables is disturbed in a certain defined group of hypertensives . . . a similar state of affairs (as yet undiscovered) could account for essential hypertension."[23]

From Normal Pressure to Treatment Pressure

Platt's position, however, suffered in the long debate with George Pickering and eventually gave way.[24] By the late 1950s, with a consistent central cause of essential hypertension remaining elusive, Pickering's proposal that hypertension might *never* be found to be a consistent pathological species—along with Irvine Page's theory of hypertension as a multifactorial mosaic rather than a mechanistically singular disease—slowly came to occupy the mainstream of medical thought.[25] Pickering replaced Platt's bimodal curve with data suggesting that the blood pressures in a given population followed a normal Gaussian distribution, with the pathological phenomenon of hypertension simply reflecting the right-hand side of the bell-shaped curve. This view came to be increasingly supported by the graded mortality curves collected by life insurance companies.[26] If the distribution of blood pressure in a population was continuous, any attempt to define a numerical limit—for example, 140 mm Hg systolic pressure, or 90 mm Hg diastolic pressure—as a dividing line between normal and pathological was bound to be arbitrary and theoretically unsatisfying. Pickering criticized overreliance on what he called "the fallacy of the dividing line" between normotension and hypertension. "The practice of making a sharp division between normal and pathologically high pressure is entirely arbitrary and is in the nature of an artifact . . . In fact, arterial pressure seems to behave as a graded characteristic: the differences between the lower pressures and the higher are quantitative, not qualitative; they are differences of degree, not of kind."[27] By 1956, Pickering noted, a number of different schema had arisen to delineate normotensive from hypertensive (table 2.1).

These thresholds were multiple and contested and originated from a variety of theoretical and methodological positions. Some were based on extrapolation from physiological experimentation, others came from studies of mortality and symptomatic disease, and others (like Gallivardin's) were meant to inform health standards of life insurance companies. Robinson and Brucer derived their thresholds from standard deviations from the arithmetical mean of a population.[28] A patient with a blood pressure of 150 over 95 might be hypertensive on one system and normotensive on another. But until the 1940s, the difference was mostly an academic point, because most patients were treated only if they were evidently ill. When specific drugs became available to lower pressure—drugs that also caused significant side effects[29]—the bounding of normal became an immediate clinical concern: when did the unknowable future danger of a patient's elevated blood pressure validate the risk of the treatment at hand?

Until the development of Diuril, any treatment that effectively lowered blood pressure entailed significant bodily risks. As a twentieth-century disease, hypertension attracted numerous twentieth-century cures, including elec-

Table 2.1. Proposed thresholds for hypertension before 1958

Blood Pressure (mm Hg)

Systolic	Diastolic	Author	Year
120	75	Gallavardin	1920
121	74	Robinson & Brucer	1939
124.7	—	Alvarez	1920
130	70	Brown	1947
140	80	Ayman	1934
140	90	Perera	1948
150	—	Thomas	1952
150	100	Hamilton	1954
160	—	Janeway	1913
160 (women)	—	Potain	1902
170 (men)	—		
180	100	Burgess	1948
180	110	Evans	1956

Source: Adapted from George Pickering, *Hypertension Manual: Mechanisms, Methods, and Management,* ed. J. H. Laragh (New York: Yorke, 1973).

trotherapy, radiotherapy, and pyrotherapy, as well as sedatives, nitrates, and psychotherapy. The first modality to generate a widespread sense of therapeutic optimism in the medical community, however, was the development of surgical therapy for malignant hypertension. The surgical approach was rooted in the autonomic-nervous hypothesis of hypertensive disease: hypertension, according to this theory, developed due to the overstimulation or excessive resting tone of the sympathetic, fight-or-flight component of the autonomic nervous system. Initial successes with experimental excisions of sympathetic ganglia led to the safer and more widely popular practice of surgical sympathectomy—the severing of the sympathetic trunk. This technique was indeed effective at lowering blood pressure and reversing the symptoms of malignant hypertension, but it entailed significant risks: spinal sympathectomy was a painstakingly long neurosurgical procedure with frequent complications, and the removal of sympathetic function produced known adverse effects. Even the improved technique popularized by Reginald Smithwick of Boston University in 1940 still required two operating sessions, separated by an interval of ten days, with multiple incisions and retropleural, retroperitoneal, and transdiaphragmatic approaches to resect the sympathetic chain from the middle of the thorax to the middle of the abdomen. The procedure required hospitalization for six to eight weeks, and although Smithwick himself performed twenty-five hundred such operations and his approach was quickly adopted internationally, its use was understandably limited to very severe cases.[30]

In comparison to the surgical approach, medical therapy in the 1940s and 1950s was relatively benign, but it too carried frequent and significant adverse effects. The first of the new postwar antihypertensive medications, the ganglionic blockers, were the pharmacological equivalent of a sympathectomy; they produced blurred vision, dry mouth, difficulty in urination, constipation, a paralytic ileus, and occasional hallucinatory psychosis. Hydralazine, known as a "neutralizer of 'pressor substance,'" inspired more confidence but at initially recommended doses also produced a series of aches and pains: headache, palpitation, tachycardia, and heartburn, as well as a lupus-like syndrome.[31] Rauwolfia compounds (e.g., reserpine), initially hailed as a safe and effective solution to hypertension, were soon found to be of more limited use and associated with an unusual side effect at therapeutic doses: a severe and largely intractable depression, for which reserpine developed a second life as a model for studying depression in animal models. Lastly, the Veratrum alkaloids showed a particularly thin margin between mild hypotensive effect and their more powerful action as emetics, greatly limiting their use.[32]

Initial investigations into these agents were the result of a cooperative effort between the wartime Office for Strategic Research and Development, the Ciba and Merck corporations, and specific academic medical centers that eventually became key sites for investigating hypertension, namely Boston University, the Cleveland Clinic, Philadelphia's Hahnemann Medical College, and the Georgetown Veterans' Administration. This small network of antihypertensive pioneers investigated a stunning number of potential compounds: Between 1947 and 1958, Edward Freis—who trained at the Boston University unit and spent the majority of his career at Georgetown—published sixty-four articles evaluating some thirty potential agents for hypertension, a publication rate of one new study every two months. Irvine Page, a central figure in the Cleveland Clinic's hypertension unit, was equally prolific. A small coterie of therapeutic advocates—including Cleveland's Harriet Dustan, Boston's Robert Wilkins, Hahnemann's John Moyer, and Marvin Moser—maintained a close circle of publication and cross-referenced support.[33] These networks, forged in the development of the early, highly toxic pharmaceutical therapies for hypertension, would later become essential to the spread of Diuril.

One other therapeutic option developed in this period deserves mention: the salt-restricted rice diet, popularized by Walter Kempner in the 1940s.[34] Although seemingly less drastic than the surgical and medical remedies, Kempner's dietetic program represented its own extreme. He organized intensive "boot camps" where hypertensive patients were to live for three months at a time, focusing their daily habits around dietetic awareness. The diet itself consisted mainly of fruit, fruit juice, and rice and was measured to contain 20 g of protein, 5 g of fat, and 200 mg of sodium. The 1956 edition of the *Merck Manual* mentioned the Kempner diet as occasionally effective but judged that "such rigid restriction can rarely be maintained and is a severe ordeal rarely warranting the effort."[35] Pickering regarded the Kempner rice diet as "insipid, unappetizing, monotonous, unacceptable, and intolerable." To remain on the diet, he added, "required the asceticism of a religious zealot."[36]

By 1958, then, many antihypertensive therapies with perceived efficacy had emerged. But Harriet Dustan pointed out in that year, "Even though modern treatment of hypertension is effective, many problems remain. We know so little about the fundamental mechanisms for the disease that our treatments are non-specific; furthermore, they are clumsy for the patients and are often associated with troublesome, and sometimes dangerous, side effects. Because we do not understand the basic mechanisms, we cannot predict the type of treatment which will be effective for a particular patient."[37]

As the question of who to treat began to redefine what was normal, the practicing physician of the 1950s encountered a confusing plurality of hypertensions. For any patient with high blood pressure, the physician was first confronted with an etiologic question: was this patient's high blood pressure a symptom of some other potentially curable process—an adrenalin-secreting tumor, local pathology of the kidney, primary aldosteronism, perhaps—or was it a condition in its own right, essential hypertension? If the latter was true, and hypertension was the primary condition, the next question was one of temporality: did this patient's high blood pressure represent an *acute* event (known as "malignant hypertension," or a "hypertensive emergency") or an insidious and *chronic* illness (known as "benign hypertension")?

The latter category, benign essential hypertension, comprised the largest population and was the center of a good deal of controversy. Depending on the mode of diagnosis, chronic hypertension could be classified as severe, moderate, or mild, a distinction that roughly correlated with diagnostic presentation of symptom, sign, or number. Consider, for example, this excerpt from a 1956 medical handbook:

A. *Severe:* Papilledema or soft exudates in the fundi, cardiac failure or disabling dyspnea, disabling coronary insufficiency, repeated cerebral thrombosis with neurological sequelae, rapidly advancing diastolic hypertension with progressive left ventricular hypertrophy.

B. *Moderate:* Signs of left ventricular hypertrophy, arteriosclerotic changes in the fundi, old cerebral thrombosis with sequelae, easily controlled coronary insufficiency.

C. *Mild:* Diastolic pressure below 125, with minimal or no objective signs of vascular damage in fundi, heart, brain, or kidney.[38]

Presented with a case of chronic hypertension with no secondary cause or evidence of hypertensive emergency, the physician asked: is this patient's blood pressure (a) severe: *symptomatic* to the patient, (b) moderate: invisible to the patient but manifest to the trained eye of the clinician through a series of subtle *signs*—such as retinal changes—of which the patient was unaware but which nonetheless demonstrated material proof of pathological processes, or (c) mild: imperceptible to both doctor and patient and visible only in the *numerical* threshold of the manometer.

By the late 1950s, hypertension as a treatable disease was largely limited to the symptomatic (malignant and severe benign forms only) and treated with potent and dangerous medications.[39] For the patient with no symptoms, there was no clear consensus on how to proceed; indeed, many medical textbooks depicted uncomplicated asymptomatic hypertension as a probability of a disease rather than a disease in itself, suggesting that "hypertension should be *suspected* if the diastolic blood pressure occasionally rises above 90 mm. of mercury. The diagnosis becomes increasingly probable the more often this value is exceeded."[40] Another textbook from the early 1950s emphasized that "too much attention is often paid to the height of the blood pressure, and not enough to the clinical picture as a whole . . . Increased blood pressure, in itself, is not a disease. It is a sign of some underlying disorder."[41] *Your Blood Pressure,* a patient-education text published in 1958, explained that high blood pressure—in and of itself, without symptoms—was not necessarily a disease: "If, as so happens, your high blood pressure was discovered in the course of a routine physical examination and you have no other symptoms, very probably your doctor will prescribe nothing at all—beyond a sensible mode of life. You may continue with your work and, as long as it is not unduly strenuous for your age, with your play. Unless you are heavy, he will say nothing about diets and reducing.

Unless you are high strung and jittery, he will give you no medicines beyond mild sedatives from time to time. Above all, he will tell you, forget your blood pressure and don't keep everlastingly having it measured."[42] Cast into this environment of diagnostic uncertainty, Diuril would prove a crucial catalyst in broadening the definition of "treatable" hypertension.

Diuril and Hypertension, 1958

Diuril's launch, for all its finely tuned promotional effort, was not sufficient to immediately overcome the general sense of diagnostic hesitation and therapeutic nihilism regarding asymptomatic hypertension. Over the course of Diuril's career, however, the threshold of the pathological became tied to explicitly pragmatic concerns of treatment efficacy and preventive benefit. Diuril was empirically and theoretically appealing to those physicians already disposed to treat the asymptomatic patient with measurably high blood pressure. All previous medications for hypertension lowered blood pressure equally in hypertensive and normotensive patients, but Diuril appeared to lower blood pressure only in individuals with high blood pressure, suggesting that chlorothiazide was a specific agent to counter the hypertensive state. Merck Sharp & Dohme marketers encouraged researchers and clinicians to see Diuril as the first true "anti-hypertensive" drug; one "Dear Doctor" letter from the company reminded physicians, "DIURIL is the only hypotensive agent with 'specificity' of action, i.e., it reduces B.P. only in hypertensive patients."[43] Although the exact mechanism was not understood, the promise Diuril held out—of a unifying mechanism legitimating the disease status of essential hypertension—was swiftly adopted.

Diuril was not more effective than other agents, but the mildness of its side effects and its availability in pill formulation made it the most palatable antihypertensive on the market. Furthermore, as an oral medication, Diuril was easy to prescribe; no hospitalization was required for administration, and no testing was required for immediate adverse effects.[44] The effect of chlorothiazide on other antihypertensive agents was equally important: in combination with ganglionic blockers, reserpine, and hydralazine, for example, chlorothiazide had the curious effect of increasing their efficacy while minimizing their toxicities. When Diuril was added to the regimen, dosages of ganglionic blockers could be halved while their efficacy in dropping blood pressures increased. So Diuril was not only a more palatable antihypertensive; it increased the palatability and efficacy of the entire class of antihypertensive drugs.

By 1958 there was a small camp of hypertension researchers who were already convinced that the broad development and use of antihypertensive drugs in mild and moderate cases of hypertension was itself a public health movement with self-evident merit. Many of these individuals, such as Cleveland's Irvine Page and Edward Freis of the Washington, D.C., Veterans' Association Hospital, were in the network of researchers who had conducted initial clinical research on ganglionic blockers, hydralazine, and reserpine. Freis had begun to argue for the mass treatment of asymptomatic hypertension well before he had ever heard of chlorothiazide. In a 1956 guest editorial for the generalist journal *GP,* he argued that the consistent findings of excess mortality in asymptomatic mild-to-moderate hypertensives—culled from actuarial data from the life insurance industry—already presented "a cogent argument for the treatment of hypertension early before vascular damage has occurred" and that the "present reluctance to use more potent and effective hypotensive agents because of their supposed 'toxicity' needs also needs to be reexamined."[45] Freis recognized that the potential for preventive chemotherapy was limited by the perceived toxicity of available cures and understood that only a limited audience would agree with his position; the editorial, he claimed, was a "think piece." By 1958, however, the advent of chlorothiazide allowed Freis to argue strongly in a review article, "It is now possible to reduce blood pressure in many mild and moderate cases with little or no discomfort to the patient."[46]

To Freis and Page and a handful of other enthusiasts, the safety profile of chlorothiazide allowed for more expansive vision and a larger audience. With so little to risk, clinicians could afford to be more liberal with treatment. Edward Freis noted, during a Merck Sharp & Dohme symposium in early 1958, that Diuril's lack of toxicity made it worth using even if the benefit could not yet be quantified: "While it may take 20 years to prove that reduction toward normal of elevated blood pressure in mild hypertension is beneficial, do it now using DIURIL alone (rarely adequate) or in combination with reserpine, veratrum alkaloids or ganglionic blockers, and reduced sodium intake."[47]

Other hypertension specialists echoed Freis's optimism: with such minimal side effects and such potential gain, why *not* treat asymptomatic hypertension? Reviewing the prognostic data on existing drugs, Henry Schroeder suggested that "even though we don't have evidence at hand now which would validate the concept, mild hypertension should be rigorously treated or managed in whatever way is necessary to effect the desired result." From a "philosophical position," A. C. Corcoran maintained, current understanding of high blood pressure as a potentially reversible condition *demanded* early treatment.[48]

Schroeder's and Corcoran's statements were made at the closing panel of the first Hahnemann Symposium on Hypertensive Disease, held in Philadelphia at the end of 1958. Ninety-one researchers in hypertension from across the country convened for the symposium, which included as participants most of the original Diuril clinical researchers as well as proponents of sympathectomy, diet therapy, psychotherapy, and other measures for the management of hypertension. Not all were as enthusiastic. Even several ardent supporters of drug therapy, such as H. Mitchell Perry, took pains to publicly taper their optimism. Perry argued that the risks of treatment justified therapy only in the symptomatic. "The other people I honestly believe we do help," he added, "but I think that we have to keep in mind that we don't have the data which supports the thesis that blood pressure reduction really does help them." Marvin Moser, himself a proponent of therapeutic activism, clarified the point by tying discussion of benefits to the risk of therapy. "Dr. Perry," he clarified, "is talking about treatment with specific antihypertensive drugs that are potentially dangerous or perhaps carry some risk to the patient. What we are trying to get at here is that before we use a drug with even minimal risk, it should be justified on the basis of improved prognosis and knowledge of the natural history of the disease."[49]

As Moser's comments suggest, some level of anxiety over iatrogenesis in the wholesale use of largely unknown novel therapeutic products was common to even the most ardent champions of antihypertensive treatment.[50] Even Freis, perhaps the most visible therapeutic enthusiast supporting antihypertensive medications, sounded a note of caution during the discussion: "There are a number of mild hypertensives that I don't treat with specific drugs which lower blood pressure, because I think they are neither safe enough nor cheap enough to warrant their use . . . Our chemotherapy is far better than it was five years ago. But it is not perfect yet. If we had a pill that was absolutely safe, that was inexpensive and that would uniformly control the blood pressure at normotensive levels with one dose of the drug each day, then I would say by all means treat all hypertensives. But since we haven't reached that stage, we still have to be selective."[51]

Freis well understood that Diuril did not represent an instant victory for advocates of asymptomatic treatment of hypertension. As arguments for expanded antihypertensive therapy were met with skepticism, supporters of antihypertensive therapy in 1958 found themselves challenged on both rational and empirical fronts. On a theoretical level, several medical authors of the 1950s dismissed a priori that a strategy of "treating the numbers" could be a successful means for controlling disease: at issue here was whether it was ever appropriate to treat an asymptomatic patient when the mechanism of disease was unknown and the medication itself had no specifically curative claims. Concerns over "treating the numbers" were cast on two levels, the first being unease at defining a disease based solely on a numerical threshold. This hesitancy was brought up in the final panel discussion of the Hahnemann conference, which is worth quoting at length:

> DR. MOYER: There was some objection yesterday about treating numbers . . . referring to the use of manometric blood pressure observations as a guide to therapy. This objection seems quite inconsistent with the practices of practically all the panelists here, because I notice that despite the fact that we say we should not treat numbers (that is, use manometric observation as a guide to therapy), we practically all do. This then makes us come around to a consideration of what the diastolic level of blood pressure is before drug therapy is indicated. May I ask Dr. Ford specifically what the numbers are that he treats relative to diastolic hypertension?
>
> DR. FORD: I'm not sure that I understand the type of patient that Dr. Freis would not treat. Anybody, regardless of age, who makes three visits to the office or clinic for recording of blood pressure, and two out of three times the blood pressure is greater than 155/95, gets treatment.
>
> DR. MOYER: This is even without evidence of vascular changes other than arteriosclerotic changes?
>
> DR. FORD: This is numbers—155/95 and two out of three visits.
>
> DR. MEILMAN: I think this discussion points up the great need for common definition and agreement. Frequently someone makes a comment that he has picked arbitrarily one level of blood pressure, maybe 140/90 or 160/100. Until we get some agreement as to what we are talking about, I don't see how we can compare treatment programs of any kind. Everyone seems to be agreed that we treat the patient with evidence of the degenerative disease; either the heart has gotten into trouble or there is some kidney damage. We all agree that we are going to treat the fellow who is already in trouble. If

normotension begets normotension, maybe it is easier to treat the patient when the disease is milder before there is overt evidence of advanced vascular damage. I realize, too, that we haven't got the ideal drug but I agree with Dr. Ford that we should treat the patient with a mild increased diastolic pressure, even before there is clinical evidence of vascular damage, which obviously is an advanced stage of the disease.

DR. MOYER: Dr. Meilman, you refer to mild hypertension, which is an ambiguous term.

DR. MEILMAN: Yes. As I pointed out, we have to make definitions first.

DR. MOYER: What is mild and what is severe hypertension, and what is the diastolic pressure that is considered abnormal? We are here to define.

DR. MEILMAN: Well, if we are going to start with numbers, I think that the numbers have been castigated too much in recent years. I often wonder how you would take care of the patient if no one had invented the sphygmomanometer. You would have a terrible time taking care of such a patient. You wouldn't treat any patients until they were in serious trouble.

DR. MOYER: May I ask what this number is then?

DR. MEILMAN: I can give you "my" number; 150/100 happens to be my number. Any patient with blood pressure above this receives antihypertensive drugs.[52]

Clinicians not only disagreed over the arbitrary and potentially misleading use of the sphygmomanometric number as a guide to treatment; they also objected to the use of the actuarial number as proof of pathologic process in the case of mild-to-moderate hypertension. Treating individual patients on a population-based probabilistic model—with full knowledge that many of the patients themselves would never manifest symptoms or complications from their condition if left untreated—was troubling when there was no solid etiologic framework of essential hypertension as a disease. As J. Estes of the Mayo Clinic noted in a 1958 review of chlorothiazide and mild-to-moderate hypertension, "Until we obtain the crucial facts about the cause of essential hypertension, or until we have a specific curative antihypertensive agent, truly adequate treatment for most hypertensive patients is out of reach."[53]

Other critics simply claimed there was not enough *experience* with antihypertensive medications in asymptomatic patients to warrant their widespread use. Arguments of experience traced out cyclical influences of drug and disease, a cycle of nihilism. Novel drugs represented risky and largely experimental chemotherapeutic agents, and risks of therapy were particularly difficult to justify in a group of patients who were not directly suffering from disease. And yet, if asymptomatic hypertension was not a treatable disease because there was no convincing evidence showing that treatment had any positive value, it was ethically difficult to justify trials of antihypertensive drugs with otherwise healthy asymptomatic patients to create convincing evidence. "Because effective antihypertensive drugs have been used for relatively short periods of time," one set of prospective investigators wrote in late 1958, "they can be adequately evaluated only in seriously ill patients with significant and predictable mortality rates during the available observation period."[54] In a major randomized controlled trial in Britain the following year, another set of investigators refused to treat any participants who presented with asymptomatic hypertension, noting that "although this possibility may nevertheless be regarded as a suitable objective for a future controlled experiment, the routine use of ganglion-blocking agents in this class of hypertensive patient seems as yet scarcely justified."[55] This cyclic negation, linking lack of evidence for treatment and lack of subjects for clinical evaluation, buttressed the rationale for therapeutic nihilism.

Expanding the Ranks of the Treatable

In the absence of convincing data, pharmaceutical marketers encouraged prescribing physicians in the late 1950s and early 1960s to feel comfortable with the broader use of antihypertensives—and to relax their iatrogenic anxieties—through an empirical plea to clinical experience and a theoretical model of hypertension as an insidious and progressive illness in which early intervention would prevent later calamity. As early as 1959, the visual rhetoric of Diuril's advertising had shifted away from the theme of novelty toward a reassuring theme of clinical familiarity (see fig. 2.2). Through its own surveys of physician relationships to pharmaceutical advertisements, the pharmaceutical industry well understood the value of what it termed "reassurance symbols" for prescribing physicians, symbols that could be used to support what marketers termed a "climate of believability about a new drug."[56] Diuril's one-year anniversary campaign had utilized the clinical ubiquity of the Diuril Man to proclaim the "weight of evidence" supporting Diuril's usage. Subsequent journal advertisements in the early 1960s continued this tone of familiarity, urging physicians to "start with what you know is right" when deciding whether a mildly hyper-

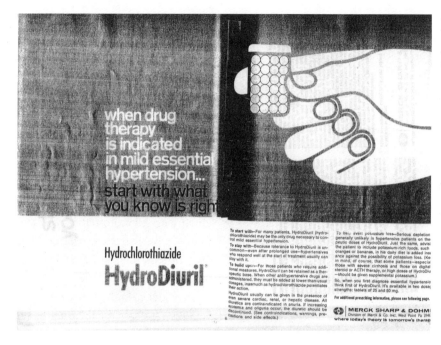

Fig. 2.2. Promoting clinical familiarity. *Source:* Diuril journal advertisement, 1963, reproduced from the *Journal of the American Medical Association.*

tensive patient should be placed on medications. As we have seen (recall fig. 2.1), by 1968 advertisements could claim that Diuril was enough of a staple in clinical practice to constitute its own form of textbook knowledge.

By 1960 Merck had become the leading advertiser in most medical journals, spending $476,000 that year advertising in the *Journal of the American Medical Association* alone.[57] This advertising, evidently intended to cast Diuril as a familiar figure to front-line physicians, was also geared to help distinguish Diuril from the subsequent thiazides and other antihypertensives that emerged shortly after Diuril. Within one year of Diuril's release, Ciba brought to market a competitor product, Esidrix (hydrochlorothiazide), which was ten times as potent as chlorothiazide. Although the increased potency did not translate into either increased clinical efficacy or safety, Esidrix would have offered significant competition had Merck's Karl Beyer not submitted a patent for hydrochlorothiazide at roughly the same time.[58] Merck Sharp & Dohme's brand of hydrochlorothiazide, HydroDiuril, was the first extension of the Diuril

brand line; subsequent combinations such as Diupres (Diuril mixed with reserpine) worked to cement the names *Merck* and *Diuril* in the mind of the prescribing physician. As other competing products entered the thiazide market,
such as Squibb's Naturetin (bendroflumethiazide), Robins's NaClex (benzthi-
azide), and Geigy's Hygroton (chlorthalidone), Merck's continued promotional efforts helped to keep Diuril's sales robust for the majority of the decade.

In addition to brokering a shift in physicians' perceptual environments
through targeted promotional materials, chlorothiazide and other antihypertensives also exerted a material influence on the natural history of hypertensive disease. Over the course of the 1950s and 1960s, as increasing numbers of
severely hypertensive patients were treated with oral agents, the epidemiology
of hypertensive mortality shifted from fatal processes internal to the disease,
such as acute hypertensive crises and hypertensive kidney failure, to fatal
processes external to the disease, such as stroke and heart attack. By the late
1960s, the most common cause of death among hypertensives was not a specific hypertensive condition but the more general pathology of coronary artery
disease.[59] This shift in mortality burden altered the rationale by which therapeutic enthusiasts could argue for the value of early treatment. Through the
1950s a deterministic model of hypertensive disease had maintained that the
condition progressed along an irreversibly degenerative course; therapeutically, the best one could hope for was to slow progression from a twenty-year
sentence to, say, a thirty-year sentence. During the 1960s, however, arguments
for preventive treatment began to shift from a fatalistic logic of degeneration
toward an activist logic of reversibility and prevention of secondary conditions.[60] Freis noted in an editorial that year that "although definitive proof is
still lacking for the treatment of the milder cases, it seems entirely reasonable,
in view of the available evidence, that reduction of the blood pressure in an
early stage of the disease will prevent the complications of hypertension from
developing, including many of the atherosclerotic complications."[61]

The increasing impact of the Framingham Study, which established hypertension as a risk factor for heart disease in the early 1960s, necessarily influenced this development.[62] As Jeremiah Stamler, one of the original Framingham investigators, argued in 1958, hypertension promised to become the first
successfully modifiable risk factor. "True, it is not yet clear whether such intervention will lower risk," Stamler admitted. "The next 5 or 10 years should answer that question. But . . . it is all too evident what will happen to high risk
men if they are left alone."[63]

Nonetheless, in the absence of conclusive data, critics of asymptomatic treatment, particularly older physicians, continued to object. The widespread prescription of drugs that offered no proven benefit, it was argued, was ethically untenable in the wake of the Nuremberg trials, constituting nothing less than a large-scale clinical trial on the unwitting general population. The venerable hypertension researchers William Goldring and Herbert Chassis wrote in 1965, "We believe we are now in an era in the empiric treatment of hypertension when a huge uncontrolled clinical-pharmacological experiment may be masquerading as a clinically acceptable therapy."[64] The recent enthusiasm for treating asymptomatic patients, they accused, was a grave example of the growing influence of the pharmaceutical industry and the naive desire of practitioners to feel they could do something in the face of a disabling disease. Goldring and Chassis challenged the existence of essential hypertension as a unitary entity; they also challenged the ability to quantitatively study drug effects and the causal relationship between blood pressure and end-organ damage. As they noted, asymptomatic hypertension was characterized by a course so variable and with treatment outcomes "so unpredictable as to preclude an acceptable comparison with an untreated control group."[65]

Goldring and Chassis also noted with some alarm the trend of published hypertension research away from basic etiology and toward antihypertensive drugs, a move they took to be representative of the pharmaceutical industry's corrupting influence on the practice of clinical research. They illustrated the trend with a review of the literature from the year of Diuril's launch to 1963, pointing out that the number of published studies on antihypertensive drug therapy had increased from 132 to 344 per year: "Current enthusiasm generously supported by the persuasive influence of pharmaceutical manufacturers, for the widespread use of antihypertensive drugs, is an example in point. After about 15 years of assorted data collecting, we believe that the alleged usefulness of antihypertensive drugs rests on conclusions drawn from notoriously uncertain statistical complications compounded by equally uncertain estimates of morbidity and mortality in the natural history of a disease of highly unpredictable course."[66] The relation of blood pressure level and coronary disease, they insisted, was equally unimpressive; arguments based on statistical difference "would appeal only to those who have more confidence in the statistical approach than the writers."[67]

Antihypertensive researchers like Edward Freis and interested pharmaceutical companies like Merck Sharp & Dohme recognized that the theoretical and empirical rationales for treatment were not sufficient to close this debate. What was needed was an epistemological tool with irrefutably demonstrative value, and at the time the large-scale randomized controlled trial (RCT) was fast becoming a gold standard of medical knowledge.[68] RCTs, however, were easiest to conduct in acute interventions that produced clear results over a short term; to show benefit for a therapeutic of risk reduction in an asymptomatic patient population, such a study would require the recruitment and long-term monitoring of a large population and would be financially and logistically challenging.

Before Diuril, it had been impossible to produce such a study, because no investigator could have convinced a large enough pool of symptomless people with blood pressure elevations to take a long-term course of a relatively toxic antihypertensive therapy.[69] As Freis recalls, however, "Diuril made it possible for the first time to control the blood pressure in a practical way, in a sufficient number of patients, so that one could then make a comparison between people whose blood pressures were successfully lowered and those whose were not. It was a stepped procedure: we couldn't do the trial before Diuril."[70] Freis had begun reading the works of Ronald Fisher and Austin Bradford Hill—two central figures in the development of the randomized clinical trial—after an influential European trip in the mid-1950s, and he recognized that the structure of the Veterans Administration (VA) care system offered an unusually viable framework for conducting long-term outpatient controlled trials. Freis's initial studies at the VA attempted to evaluate the long-term preventive benefit of antihypertensive medication while also comparing various antihypertensive agents against one another; he soon found that these two goals were difficult to address in a single study.[71]

What would later be known simply as "the VA study" began to admit research subjects in 1964 and assigned more than five hundred asymptomatic hypertensive patients, recruited from VA hospitals across the country, to receive either an active oral combination of hydrochlorothiazide-reserpine-hydralazine therapy or a placebo. Among patients with diastolic blood pressures in excess of 115 mm Hg, the differences in mortality between treated and untreated patients appeared so quickly that the "severe asymptomatic" placebo arm of the study was discontinued after two years. By 1967 data indicated a statistically significant effect in the prevention of hypertensive complications in patients with diastolic pressures over 115 mm Hg, and by 1970 a significant difference was recorded for patients with diastolic pressures from 105 to 114 mm Hg.[72] The following year, the VA study earned Freis the coveted Lasker award with a citation recognizing the VA trials as "the definitive study and demonstration of the fact that even *moderate* hypertension is dangerous, and should, and *can* be treated successfully."[73]

Yet the VA study should not be understood merely as the triumph of a more powerful technique of therapeutic research. It also represents an early moment of realization—on the part of pharmaceutical manufacturers—that large-scale postmarketing research was a powerful and essential marketing tool. Eugene Kuryloski, then director of marketing for MSD, recollects: "Another thing that was done was to develop clinical studies all over the place . . . Through our contacts with the government, we got them to agree to do a large five-year study on Diuril. We agreed to supply them all the Diuril and placebos they wanted. They were going to do cross-over studies, and it was going to be a long-range study. But it paid dividends."[74] Merck swiftly produced a set of advertisements promoting the results of the VA study as a justification of early and widespread use of antihypertensives; these promotional materials appear to have been well received among cardiologists and primary care physicians. A Merck marketing executive wrote Freis in May of 1973: "I have wonderful news to report on the journal ad based on the VA Study. Last week we learned that it broke all market research records for reader impact. The researchers told us the ad scored a new high for an 'impact score,' a measure used to indicate reader interest and attention to a journal ad message. It was quite an accomplishment because this type of testing has been underway for about four years, and more than 600 ad messages have been researched."[75] Although the material made for good advertising campaigns, these were not the only dividends Kuryloski was referring to. The VA study would prove far more useful than mere ad copy.[76]

Symptoms Lost, Guidelines Gained: Toward the Joint National Committee and Beyond

In his speech accepting the Lasker award in November 1971, Freis recommended that a national body be established to make the increased detection and treatment of hypertension a public health priority.[77] He received hundreds of letters of congratulation from eminent physicians throughout the country; at least one of them, a professor of surgery at George Washington University, took up his cause bodily, noting "Congratulations on the Lasker Award! I take Diuril every day."[78] Perhaps the most significant reply, however, bore the name of the award's benefactor: Mary Lasker wrote shortly after the award ceremony to congratulate Freis personally and to note, "We think that a great deal more can be done in promoting the treatment of moderate hypertension, and hope possibly to be in touch with you about more public information programs."[79]

Lasker had already been planning to use the occasion of Freis's award to gather momentum for a national hypertension education program.[80] For Mary Lasker was not merely the name behind the Lasker Prize; she was, by 1971, a formidable force in the center of Washington social and political circles with a long-standing commitment to increasing federal funding in the area of chronic disease, especially for mental health, cancer, and the prevention of heart disease. Lasker and her tight crowd of intimates, including the noted Texas heart surgeon Michael Debakey and journalist-turned-lobbyist Mike Gorman, had to a large degree been responsible for the massive increase in federal appropriations to the National Institutes of Health (NIH) that allowed the founding of the National Institute of Mental Health, the expansion of the National Cancer Institute, and the expansion of the National Heart Institute into the National Heart, Lung, and Blood Institute (NHLBI).[81] Assisted by the efforts of the Lasker Foundation, Freis's VA study would become the central evidence justifying the creation of the National High Blood Pressure Education Program (NHBPEP) in September of 1972 within the NIH's National Heart, Lung, and Blood Institute.[82]

The Lasker Foundation continued to work to publicize hypertension once the National High Blood Pressure Education Program had been established. In August of 1972, after Edward Freis appeared on the *Today Show*, Mary Lasker received a memo from her press secretary, Ruth Maier, documenting the event as "the direct result of our work starting last November, to arrange to get Dr. Freis interviewed on the 'Today Show' about moderate hypertension."[83] The transcript Maier forwarded to Lasker reflected the continuing problems of communicating the importance of the diagnosis and treatment of an asymptomatic disease to a public audience:

FRANK MCGEE: Well how can a person know that he has even moderate hypertension, unless their doctor tells them while taking their blood pressure test or is there any other way?

DR. EDWARD FREIS: There are no symptoms.

MCGEE: No symptoms?

FREIS: A person feels perfectly well. The only way that he would know is if he had his blood pressure taken. And therefore I think it's very important that people have regular checkups, particularly those who have a family history of hypertension. They should see a physician and have their blood pressure checked. This is the only way they can find out.

MCGEE: How do you define moderate hypertension? Is it something that most of us could understand?

FREIS: Well yes, there is a normal level of blood pressure, which is about a hundred and twenty over eighty. And then progressive elevations above that. But the higher level then, the more severe the hypertension. Now moderate hypertension would be defined as a blood pressure of about a hundred and seventy to two hundred over a hundred to a hundred and fifteen.[84]

Freis's use of the term "moderate hypertension" reflects the degree to which the category had changed by the early 1970s. Whereas earlier the distinction between severe, moderate, and mild had reflected the difference between symptom, sign, and number, by the early 1970s all forms were defined in terms of numbers. Educating Americans to learn their numbers and be concerned about them demanded a large-scale and well-coordinated institutional effort.

The national education program was designed as a federal interagency collaboration that involved professional and voluntary groups as well as governmental bureaus. To continue its participation in the policy process, the Lasker Foundation founded a shadow organization, one of the first health lobbying organizations to be formed around a specific chronic disease entity, called Citizens for the Treatment of High Blood Pressure, with Mike Gorman at its head. Gorman later recalled, "As the outlines of the original education program began to unfold, a group of us who had been involved in this noble conspiracy decided to hold a strategy meeting and see what we could do about developing a parallel Citizens organization which could not only work closely with federal and state governments, but could generate a significant amount of activity in the private sector."[85] Citizens (as the group informally referred to itself) was carefully constructed to include most of the important players of the NHBPEP, such as the secretary of Health, Education, and Welfare (Elliott Richardson), the head of the NIH (Theodore Cooper), and the head of the NHLBI (Robert Levy), as well as influential hypertension researchers such as Harriet Dustan and Marvin Moser and prominent Washington figures such as Lady Bird Johnson.[86] Citizens would become an integral component of the program's Coordinating Committee, which worked to unify stakeholders from public, private, and professional domains.[87]

Successful public-private collaboration would be a particular strength of the national education program; its ability to flexibly articulate public and private stakeholders under a nominally federal program would subsequently be held up as an ideal model of Nixonian "New Federalism."[88] As Cooper wrote to Richardson (by then secretary of defense) in 1973, the first NHBPEP National Conference on High Blood Pressure Education elicited a strong show of support from the pharmaceutical industry. Among other efforts, "Smith, Kline, and French Pharmaceutical Company has agreed to sponsor a trial education program on high blood pressure for physicians utilizing information developed by our task force and effected through their field representatives (salesmen). This effort is one of the first, if not the first, attempt to coordinate physician education efforts between the Government and pharmaceutical industry."[89] The following year, Merck Sharp & Dohme published *The Hypertension Handbook,* bearing the seal of the National High Blood Pressure Education Program on its cover.[90] Merck Sharp & Dohme; Smith, Kline & French; Ciba; and the Pharmaceutical Manufacturers of America all began nationwide advertising campaigns in the medical and popular literature that prominently displayed the program's logo and quoted Freis, Richardson, Cooper, and the VA study to mobilize greater detection and treatment of hypertensives.[91]

Perhaps the single most durable activity of the NHBPEP in promoting the detection and treatment of what it promoted as the "silent epidemic" of hypertension was the series of conferences that produced, in 1977, the first *Joint National Committee (JNC) Report on Detection, Evaluation, and Treatment of High Blood Pressure in the Treatment of Hypertension*—one of the first broadly binding sets of clinical guidelines in contemporary medical practice. The first JNC report was published in the January 17, 1977, issue of the *Journal of the American Medical Association* and was quickly republished in many other medical publications. Mike Gordon wrote to Mary Lasker after the unveiling of the JNC guidelines at a NHBPEP coordinating meeting, "This is not just a marshmallow study . . . it gets down to specific recommended drugs."[92]

Partly as a result of the implementation of the JNC guidelines, by the early 1980s hypertension had become the single most common reason for visits to a primary care physician.[93] Over the next three decades, the JNC convened seven times, each time revising the thresholds of treatment to promote the broader definition of treatable hypertension. In 1984 the JNC-III report lowered the diastolic blood pressure required for diagnosis with hypertension from 95 to 90 mm Hg, which almost doubled the population of Americans considered to be hypertensive. In 1993 JNC-V lowered the borderline between mild and moderate hypertension from 105 to 100 mm Hg diastolic and reclassified any aberration over 140/90 as grounds for antihypertensive therapy; the most recent report, JNC-VII in 2002, lowered the threshold for treatment again, to 130/80 in patients with diabetes or renal disease.[94] While each of these changes helped to reduce the risk for heart disease and stroke for at-risk populations, each successive lowering of threshold also generated substantial increases in the population of daily consumers of chronic antihypertensive pharmacotherapy.

Conclusion: Diuril and Hypertension

Although their careers ultimately took them in different directions, in 1978 Elliott Richardson and Theodore Cooper both received their own Lasker awards for their role in the creation of the National High Blood Pressure Education Program.[95] After leaving his position as Secretary of Heath, Education, and Welfare, Richardson became secretary of defense and then attorney general, secretary of commerce, and ambassador to Great Britain. In counterpoint to Richardson's impressive career in government, Cooper traced an equally successful career in the private sector, leaving the helm of the National Institutes of Health to become CEO of the Upjohn pharmaceutical corporation, illustrating what was later called the "revolving door" between government and industry. Taken together, Cooper and Richardson biographically embody the sort of public-private collaboration that enabled both the influence of the NHBPEP and the spread of antihypertensive treatment more generally.

As we have seen, Diuril was a crucial catalyst in the establishment of the specific therapeutics of prevention. Diuril provided a specific and palatable preventive agent that validated the detection and management of a cardiovascular risk factor. The transformation toward preventive treatment was neither immediate nor complete: for many years, the decision to treat or not to treat was left as a matter of clinical judgment, while lack of data and lack of consensus continued to enable the individual clinician to choose between different models of hypertension: a degenerative disease or a process amenable to early intervention. Even the VA study, the prototypical postmarketing study for preventive pharmaceuticals, did not silence all critics.[96] It did, however, deftly change the site and means of engagement so that the everyday practicing physician gradually came to adopt numerical thresholds as a principal basis for diagnostic and therapeutic decision-making. Moreover, the VA study became a crucial plank in the hybrid public-private mobilization that promoted hypertension as a public health priority to physicians and potential patients. The promotional body (NHBPEP) and the set of clinical guidelines (the JNC reports) that the VA study made possible have since become the model for the public health mobilization of other asymptomatic risk factors.

The narrative of Diuril and hypertension is a story of successful convergence of epidemiological and therapeutic developments. The result was that a population at risk came to be visualized as both a public health priority and a viable long-term market for goods, encouraging public-private collaboration in disease promotion and rationalization of practice patterns. In addition to its significance as a profitable consumer good, Diuril as a therapeutic agent prompted new ways of thinking about hypertension and created the possibility of long-term randomized clinical trials in symptomless subjects. Before Diuril, limitations in therapeutic possibility had themselves constrained the ethics and logistics of clinical trial research and were therefore incorporated into prevailing conceptions of pathology and normality. Diuril helped to transform this cycle from a negative, mutually nihilistic relationship to a positive, mutually reinforcing and potentiating one.

The historical encounter between Diuril and hypertension has therefore been mutually defining, an inelastic collision in which both entities were changed irreversibly. In subsequent decades the extended preventive pharmacotherapy of high blood pressure would be ratified as one of the most cost-effective achievements of late-twentieth-century preventive medicine. As the decision to treat became reduced to a numerical threshold, informed by large-scale trials and shaped by guideline and committee, hypertension became the example that other conditions of risk would emulate.

Part II / Orinase and Diabetes, 1960–1980

CHAPTER THREE

Finding the Hidden Diabetic

Orinase Creates a New Market

> There are diabetics enough to go around. No doctor needs to lack for them. If he is not satisfied with the number of his cases, extra attention paid to those patients most susceptible to diabetes will disclose new instances previously overlooked. —ELLIOTT P. JOSLIN, 1931

In 1961 Milton Moskowitz, the editor of *Drug and Cosmetic Industry* and a frequent commentator on developing practices in pharmaceutical marketing, prepared a feature article tracing the successes of Diuril, entitled "DIURIL Creates a New Market." Searching for an example that illustrated the significance of Merck's Diuril campaign, he settled on Upjohn's new diabetes drug, Orinase: "The Upjohn Co. is now chalking up an annual volume of $30,000,000 in Orinase, the oral antidiabetic it introduced several years ago. Orinase was the first product of its kind. Previously, the principal therapy was insulin, a market dominated for many years by Eli Lilly and Company. Insulin sales have contracted—but not substantially. It would seem that Orinase's introduction has expanded the total market by bringing under medical care diabetics who formerly were not treated."[1] Diuril and Orinase, Moskowitz argued, were two examples of a new form of pharmaceutical marketing that refused to accept the incidence of disease as a fixed market or a zero-sum game. Any disease was a potential market for a drug, but chronic diseases such as diabetes and hypertension were growth markets that could continue to expand—as long as the screening and diagnosis could be pushed further outward to uncover more hidden patients among the apparently healthy. In the infinitely expandable universe of chronic conditions, in the logic of preventive pharmacology, Moskowitz saw unlimited growth capacity for the pharmaceutical industry.

Like Diuril, Orinase catalyzed a shift in the basic conception of chronic disease from a model of inexorable degeneration to a model of surveillance and early detection. Both drugs fueled a movement to make the screening and treatment of "hidden patients," or those unaware of their own pathology, into a public health priority. Both represented more palatable alternatives to inconvenient and painful therapeutics. And yet the story of Orinase's relationship with diabetes constitutes a much different narrative from that of Diuril and hypertension. Unlike hypertension, which was largely a disease of the twentieth century, diabetes had been a stable category for centuries. When Orinase's marketers tried to promote a product that promised to simplify the treatment of diabetes and extend the boundaries of the condition, they found themselves simultaneously aided and foiled by this historical inertia. And unlike the domain of hypertension, which expanded as a single numerical threshold was lowered, the definition of diabetes grew in concert with another condition, a flexibly defined precursor state known as prediabetes. We need to understand the relationship between diabetes and prediabetes, and the role of Orinase (tolbutamide) at the interface between the two, to comprehend the pharmaceutical articulation of risk in contemporary therapeutics.

But first, a brief history of diabetes as a symptomatic disease.

A Disease in Motion

Diabetes had become a site for theoretical debate over the arbitrary division of health and disease well before Orinase was developed. Claude Bernard singled out diabetes in his 1865 *Introduction à la médicine expérimentale* to illustrate how difficult it was to demonstrate any exact boundary between health and disease once the body was understood in terms of physiological chemistry.[2] Fifteen years before the launch of Orinase, the philosopher of disease Georges Canguilhem used the example of diabetes to demonstrate that the value distinction between pathology and normality was fundamentally arbitrary.[3] However, neither Bernard nor Canguilhem argued that the arbitrariness of the distinction in any way undermined the status of diabetes as a disease.

Diabetes is one of the earliest named diseases in history. Symptomatically described in the Ebers Papyrus some thirty-five hundred years ago and mentioned in the subsequent writings of Galen and Celsus, diabetes was named after the Greek term for siphon—in reference to its most characteristic symptom of copious urination—by Arataeus of Cappadocia in the first century AD. It was not until the late seventeenth century that diabetes with sweet urine (diabetes mellitus) was formally differentiated from other conditions of frequent urination, and in the late eighteenth century the English physician Matthew Dobson first demonstrated that the sweetness of diabetic urine was due to sugar in the urine, termed glycosuria. The clinical sign of glycosuria was thus added to the symptomatology of polyuria (frequent urination), polydipsia (frequent thirst), polyphagia (frequent hunger), and autophagia (wasting) that had characterized the disease in clinical practice. Patients presenting with the full constellation of symptoms were thought to have a poor prognosis and typically were treated on a symptomatic basis.[4]

Dobson's work also suggested that diabetic patients tended to have an excess of sugar in the blood. It was the nature of this excess—clinically termed hyperglycemia—that led Claude Bernard to his arithmetic musings on the nature of pathology and physiology. Bernard's research into carbohydrate metabolism demonstrated that some level of sugar was always found in the blood of living organisms: indeed, the absence of sugar in the blood was inconsistent with life. Defining the point that separated euglycemia (ideal blood sugar levels) from hyperglycemia was difficult. Bernard insisted: "There is only one glycemia, it is constant, permanent, both during diabetes and outside the morbid state. Only it has degrees: glycemia below 3 to 4% does not lead to glycosuria; but above that level glycosuria results . . . It is impossible to perceive the transition from the normal to the pathological state, and no problem shows better than diabetes the intimate fusion of physiology and pathology."[5] Sugar in the blood was a continuous variable with a bell-shaped distribution in the human population. But the presence or absence of sugar in the urine was not a continuous variable: any person could be categorized to have either one or the other. Bernard postulated that the kidney acted as a physiological threshold, imposing its own arbitrary line on the graded presentation of blood sugars to determine what level of blood sugar resulted in glycosuria. Like the water level of a river rising above a dam, sugar levels in the blood would spill over into the urine only once they surpassed this level. This level, though, was hard to define with any exactness outside of the symptom itself. Bernard challenged physicians of his time to see beyond the kidney to the more fundamental logic of carbohydrate metabolism.

Bernard's research on carbohydrate metabolism supported a liver-centered model of diabetes and helped lay the theoretical and methodological foundations for experimental medicine and subsequent characterizations of scores of hyper- and hypo-physiological disorders. Nonetheless, the core philosophical questions he raised regarding the numerical definition of disease had a negligible influence on diabetes in clinical practice.[6] For the rest of the nineteenth century and a good part of the twentieth, most clinicians were concerned about treating diabetic patients who presented with symptomatic complaints rather than broadly screening physiological parameters to detect the silent physiological deviations observed in Bernard's laboratory. Even for acutely symptomatic patients, it appears there was little that could be done in the late nineteenth century to extend life or ameliorate suffering aside from strict adherence to a starvation diet.[7]

The discovery of insulin—the paradigmatic miracle drug of the early 1920s —greatly changed the diagnostic, prognostic, and therapeutic calculus of diabetes mellitus.[8] The isolation of this essential hormone won Frederick Banting, Charles Best, J. J. Macleod, and James Collip the Nobel Prize in 1923 and dramatically transformed the theoretical understanding of diabetes as well as the clinical course of the disease and the lives of its victims. The transformation was particularly visible among juvenile-onset, or "severe," diabetics, whose life span upon diagnosis had typically been measured in single digits before insulin. After insulin, a diabetic child could be expected to become a diabetic adult: a newly natural process that had previously been seen as an impossibility. In spite of tremendous publicity surrounding the new "cure" for diabetes, however, insulin was no cure: although these children lived, they lived diabetic lives. In a shift that has been termed a bittersweet transformation, insulin delivered diabetic children not from sickness into health, but rather from disease into attenuated disease with newly revealed chronic manifestations.[9]

"Insulin has not only prolonged the life of the diabetic," the celebrated diabetologist Elliot Joslin wrote in 1931, "but by doing so it has disclosed facts about diabetes hitherto unknown, because the patients died so soon."[10] Along with longer life spans, the postinsulin era yielded a harvest of previously undescribed diabetic conditions. Diabetic eye disease, formerly a rare and noteworthy occurrence, now became commonplace. By 1936 Paul Kimmelstein and Clifford Wilson had described a uniquely diabetic form of glomerulosclerosis, a kidney lesion that led to renal failure and death.[11] Coupled with neurological complaints (neuropathy), these eye and kidney complications formed the "diabetic triopathy" of chronic complications that constituted the initial costs of living longer lives with diabetes. In later years, the long-term diabetic would also suffer from increased susceptibility to infections, poor wound healing, and vascular disease.

Although insulin produced its most memorable images in the snapshots of diabetic children restored from skeletal emaciation to health and plumpness, it generated more pressing public health dimensions as a growing population of adults living with diabetes found themselves plagued by these degenerative changes. The decade following the introduction of insulin saw a paradoxical increase in the prevalence of diabetes, particularly in the population diagnosed as adults. By 1956 the category of late-onset or "overweight diabetics" made up the greater part of the epidemiology of diabetes mellitus. Estimates suggested that up to 80 percent of all diabetics fell into this category.[12]

The postinsulin era also brought a new set of social challenges for persons living with diabetes. Even after the introduction of newer insulins with gentler properties and longer half-lives, the diabetic's life required a swift and total indoctrination into a demanding lifestyle with meticulous labor practices of calorie calculation and insulin self-administration. Becoming a diabetic also involved transforming one's own bathroom into a diagnostic laboratory to perform regular self-surveillance of urine sugar with the Benedict's tests, which involved test tubes and an open flame.[13] In addition to this burden of self-care practices, individuals known as diabetics faced a pronounced social stigma, especially in terms of employment and insurance discrimination. These shared experiences of diabetic patienthood worked to produce a context in which patients diagnosed with diabetes came to strongly identify as "diabetics" socially as well as medically.

Although it is tempting to view the growth of this diabetic community in relation to contemporary patient empowerment movements, the identity of the American diabetic was as much a product of paternalistic "top-down" injunctions as of any authentic "bottom-up" patient populism. This distinction is well illustrated by comparing the British and American diabetic associations. In 1938 a group of well-heeled diabetics founded the British Diabetic Association—with H. G. Wells as the best-known charter member—to defend and promote the interests of diabetic patients. In contrast, the American Diabetes Association (ADA), was founded by a group of clinicians and researchers professionally interested in diabetes. The two organizations appeared analogous, but the ADA was explicitly an organization of *diabetologists* rather than persons living with diabetes, and it took as its primary concern a paternalistic responsibility toward the diabetic patient.[14] This managerial sensibility was perhaps best expressed in the clinic of Elliot Joslin, whose recommendations reflected a highly moralized sense of disease and illness tightly connected to discipline and right living.[15]

Although many diabetologists had been afraid that the easy availability of a "miracle drug" would cause diabetics to slack off on their diets, insulin supplemented rather than supplanted the moral architecture of diabetes care. In the postinsulin era, the needle came to characterize and bound the discipline of the diabetic life: taking one's insulin and taking care of one's diabetic self was a full-time job. Given the consequences of illness identity and management, it is hardly surprising that for the first half of the century, the diagnosis of diabetes was largely limited to the symptomatic.[16]

Oral Hypoglycemics from Hoechst to Kalamazoo

In addition to the pain of the injection and the labor of self-surveillance that marked diabetic life in the postinsulin era, any regular usage of hypodermic syringes had, by the 1920s, acquired a decided social stigma.[17] Immediately after the release of insulin, many doctors and patients alike were eager to find an acceptable oral alternative that might free diabetics from the needle. Although insulin, being a digestible peptide, was not effective in oral form, the drug's codiscoverer, Nobel laureate Charles Best, was confident that an oral form of the hormone would be made available in the near future.[18] Indeed, the search for an effective oral antidiabetic agent had preceded the discovery of insulin. A variety of extracts from bacteria, yeasts, and vegetable substances had been tried as oral agents with little success; the drugs that were the most effective at reducing blood sugar also tended to be the most toxic.[19] The most promising oral hypoglycemics in the first half of the twentieth century were the synthetic guanidine derivatives, most significantly Synthalin (decamethylene diguanidine), which was reported in the early 1920s to have a pronounced hypoglycemic action and brought to market with much acclaim before reports of liver toxicity severely curtailed its use.[20] Several of these compounds had come and gone by the time of World War II, but not one had made its way into widespread usage.

The new generation of oral hypoglycemic drugs did not come, as Diuril did, from the research laboratories of the American pharmaceutical industry. Orinase and the other sulfonylureas emerged from the venerable pharmaceutical houses of Europe in a context of epidemic infectious disease and global war. In the years following Gerhard Domagk's 1937 introduction of Prontosil—the first antibacterial sulfa drug—a wide variety of related compounds were put into clinical trials by French, German, and Swiss pharmaceutical companies for potential use against infectious diseases. One of these samples, while being tested by M. J. Janbon, professor of pharmacology at Montpellier University as an antityphoid agent, was found to produce blackouts, convulsions, and coma, a set of side effects not observed in any other sulfa drug. Janbon's colleague at Montpellier, Auguste Loubatieres—a physician performing experiments with protamine zinc insulin—recognized these reactions as consequences of lowered blood sugar. The pair collaborated on a set of studies evaluating the hypoglycemic action of these particular sulfa derivatives, now called the sulfonylureas.[21]

Janbon and Loubatieres's initial studies were performed in early 1942, a difficult year for French medical science, and Loubatieres's research program was disrupted when German forces extended their occupation of France to Montpellier. Though the two scientists published several times in the Francophone medical literature, subsequent development of these agents for therapeutic use took place inside of the great German pharmaceutical conglomerates.[22] Then ensuing clinical research in the Chemische Fabrik von Heyden in Dresden on the hypoglycemic properties of a related sulfonylurea was disrupted by the subsequent defeat and partition of Germany. Not until 1952, after a drug sample was smuggled from Dresden (then East Germany) to C. F. Boehringer (in West Germany), was the compound known as carbutamide (BZ 55) further developed. It underwent clinical trials in diabetic patients at Berlin's Auguste Viktoria Hospital in 1954. Two years later, Boehringer brought carbutamide to market under the trade name Nadisan. By that time, a competitor, Hoechst AG, had also developed a hypoglycemic sulfonylurea, D860, or tolbutamide, which was brought to market in the same year under the trade name Rastinon.[23]

Before the war, Hoechst had been part of the sprawling I. G. Farben conglomerate, which was closely tied to the Nazi state and formally dissolved with the collapse of the Third Reich. In the aftermath of the war, American pharmaceutical firms—scrambling for the diamonds they felt must be hidden in the ashes of I. G. Farben—eagerly sought licensing agreements with its splintered and financially crippled remnant companies. In the midst of this activity, a representative of the Kalamazoo-based Upjohn Company succeeded in making favorable contacts with executives of Hoechst in 1949 and obtained a co-operative arrangement for cross-licensing in 1950. Orinase was the first fruit of this agreement.[24]

In September of 1955, the assistant medical director of Upjohn received word of Hoechst's development of D860 as a promising hypoglycemic agent. A sample was sent to Kalamazoo, where a small research plant was established, and within two months the experimental drug was sent out for widespread clinical investigation. The time scale for investigation was tight, because Upjohn had learned that the Eli Lilly Company had recently begun clinical trials for carbu-tamide in the United States to evaluate whether it should proceed with a similar licensing deal with Boehringer. The medical director of Upjohn, Dr. Earl Burbridge, met at that time with representatives from Boston's Joslin Clinic and New York's Mount Sinai Hospital, to arrange for large-scale clinical trials in those institutions along with 15 other sites. Over the next two years, 12 million tablets were distributed through this network, and five thousand patients received tolbutamide in one of the largest-scale clinical trials yet conducted for a novel therapeutic agent.[25]

Although Orinase had initially been promoted within Upjohn and in the popular media as an "oral insulin," it was clear by early 1956 that tolbutamide and insulin did not have the same level of function in lowering blood sugar.[26] Unlike insulin, tolbutamide had no efficacy in the diabetes of young children or in "surgical diabetics," those whose condition developed after surgical removal of the pancreas. Orinase seemed to work best in older, overweight patients who had only recently been diagnosed as diabetic—the type of patient whose disease had traditionally been managed with diet. In April of 1956, the American Medical Association issued a warning against "indiscriminate use of new sulfa-like drugs . . . taken by mouth by diabetics as a replacement for insulin," noting that they had little, if any, effect on child-onset diabetes and that patients who appeared to respond best to the drugs were aged, obese persons recently diagnosed with mild diabetes.[27]

This difference in drug response ultimately proved significant in the reclassification of diabetes into two types: type I and type II. Although some division of diabetes into severe and mild forms (also known as "thin diabetes" and "obese diabetes," respectively) had been discussed much earlier, the elucidation of how tolbutamide acted offered a more mechanistic basis for the differences between these categories.[28] This new view was prominently announced at an "Orinase Symposium" held at Upjohn's Brook Lodge and subsequently published as a special edition of the journal *Metabolism* in which seventeen teams of clinical researchers reported favorable use of Orinase in five thousand patients.[29] Distinguishing "Orinase-sensitive" from "Orinase-responsive" diabetes would then become central to Upjohn's promotional strategy for the drug. Two of Upjohn's investigators reported at the Brook Lodge conference: "It is fashionable to regard diabetes mellitus as a single disease in spite of clinical and experimental data to the contrary. The thin, weight-losing, ketotic patient who is insulin deficient and who needs insulin to survive appears to us to have little in common with the obese, weight-gaining, nonketotic patient, who is not insulin deficient and who does not need insulin to survive. *The fact that one responded to Orinase and the other did not is further evidence that they have different metabolic disorders.*"[30] Furthermore—and central to the marketing of Orinase—among the larger population of mature-onset, nonketotic, largely overweight diabetic individuals, Orinase seemed to work particularly well in lowering blood sugars.[31]

Concerned that the new drug might be dismissed as a shabby substitute for insulin that worked only in mild cases, Orinase's researchers and marketers instead promoted the type of diabetes in which Orinase did work as an independent disease classification. The term "Orinase-responsive diabetes" circulated through the clinical literature, with reference to an "Orinase response test" as a crucial diagnostic step separating the classes of diabetic patients. Clinical research demonstrated that many Orinase-responsive patients could achieve good blood sugar control with Orinase.[32] Mount Sinai Hospital and the Joslin Clinic were the two most important sites for this early clinical research, and they also generated some of the most visible proponents for oral hypoglycemics. Mount Sinai's Henry Dolger and the Joslin Clinic's Alexander Marble, Howard Root, Robert Bradley, Peter Forsham, and Rafael Camerini-Davalos were some of Orinase's strongest supporters in subsequent years. Like most academic medical centers, the Joslin Clinic did not limit its clinical research arrangements to one pharmaceutical company; as the clinic was actively testing Orinase, it was at the same time testing the drug's potential competitors: Lilly's carbutamide (BZ 55); and later Pfizer's Diabinese (chlorpropamide).[33] In 1956 Orinase's currency increased when Joslin trials of the Lilly drug demonstrated mounting evidence of toxicity. After eight deaths attributed to the drug, Eli Lilly pulled it from trials and halted development.[34]

Suddenly the front-runner in the race to produce an oral hypoglycemic, Upjohn concluded by August of 1956 that enough results had accumulated to file a formal new drug application (NDA) with the Food and Drug Administration. The NDA for Orinase was the most monumental application of its kind that Upjohn had ever composed; when submitted, the Orinase application totaled 10,580 pages in twenty-three volumes with a total of 5,786 tabulated case reports.[35] The magnitude of these clinical trials indicated the large amounts of data required to render visible the relatively small improvements provided in less severe forms of diabetes. Processing of the immense data set was facilitated by IBM punch cards, in one of the earliest uses of the computer in pharmaceutical clinical trial data analysis.[36]

In the months between submission and approval, Upjohn mobilized its researchers in a continuing series of symposia. As was the case with Diuril, the New York Academy of Sciences was a crucial kick-off venue for Orinase promotion, and the Upjohn-NYAS event received day-by-day coverage in the *New York Times* and other newspapers.[37] The company newsletter bragged that the conference featured twenty-seven "top-flight participants," including Rachmiel Levine (later head of the American Diabetes Association), sulfonylurea discoverer A. L. Loubatieres, and the Joslin's Alexander Marble.[38] During the conference Henry Dolger of Mount Sinai reported that 90 percent of the 500 adult diabetics given Orinase were now managed successfully. The new Joslin data was more conservative but still showed improvement of the 420 adult patients treated with Orinase.[39] Not a single significant toxic reaction had been noted in either case series.

After FDA approval was received, Upjohn marketers planned to launch the new drug in June of 1957, the week following the annual meetings of the American Diabetes Association. An internal report before launch described the importance of the drug to the rest of the company: "This is no doubt the most important drug, both from the standpoint of medical interest and sales potential, that The Upjohn Company has ever studied. Orinase has firmly identified the Company in the field of diabetes. As Dr. Upjohn has pointed out, our problem is to remain alert in the field so that our eminence will not be short-lived."[40]

A More Comfortable Therapeutic

During the early development of Orinase, Upjohn's medical division provided limited hints to the rest of the company—and to the financial community—that it had something promising in the pipeline, but these discussions had been characterized by a vague and restrained tone.[41] Once the NDA was filed, however, Upjohn swiftly worked to mobilize its sales and marketing teams. In the context of failed oral hypoglycemics of prior decades, and in the limited setting of the adult-onset "mild diabetic," the company's leaders recognized that the product would demand a concerted marketing effort.

"This is probably the greatest challenge you will ever have," a circular to all Upjohn salesmen read. "The educational job you must do is tremendous in scope. Good Luck."[42] The challenge was particularly difficult in that the oral drugs faced an established "miracle drug"—insulin—whose market had been dominated by Eli Lilly for more than thirty years. Upjohn had not previously been associated with diabetes care, and although the failure of Lilly's own oral drug, carbutamide, had bolstered Orinase's prospects, it also rekindled the suspicion of oral agents still lingering from earlier toxic agents such as Synthalin. In the calculus of novel drug introduction, specialists typically tried riskier, unknown regimens first, on more severely ill patients. And here, too, Orinase was at a disadvantage, having been demonstrated *not* to be effective in severe diabetes, but only in milder cases of the disease; the diabetes of these patients tended to be already well controlled by diet or insulin therapy, and they were often treated by generalists rather than diabetes specialists. Rachmiel Levine summed up these challenges in the first Brook Lodge symposium on Orinase: "If insulin were not available, if it were not the ideal physiological agent that it is, or if good control of the diabetes in the overweight group could not be achieved by dietary means, there would be justification for a prompt wider clinical adoption of the sulfonylureas." However, given the present context, Levine advised that "the wise course would appear to be that of making haste slowly."[43] Assessing the marginal benefits of the drug and the widespread desire for an "oral insulin," he counseled Upjohn to be cautious with initial steps.[44]

Upjohn marketers appear to have taken his advice seriously: they cautioned their salesmen not to overpromote the drug. The training of the Orinase sales force began in late 1956 with a four-part series of articles detailing (1) the pathogenesis of diabetes and the mechanism of insulin, (2) the epidemiology and classification of diabetes, (3) the diagnosis of diabetes, and (4) the prognosis and therapeutics.[45] Circulars and letters to salesmen explicitly admonished that Orinase was "*not* an oral insulin" and that it did not work in all diabetics. The company also sent out a "double warning" to physicians that year, stating that Orinase was still neither proven effective nor proven to be safe.[46]

Sales representatives were told that the foremost advantage of the drug lay not in enhanced efficacy or safety in comparison to insulin, but rather in its convenience. A hint was present in the name of the drug itself: "Orinase" brought to mind orality. Early promotional materials emphasized the pill itself, depicting an Orinase tablet resting in a woman's manicured hand. Orinase's foremost selling point to diabetics and their doctors was the promise of replacing the needle with a more comfortable model of therapeutics, and early publicity for the drug focused on the cruelty of the needle (see fig. 3.1). As an in-house circular noted: "In 1921 the discovery of insulin injections revolutionized diabetic treatment. Since then, diabetics live normal though uncomfortable lives, so long as they follow physician's instructions. Now Orinase, a medicine taken by mouth instead of daily injection, can make life more comfortable for most (though not all) diabetics. Though a true cure for diabetes is yet to be found, Orinase is an important milestone—a tremendous advance in the management and treatment of the disease."[47] "Last week," *Time* reported in early 1956, "reports in two scientific journals gave promise that some day many of the 1 million known diabetics in the U.S. may throw away their hypodermic needles in favor of an insulin substitute taken in pill form." The *New York Times* added that "the development of Orinase, a tablet capable of lowering blood sugar, has brought freedom from the needle to 300,000 of the nation's diabetics." A year after Orinase's release, *Time* noted that "hundreds of thousands of diabetics all over the world . . . have rejoiced at their new-found freedom from the need for daily needlework." A subsequent *Time* article in 1959 similarly enthused that Orinase had liberated thousands of "diabetes victims" who until recently "were slaves to insulin and the needle."[48]

The marketing of convenience was a delicate subject for an industry whose public standing was based on producing life-saving contributions, and it was not an entirely successful strategy with diabetologists. The week of Orinase's launch, Henry Ricketts, former president of the ADA, was quoted in *Newsweek* suggesting that many physicians felt a deep ambivalence toward tolbutamide. Since the treatment of diabetes was a long-term affair and since the drug, however exemplary it might have been in clinical trials thus far, had been tried only for a few years, Orinase placed general practitioners in a double bind. On the one hand, it hardly seemed fair to withhold oral medications, given how much desire there was among the diabetic population for an oral drug; on the other hand, he warned, family doctors should be careful about switching patients

Good-by needle: Siegfried Fleischer, 79, has given himself 10,000 injections of insulin in his 55 years as a diabetic. Now he can use a new drug, tolbutamide, in pill form.

GOOD NEWS FOR DIABETICS

Now, a new pill may free thousands of diabetics from their
lifelong slavery to the hypodermic needle.

Fig. 3.1. Freedom from the tyranny of the needle. *Source:* "Good News for Diabetics: ORINASE* the Long-Sought Oral Antidiabetic," *Upjohn News,* January 1957, 160, Upjohn Collections, Kalamazoo Public Library, Kalamazoo, MI.

from insulin to Orinase simply to make life easier, noting ominously that "some diabetics may pay for their convenience later on."[49]

While Ricketts emphasized concern over hidden adverse effects that might not materialize until years down the road, other diabetologists worried about adverse effects that were behavioral rather than chemical in nature. During the week of Orinase's launch, at the ADA conference, Philadelphia's Garfield Duncan warned that the ease of Orinase was itself a risk; he feared that the "temptation to give these drugs to diabetic patients who are obese rather than reduce weight by food restriction would encourage the maintenance of obesity." Mount Sinai's Henry Dolger also warned that the convenience of tolbutamide might tempt diabetics unsuited to oral therapy, noting that even during the experimental use of the drug at Mount Sinai he had discovered three patients who "deliberately misled us regarding their symptomatology when we were reducing the insulin . . . These are patients whose eagerness to go off insulin is such that they will give false statements as to their symptoms."[50]

Recognizing that excessive focus on convenience was hampering its efforts to market Orinase as a legitimate new therapy, Upjohn shifted its promotional materials slightly in 1959 to promote an image of Orinase in terms of superior *safety* as well as convenience. Adapting a visual metaphor from engineering into a clever visual rhetoric of molecular mechanism, this new "euglycemic" campaign portrayed the molecular structure of Orinase as a machine stabilized by a spinning methyl-chain spur, analogous to the governor that ensures the constancy of rotation in an internal-combustion engine (see fig. 3.2). The euglycemic campaign emphasized that Orinase was more than a mere convenience. At the beginning of 1959, Orinase's product coordinator, R. M. Royle, laid out for salesmen the "seven wonders of Orinase":

1 Orinase is effective treatment for a high majority of maturity onset diabetics
2 Orinase provides a smoother and better quality of control
3 Orinase has an extremely low incidence of side effects
4 Orinase-stabilized patients are virtually free of hypoglycemic reactions
5 Orinase is a "euglycemic" rather than a "hypoglycemic" agent
6 Orinase has no known toxicity
7 Orinase makes for a more normal life for the diabetic and his family.[51]

This last item, Royle insisted, had "greater meaning than the obvious fact that it is easier to swallow a tablet than to be tied to the routine of daily injections."[52] Beyond convenience, the tablet as a mode of administration enabled patients to travel without concern for the availability of their insulin supply, made it easier to indoctrinate newly diagnosed patients into the habits of diabetes self-care, and freed senile, blind, arthritic, and otherwise disabled individuals from depending on family members or caregivers for regular injections. As Mount Sinai's Henry Dolger suggested, for many patients Orinase was "more than a convenience; it improves the whole life situation."[53]

In the eyes of Upjohn's competitors, however, convenience was still the name of the game. After one year as the only oral hypoglycemic on the market, Orinase was joined in 1958 by Pfizer's Diabinese (chlorpropamide), a drug intended for once-daily dosing whose first-year promotional effort alone involved $1 million of advertising. Diabinese was introduced to physicians via the

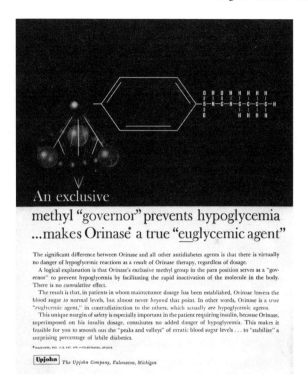

Fig. 3.2. The "Euglycemic Agent." *Source:* Orinase journal advertisement, 1959.

largest single ad yet placed in a medical journal: a twenty-four-page insert in the *Journal of the American Medical Association*. This ad and the three-page ads in eleven other national medical journals delivered the central claim of Diabinese promotion: that it was even more convenient than Orinase.[54]

In spite of this competition, Orinase held onto its market share. By 1958 an estimated 2 million Americans were diagnosed with diabetes; another 1 million were thought to be undiagnosed; of the known diabetics, 320,000 were using Orinase.[55] When the Upjohn Company went public a year later, Orinase represented an annual volume of more than $15 million out of a total $50 million diabetes market.[56] In that year more than 400,000 diabetics consumed some 243 million tablets of Orinase. At the time of the Upjohn IPO, Wall Street analysts saw in Orinase potential for growth that made investment in it seem particularly sound: "[Orinase] is considered to be the most significant development in diabetes therapy since Drs. Banting and Best discovered insulin in the

1920s. It has already freed thousands from the use of insulin needles. And many more will come to rely on it since there are 1.6 million known diabetics in the 40-and-over age group where use of Orinase has been most successful."[57] Wall Street was sanguine about the market potential for known diabetics. But Upjohn would soon target an even more expansive market: diabetics who did not yet know they were diabetics. As the firm set out to explore these outer orbits of diabetes, its financial growth from Orinase appeared limitless. In the words of Orinase's marketing manager, R. M. Royle, the "Orinase Epoch" was just beginning.[58]

Orinase and the Hidden Diabetic

Having established a valid therapeutic indication in patients already known to have diabetes, by late 1959 Upjohn marketers were able to shift the Orinase campaign into a second stage that actively recruited more members to the diabetic population. Millions of "hidden diabetics" roamed the populace undiagnosed and untreated, many of them without evident symptoms. By definition, most of these hidden diabetics had a mild, adult-onset form of disease for which Orinase could claim to be an ideal treatment. The development of the "hidden diabetic" as a potential market would become a crucial catalyst in the promotion of widespread diabetes detection efforts.

Screening for undiagnosed diabetic children had become a public health concern long before the introduction of Orinase. By the 1930s, screening for sugar in the urine was a common event in many children's gymnasiums and summer camps.[59] Broader screening in the adult population, however, had been a more elusive project. In 1947 Leo Krall and Hugh Wilkerson (of the U.S. Public Health Service) analyzed blood and urine samples from nearly three-quarters of the population of Elliot Joslin's hometown of Oxford, Massachusetts, providing the most comprehensive portrait of a community's carbohydrate metabolism yet produced. The Oxford data suggested that 1 million Americans had undetected diabetes, giving rise to the concept of "a million hidden diabetics." Supporters of diabetes as a public health problem used this figure to concretize funding appeals and to mobilize efforts for widespread screening. The American Diabetes Association began screening efforts in 1948, declaring the first National Diabetes Week and beginning its first national diabetes detection drive in October of 1949. The ADA's objective to find the "million unknown diabetics in the United States and Canada" referenced the Oxford findings directly.[60] In the same year, the ADA founded a new journal, *ADA Diabetes Forecast,* intended as a promotional and educational tool to aid early diagnosis of diabetes by primary care physicians.[61] Diabetes Week was soon jointly promoted by the American Pharmaceutical Association and the American Medical Association.

These early drives used urine sugar tests as a screening tool both easy to perform and indicative of a decidedly pathological feature, glycosuria. But the early screening efforts were of very limited scope, partly because all positive urine tests needed to be checked against a venous blood draw. As some of the principal investigators recalled, "there was no easy way to screen people for the disease . . . you had to collect samples of blood and urine, and that simply wasn't practical for large populations."[62] Analysis of samples was further complicated by limited manpower and laboratory facilities: by March of 1950, of the 304,851 tests conducted, less than 65,000 had been analyzed and processed.[63] As the authors of the Oxford study had noted, for large-scale screening, simpler methods were needed.[64]

Following Orinase's launch, diabetes detection efforts received an infusion of emphasis and funding from Upjohn and subsequent producers of oral hypoglycemics. In 1958 the American Diabetes Association could boast that forty-two local diabetes drives were in progress in cities including Boston, Los Angeles, Philadelphia, and Atlanta; the ADA distributed 1.5 million Dreypack urine test kits in 1958.[65] Manufacturers of insulin had previously had little reason to fund diabetes detection programs. "Hidden diabetics" tended to be mild diabetics, the kind best managed through diet and exercise, and therefore not immediately likely to expand the market for pharmaceutical products. Among other things, asymptomatic patients were unlikely to submit themselves to a lifetime of insulin injections. Orinase altered this logic. To newly detected, symptomless diabetics, Orinase offered a solution that many found preferable to diet or insulin.

A burst of popular media encouragement for the detection of hidden diabetics followed the launch of Orinase, and there are several reasons to believe that this was no coincidence. One genre of news and magazine articles announced the breakthrough developments of oral hypoglycemics to readers of popular magazines; such literature almost invariably mentioned the "hidden million" diabetics that such agents would help to treat. Another parallel type of popular magazine articles detailed diabetes detection efforts, and almost all of these articles called the availability of oral agents a crucial link in the argument for expanded detection programs.[66]

The popular science writer and Diuril promoter Paul de Kruif exemplified the connection between the two genres. In a 1958 article for *Today's Health* entitled "A Million Hidden Diabetics," de Kruif asked, "How can it be that despite the efforts of 2400 doctor-members of the American Diabetes Association, dedicated to detecting it, despite our alert family doctors, so many Americans live in blissful ignorance of the fact that they harbor undetected diabetes?"[67] To the reader's surprise, de Kruif then stated that he himself was a member of a diabetic family—hence a "susceptible"—and revealed in a surprisingly personal narrative that he had been diagnosed with symptomless diabetes at a recent screening examination. As personal as the motivation for writing the first article may have been, de Kruif's second diabetes article that year was likely connected to Upjohn's publicity office in much the same way that his Diuril article (published in the same year) had been connected to Merck's public affairs department. De Kruif's celebration of Orinase, titled "New Day for Diabetics," was published in *Reader's Digest* and subsequently distributed by Upjohn. The Upjohn Company even placed advertisements in the *New York Times* to encourage the populace to read the issue of *Reader's Digest* containing de Kruif's article.[68]

By late 1960 the Upjohn Company publicly announced that it was providing financial and material support to the ADA and its affiliate chapters to expand their diabetes detection drives. In addition to financial assistance, Upjohn explicitly converted its Orinase promotional apparatus into a diabetes detection apparatus. According to an in-house publication, "Since the drugstore is a very busy public establishment and is an ideal place for the use of publicity posters and leaflets, it was suggested that Upjohn salesmen could distribute [diabetes detection] publicity materials in the drugstores of their territories."[69] The following year, salesmen were asked, "What can we do to aid in the detection of the presently estimated 1,500,000 undiagnosed diabetics?"[70] Besides detailing Orinase, Upjohn sales representatives were instructed to urge physicians they visited to increase their screening of diabetic populations. Upjohn's Public Affairs Committee—in concert with the Conference of State and Territorial Directors of Public Health Education—produced a twenty-three-minute film entitled *Diabetics Unknown,* released in 1961, in which undiagnosed and diagnosed individuals were interviewed regarding their knowledge of the natural history, diagnosis, screening, and treatment of diabetes.[71]

Upjohn was soon joined by other manufacturers of oral antidiabetics interested in increasing the ranks of the treatable. In the early 1960s, with Lilly's oral agent Dymelor on the market, the medical director of Eli Lilly and Company, Franklin B. Peck Sr., assumed presidency of the ADA and oversaw its transformation into a "voluntary organization" that would allow lay membership and broaden its fund-raising efforts. By 1961 Peck's stewardship of the ADA had swiftly led to an expansion of diabetes detection and awareness programs. New press kits were created and a budget was provided for radio and television spots. Postage meter insignia were generated with the slogan "Be Alert—Be Tested—Be Sure—Check Diabetes" and adopted by ADA affiliates. Among other institutions, Upjohn featured this postmark on thousands of its letters and used it prominently as cover art for a 1960 issue of its in-house journal, *Upjohn Overflow*.[72]

But manufacturers of therapeutics were not the only financial backers of detection by the early 1960s. The hidden diabetic—and anyone else who might be persuaded to take a urine test—quickly became a target for the developing diagnostic test industry. Having identified this market, chemical companies began to mass-manufacture screening devices for sugar in the urine. The Dreypack, produced especially for the American Diabetes Association, required only that a single strip of paper be dipped into urine and then mailed to a central laboratory for analysis. In 1963 the market leader in the diagnostic test industry, Ames Company, brought to market the Clinistix, a urine sugar test that diabetics could use to test themselves in their own homes; a 1963 advertisement in *Today's Health* promoted the Clinistix "Wanted: 1,400,000 undetected Diabetics."[73] By 1964 Ames had developed a product for convenient widespread blood sugar monitoring, the Dextrostix, which made its promotional debut in a two-page spread in the journal *Diabetes*. A finger was shown with a single drop of blood falling from its tip, and the caption read, "A One-Minute Test for Blood Sugar."[74] As finger-stick blood glucose exams found widespread use in diabetes screening outside of the clinic, more intensive tests such as the two-hour glucose tolerance test (GTT) became the gold standard for clinical diagnosis.[75]

In this context, the hidden diabetic became a key figure in Upjohn's expanding promotion of Orinase. A 1964 Upjohn ad campaign titled "diabetes: detection . . . diagnosis . . . treatment," featured a montage of photographs depicting a two-hour glucose tolerance test. The connection between the detection of hidden diabetics through aggressive detection efforts and the management of patients with Orinase therapy was explicit. The caption said: "After the glucose tolerance test . . . time for oral therapy . . . it's time for Orinase."[76] Up-john also began to distribute a monthly journal to physicians, entitled *Diabetes: Research, Detection, Therapy,* which encouraged physicians to screen more of their patient populations for diabetes and included information about novel screening techniques and oral regimens for asymptomatic diabetics.[77]

The following year Upjohn produced and distributed a film, *Finding the Hidden Diabetic,* hosted by the director of the Cleveland diabetes detection program, Gerald Kent.[78] Other speakers in the film included Alexander Marble of the Joslin Clinic, Glen W. MacDonald of the U.S. Public Health Service, and Samuel L. Andelman, the commissioner of health of the city of Chicago. After describing the magnitude of the epidemic of undiagnosed diabetes and the need for early detection and treatment, the film provided models for diabetes detection programs and emphasized the importance of guiding individuals from screening programs into full clinical diagnosis and long-term antidiabetic therapy.[79]

As the film explained, the experience of being diagnosed with diabetes was undergoing rapid changes. The epidemiology of the disease shifted to include more mild cases: 75 percent of diabetics contacted in a 1960 U.S. Public Health Service survey reported no chronic limitation of function.[80] By 1965 the estimate of the population of hidden diabetics had risen to 2 million, out of a total diabetic population of 4 million.[81] Detection efforts, it seems, were locked into a Xeno's paradox with reference to their quarry; as detection effort shifted to earlier stages of diabetes, the numbers of the estimated "undetected" grew rather than shrank. Far from being perceived as a failure of detection, the ever-receding target of the hidden diabetic population served as moral imperative for the funding of detection drives, increased diagnosis of persons with diabetes, and expansive prescription of Orinase and other oral antidiabetic drugs.

And yet to advocate the detection and pharmacological treatment of hidden diabetics required a confidence that the treatment of the asymptomatic would indeed do some good. Cleveland's Gerald Kent noted in the Upjohn film *Finding the Hidden Diabetic:*

It is mainly a fourfold problem:
(1) Why should we treat diabetes early?
(2) Two, what methods should we use to detect diabetes in this early stage?
(3) Three, what in the world do we do with the thousands of people that we find in early diabetes in the way of treatment?

(4) Four, a very complex psychological problem, which has to do with how we convince people who do not feel sick, do not feel that they are patients, how do we convince them to get into medical channels under medical supervision where a treatment may be instituted if necessary?[82]

Kent's fourth question was perhaps the most difficult to answer, because the early detection of mild diabetes in the 1950s and 1960s came with substantial costs to the newly revealed diabetic. Once labeled as diabetics, patients found themselves ineligible for life insurance and were charged double premiums for health insurance.[83] Persons with diabetes could not serve in the military. They were actively excluded from working in most corporations with a payroll of more than four hundred, which tended to systematically require company physical examinations that would "weed out" infirm employees. Even in the federal government—one of the earlier foci of fair employment—only twelve hundred positions were listed by the U.S. Civil Service Commission as potential positions for persons with diabetes. Diabetics were still barred throughout the 1950s from jobs requiring the operation of a motor vehicle and moving machinery, and they were not allowed to work "above ground level" in any federal facility.[84]

Within the clinical literature, the prospect of treating a symptomless population of diabetics reactivated a debate on the nature of diabetes management that went back at least to the 1930s. On one side were traditional advocates of "tight control"—embodied in Elliott Joslin and the Joslin Clinic—for whom the insulin therapy had been framed in a disciplinary logic that equated physiological self-control with a more general philosophy of ascetic self-determination. Joslin physicians and patients were instructed that the maintenance of sugar-free urine and, later, of blood sugar levels within normal values, were the clearest guides to successful management: in Elliott Joslin's morally laden terms, "tight control" was superior to "loose control." Joslin's opponents, most notably Cornell's Edward Tolstoi, argued that experience with newer, long-acting insulins indicated that as long as symptoms remained under control, diabetics were better served by a "free control" than by the "strict control" of Joslin and company. Insistence on sugar-free urine in an otherwise asymptomatic patient, Tolstoi claimed, might actually be harmful to patients; it might carry an increased risk of hypoglycemic accidents.[85]

Other prominent endocrinologists echoed Tolstoi's challenge, noting that blood sugar measurements in and of themselves were insufficient to determine the long-term well-being of the diabetic patient. "Can one say that diabetes has been cured simply because tests for blood sugar made after treatment give negative results?" asked Michigan's Jerome Conn. He added, "One would not say that heart disease is cured when signs of decompensation disappear."[86] Herman Mosenthal, president of the ADA in 1941, argued strenuously that blood sugar was of no concern to the diabetic if the urine was sugar-free. "A high blood sugar," he noted, "without glycosuria, is in all probability of no significance." In the same year, Doctors Julius Boyd, Robert L. Jackson, and James H. Allen set forth the contrasting ideal of "physiologic control" as the extension of the Joslin creed: "A physiologic level of control is one which would avoid any degree of hyper- or hypoglycemia or glycosuria . . . Presumably if one could accomplish this, all conceivable disturbances of the body due to diabetes would be avoided."[87]

The debate over tight versus loose control (in the Joslinites' terms), or strict versus free control (as Tolstoi's followers framed it), was at heart a debate about the feasibility of prevention in diabetes care. The entire purpose of detecting populations of hidden diabetics, from a public health standpoint, lay in the possibility of checking the destructive progression of the disease. Upjohn had close ties with the Joslin school and was interested in promoting the broader preventive use of oral hypoglycemics in asymptomatic patients. Joslin's supporters argued that tight control of diabetes had merit not only in preventing specifically "diabetic" complications but also in preventing the more general sequela of cardiovascular disease, and Upjohn circulars carried their message.[88] One circular to Upjohn salesmen argued: "It is hardly necessary to make a case for early detection of diabetes since effective treatment is of necessity preceded by diagnosis of the condition. Furthermore, one can hardly read an article in the medical literature on the subject of diabetes which does not conclude with a plea for the institution of effective control at the earliest possible time. Indeed, our own promotion of Orinase has been spurred by a sincere conviction that in the fully responsive patient, Orinase-released insulin and diet provide more physiologic therapy and superior control than any other regimen."[89]

Moreover, although most diabetics were older patients, Upjohn's promotional materials began to urge the screening of younger populations among whom a greater preventive effort could be mounted. In the Upjohn promotional film *Finding the Hidden Diabetic,* Joslin's protégé Alexander Marble testified to the advantages of detecting mild diabetes at younger ages: "In young persons with abnormal glucose tolerance tests without symptoms of diabetes—without overt diabetes—if given a sulfonylurea compound or other oral hypoglycemic agent . . . the glucose tolerance may be improved as time goes on and in some instances, as long as the drug is continued, they actually revert to normal. This is a new area, not to be confused with standard treatment at the present time, but we may be optimistic about the possibilities. It certainly is a very worthwhile program with potential to prevent the onset of true diabetes."[90] The "new area" suggested by Marble reached beyond the hidden diabetic to suggest other undetected and potentially treatable precursor states of diabetes. Such populations would prove fertile ground for Orinase's market expansion.

Diabetes and Prediabetes

Echoing Claude Bernard's concerns a century earlier, Gerald Kent ended his narration of the film *Finding the Hidden Diabetic* with the admonition that "we must no longer think of sugar in the urine as being diabetes: we've got to think about the dynamic disease."[91] Without the categorical distinction of sugar's presence or absence from the urine as a diagnostic yes-or-no, how should one draw boundaries between elevated blood glucose and the disease state of diabetes? In the period of Orinase's spreading usage, this distinction became mediated through the mobile and malleable category of the prediabetic.

The origins of prediabetes lay in an optimistic attitude, after the development of insulin, that someday it might be possible to prevent not merely the complications of diabetes but the disease itself. Joslin wrote in the 1930s of the identification, management, and close scientific study of diabetic "susceptibles" as the most promising outlook for the prevention of diabetes.[92] Charles Best, in the keynote address of the first annual meeting of the American Diabetes Association in 1941, echoed Joslin in suggesting that a better technology of detecting susceptibles might lead to a policy of prevention. "Early diagnosis is a clinical problem about which much needs to be done," Best argued. "One of the great problems for clinicians and experimentalists is to attempt, perhaps by entirely new means, a way to detect the patients who are on the verge of diabetes."[93]

When the diagnosis of diabetes had required sugar in the urine, the detection of any populations "on the verge of diabetes" was impossible. But in the post–World War II era, as blood glucose levels took on an increasingly important role in diabetes diagnosis and screening, more people were receiving borderline test results regarding their glycemic status. Along with relatives of known diabetics, these borderline hyperglycemic individuals could be conceived of as a group of people at risk for diabetes. Various terms were used to describe this population, including *protodiabetic, chemical diabetic, latent diabetic, stress diabetic,* and *prediabetic.* The initial classificatory systems were multiple and overlapping, but they shared a common goal: to describe the populations who, though not clearly diagnosable with frank diabetes, should be considered susceptible to diabetes.

Before 1939 one could not receive a diagnosis of diabetes without evidence of sugar in the urine. At this time, patients with an abnormal glucose tolerance test who had no overt glycosuria were considered prediabetics. As Michigan's Jerome Conn noted that year, "One considers the so-called prediabetic state that in which the patient shows a diabetic type or tendency toward a diabetic type of dextrose tolerance curve but does not have sufficient rise in the sugar content of the blood to cause spontaneous glycosuria. When a patient has spontaneous glycosuria and a diabetic type of curve, he is called diabetic."[94] Conn argued that patients should properly be considered diabetic on the basis of their blood sugar findings alone, substituting the term *mild diabetes* for the condition recognized formerly as the prediabetic type. This transformation was neither instantaneous nor complete; in a panel discussion on the natural history and identification of diabetes as late as 1955, considerable disagreement still existed over whether an individual without sugar in the urine could be considered to have true diabetes.[95] By the early 1960s, however, the territory of prediabetes had shifted, and those formerly diagnosed as prediabetics were now diagnosed as chemical, mild, or early diabetics.

In their place, new populations of prediabetics were described: patients whose carbohydrate abnormalities were evidenced only under bodily stress. Perhaps the best-known example of this phenomenon was the "prediabetes of pregnancy," a well-described event in which the endocrinologic stresses of pregnancy yielded a symptomatic diabetes that typically resolved following childbirth.[96] After studies appeared suggesting that women who developed prediabetes during pregnancy were far more likely to develop overt (nonpregnant) diabetes later in life, the prediabetes of pregnancy was recast as an early warning sign of an underlying, incipient diabetes. Such findings led W. P. U. Jackson to develop an "iceberg theory" of prediabetes as a latent condition made visible by stresses such as pregnancy, infection, or other shocks to the system (see fig. 3.3): "We do not know the cause of diabetes, nor do we understand its very real relation to growth and size; but apparently diabetes in the latent

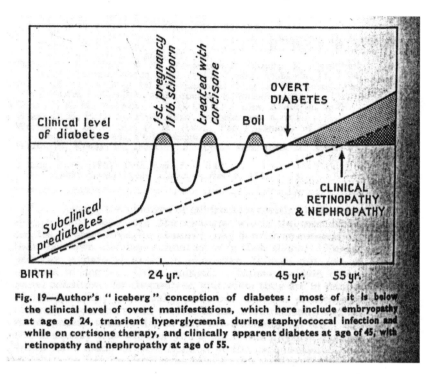

Fig. *3.3.* The iceberg model of prediabetes. *Source:* W. P. U. Jackson, "A Concept of Diabetes," *Lancet,* September 24, 1955, 630.

form ('pre-diabetes') remains with the victim years or decades before he or she becomes overtly diabetic. During this time pregnancy, corticotropin therapy, acquisition of Cushing's syndrome or of acromegaly, a staphylococcal infection, or overeating, may uncover the individual already predisposed to diabetes. The previously adequately compensating pancreas cannot stand the additional stress completely, and evidence of the latent diabetic state is brought to the surface, divulging its menacing presence."[97]

Following the synthesis and widespread availability of the stress hormones ACTH and cortisone in the late 1940s and early 1950s, pharmacological stress could be used as a diagnostic instrument. Shortly after the release of Cortone in 1949, diabetes-like responses to glucose tolerance tests in apparently normal populations were noted after a dose of cortisone had been administered.[98] Steven S. Fajans and Jerome W. Conn, of the University of Michigan, devoted much of the 1950s to the study of these "stress diabetic" variants of prediabetes.

By 1958 their research had produced a standardized protocol for a stress glucose tolerance test primed with cortisone (cortisone-GTT). The use of this tool to study prediabetes resulted in Conn's prestigious Banting Lecture of the 1958 meetings of the American Diabetes Association. Conn described the development of a prediabetes-detecting instrument as a central goal for the public health approach to diabetes. "When we can detect the prediabetic with reasonable certainty," he predicted, "only then can we justify the use of therapeutic or prophylactic measures designed to prevent the disease."[99]

At the time of the Banting Lecture on prediabetes, Orinase had been on the market for a year, and Upjohn's executives saw promising possibilities in the expanded treatment of prediabetic states. Fajans and Conn were quickly included within the network of Upjohn-funded research, and by 1959 they announced the predictive utility of their cortisone-primed glucose tolerance test at an Upjohn-Hoechst-sponsored forum at the New York Academy of Sciences.[100] By late 1960 Upjohn salesmen were informed that in addition to the overt diabetic and the hidden diabetic, they should begin to pay attention to "a third group of persons who are latent diabetics, or prediabetics." Whereas the number of hidden diabetics was at that time estimated to be 1.6 million, the ADA had estimated the prediabetic population in the United States to be much larger—over 5 million. An article in the *Upjohn Overflow* noted, considering Orinase's possible market, "It is interesting to speculate on . . . prophylactic measures in this vast group of potential diabetics."[101] At Upjohn's 1962 Brook Lodge conference entitled "Tolbutamide Therapy after Five Years," Fajans and Conn presented several sets of data supporting the use of tolbutamide in asymptomatic patients with elevated GTT, concluding that "it is possible that use of sulfonylurea compounds may be prophylactic in the very earliest stages of diabetes in the truly prediabetic individual."[102] In the aftermath of the conference, Upjohn company newsletters noted that prediabetes was "of extreme interest for future study of Orinase."[103]

Upjohn briefly attempted to extend the Orinase product line into a total system of patient detection and treatment for prediabetes. An article in *Today's Health* explained that while the cortisone-GTT test had been promising, a new "tolbutamide test" had recently emerged that might provide a superior method for detecting precursor states to the disease: "This test uses an intravenous injection of tolbutamide following the administration of cortisone and is extremely sensitive and reliable. An abnormal tolbutamide response, in cases where the response would be normal without the stress induced by cortisone, indicates a pre-diabetic state."[104] The Orinase project team at Upjohn seized on this new possibility for their product and quickly worked to position Orinase Diagnostic as a new product in the Orinase line and a new entry in the diagnostic test market.[105] By 1962 Upjohn salesmen were trained to enumerate for physicians the reasons why Orinase Diagnostic was superior in sensitivity and specificity to the glucose tolerance test.[106] In addition to being a potentially profitable product in itself, Orinase Diagnostic generated a direct flow of patients into subsequent long-term Orinase therapy. This line of flow was explicitly laid out for Upjohn salesmen in the company newsletter: "Since diabetes mellitus is caused by a lack of or an insufficient supply of effective endogenous insulin, the Orinase Diagnostic test becomes a more logical, more rational tool for the accurate diagnosis of mild diabetes. With an early diagnosis, the excellent control possible with Orinase tablet therapy is more likely to reduce or halt the degenerative changes of diabetes."[107] In 1966 an oral form of Orinase Diagnostic joined the intravenous version on the market, but by the late 1960s enthusiasm for the product had declined and Orinase retracted to its more focused role as a therapeutic agent alone.[108] Nonetheless, the brief career of Orinase Diagnostic reminds us that the influence of pharmaceuticals on disease definition is not merely a product of their therapeutic action.

The failings of Orinase Diagnostic were more than balanced by the contributions that the various forms of prediabetes—and their steady transformation into treatable diabetic states—made to the antidiabetic market over the course of the 1960s. A confusing array of terms had proliferated, and *latent diabetes, early diabetes, mild diabetes, stress diabetes, chemical diabetes,* and *protodiabetes* all shared an uneasy existence with each other and with the term *prediabetes.* The Joslin-trained Rafael A. Camerini-Davalos, who would go on to stake his career in the field of early diabetes, devised a linear taxonomy of multiple prediabetes in 1963 as stages on a timeline of disease progression.[109] *Overt* diabetics were fully symptomatic, with abnormal blood sugars and characteristic clinical presentations. *Chemical* diabetics had normal resting or fasting blood sugars but tested positive in glucose-response tests (GTT). Those with stress-induced diabetic test results were no longer prediabetic but instead were said to have *latent chemical* diabetes, while the category of *prediabetes* had been relegated to those with neither symptomatic, hematologic, nor stress-induced signs of abnormality but who, for reasons of family history, were thought likely to have increased risk for disease—a category formerly known as *protodiabetics* or merely *susceptibles.* Subsequent taxonomies of diabetes, such as that of

James Moss, explicitly emphasized the prophylactic value of treating these multiple stages. Moss's schema symbolized preventive therapeutics with the image of a thin white retrograde arrow pushing back the tide of a much larger, progressive arrow along the axis of deterioration.[110]

The clinical importance of prediabetes rested on the premise that prediabetes, unlike overt diabetes, might in fact be reversible.[111] Buttressed by new evidence on the possibility of reversing prediabetes, Orinase promotional materials from 1965 onward insisted that the drug had changed the goals of diabetes management from passive reaction to active prevention. The following circular was provided to physicians:

> Initially accepted as simply a drug of convenience, Orinase (tolbutamide) quickly spearheaded research into many previously unprobed mysteries of diabetes . . . the scope of therapeutic attention was thus broadened to include not only better symptomatic control but possible reversal of the certain progression of the disease . . . These goals, which six years ago were unattainable, are more feasible today. However, their fulfillment is largely dependent upon the earliest possible detection of the disease and initiation of therapy with Orinase (tolbutamide) . . . If one can speculate about the possibility of a patient's reverting to a previous stage in the progression of the disease or even delaying the normal progression, the true significance of Orinase (tolbutamide) therapy can be appreciated.[112]

The transformative, expansive influence of Orinase on diabetes was translated into the visual rhetoric of Orinase's 1965 logo (see fig. 3.4). Gone were the hand holding the pill and the spinning methyl governor stabilizing the function of its molecular structure. In its place was a single pill at the center of an eccentric or conical series of concentric curves, each expanding outward, each successively larger circle marked with a year: 1957, 1958, 1959, 1960, 1961, 1962, 1963, 1964, 1965. Whether the ever-larger spheres expressed the volume of prescriptions written, pills sold, or patient populations treated, or the expanding boundaries of diabetes itself, it was all the same thing: the image expressed wordlessly the central property of the drug as an agent of growth with an ever-expansive trajectory.[113]

That the greater part of these diabetics would have been considered prediabetics a decade earlier illustrates the dynamic nature of prediabetes and its importance in redrawing the boundaries of diabetes as a disease state. Jerome Conn had said as much in the Banting Lecture in 1958, when he first introduced the term to many of his colleagues: "Thus defined, the prediabetic state be-

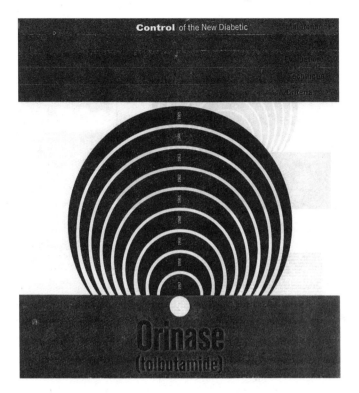

Fig. 3.4. Concentric spheres of Orinase. *Source:* "Orinase (Tolbutamide): Control of the New Diabetic," C38(a)I Upjohn Co., Kremers Files, American Institute of the History of Pharmacy, Madison, WI.

comes a target which will continue to move toward the goal of eventual complete understanding. What may be regarded as prediabetic today will later be shown to be a clear manifestation that the disease is present . . . What I wish to say is that the prediabetic state must be defined in terms of both time and objective manifestations, each newly discovered manifestation taking us further back into the prediabetic state."[114] Just as the concept of prediabetes had mediated, in the late 1950s, a shift in the diagnosis of diabetes from urine testing to blood testing, it helped catalyze the shift in the early 1960s toward increased glucose tolerance testing, and later still it brokered the increased usage of cortisone-primed glucose tolerance testing. As the definition of prediabetes receded ever earlier in the progression of the disease, via a series of increasingly sensitive diagnostic technologies, the expanding definition of "true diabetes" pursued it at every step.

Prediabetes can be understood as a mobile and efficient means of translating diabetes into an infinitely expandable market, a vision that coincided exactly with a public health goal of broader and earlier preventive measures to combat an ancient scourge. By equating the linear gradient of physiological parameters with the temporal progression of disease, the concept of prediabetes invested borderline test results with a sense of pathophysiological urgency.

Conclusion

Mild, hidden, chemical, early, or latent, the expansive taxonomy of adult-onset diabetes had by the end of the 1960s provided for an ample population of newly diagnosed diabetics, and by 1968 the National Health Survey revealed that over three-quarters of all known diabetics were being treated with oral hypoglycemics.[115] In that year alone, more than 3 million new prescriptions were written for oral antidiabetics—more than ten new oral agent prescriptions for every one new insulin prescription—and Orinase grossed over $40 million in sales.[116] In 1968 the average person living with diabetes had been diagnosed on the basis of blood sugar readings, offered a choice of oral antidiabetic agents, and told that in exchange for taking a pill a day for the rest of her life, she could reasonably hope to reduce her chances of diabetic complications, heart disease, and stroke. For the populations of asymptomatic diabetics detected and dosed, Orinase was intended to be a lifetime partner. As one Orinase salesman noted in a letter sent back to his supervisors with a copy of a prescription written for six hundred Orinase tablets, "the M.D. told the patient that he was going to be on Orinase for life and he might as well buy in quantity (. . . love that doctor)."[117]

This transformation in diabetes practice might superficially resemble the transformation in hypertension practice after Diuril, but there are significant differences. First, the therapeutic innovation represented by Orinase was not initially connected to claims of improved safety or efficacy, but instead emphasized palatability and ease of administration. This marketing of convenience encountered resistance and soon took on additional dimensions of enhanced safety (in terms of decreased hypoglycemic episodes) and increased efficacy (through better therapeutic compliance). Orinase's larger impact, however, came from its role in making possible the logic of diabetes screening and mobilizing the various categories of prediabetes that would come to be folded into the general category of diagnosable and treatable diabetes. Metaphorically, the shift from the needle to the pill encapsulated a general transformation of the therapeutic landscape of the 1950s and 1960s, as the injectable "miracle drugs" of the 1930s and 1940s like insulin and penicillin gave way to the tablets and capsules that were the wonder drugs of the post–World War II era. The daily integration of oral medications into outpatient life was essential to building a feasible pharmacopoeia of risk reduction.

Moreover, there was simply no analogue to prediabetes in the mild-to-moderate hypertension debates of the 1950s and 1960s.[118] What was in hypertension a single threshold separating the hypertensive from the normotensive became in the case of diabetes a territory all its own: a mobile buffer state that interposed itself between the normal and the pathological and in so doing actively stabilized both categories. Though never explicitly a disease, prediabetes allowed an articulation of an activist and expansionist mode of disease prevention and broader treatment. As a buffer, its two interfaces were both active: the interface between normal and prediabetes offered a demarcation of risk, while the interface between prediabetes and diabetes constituted a boundary of experimental therapeutics versus proven therapeutic indications. Prediabetes thus provided an epistemological two-step crucial to the mechanics of market and disease expansion.

This prediabetic state did not evaporate after serving its mediating purpose in the 1960s. Rather, definitions of prediabetic populations have continued to expand. The most recent instantiation of the prediabetic state is the contemporary category of "insulin resistance," sometimes explicitly referred to as a form of prediabetes.[119] As with previous categories of prediabetes, contemporary descriptions of insulin resistance denote a prepathological state that is defined in practice by borderline test results, a population whose blood glucose levels—whether fasting or following a glucose tolerance test—lie in the region just below the numerical thresholds for frank diabetes. Recently, a large multisite clinical trial, the Diabetes Prevention Program (DPP), has demonstrated that populations diagnosed with insulin resistance and placed on a regular regimen of Bristol-Myers Squibb's newer oral antidiabetic Glucophage (metformin) demonstrated a 31 percent reduction in three-year incidence of development of diabetes relative to placebo.[120] These results have been ratified with subsequent studies, and the pharmacological treatment of insulin resistance is steadily gaining favor in clinical practice.[121]

It is possible that within a few years, insulin resistance as a category will cease to exist and will instead be known as another variant of mild diabetes. If the results of the DPP are upheld, the screening, detection, and treatment of insulin-resistant populations will offer a further step toward Elliot Joslin's 1931 call for large-scale measures for the prevention of diabetes, but they will also result in another vast increase in antidiabetic prescriptions, currently judged at about $750 per person per year.[122] They also suggest a further question: if the difference between insulin resistance and diabetes is now strictly a quantitative difference in test results, how long will it be before another study demonstrates the benefit of treating the next population with borderline laboratory values below the current threshold for insulin resistance? What logic can oppose the progress of this expansive pharmacotherapy of prevention?

It would be wrong to take away the impression that the expansive trajectory of treatable diabetes traces an unbroken line from the launch of Orinase to the successes of Glucophage. Lingering disputes over the value of glycemic control in asymptomatic individuals plagued the community of diabetologists in the 1960s much as lingering concerns over the value of normalizing blood pressure in moderate hypertensives had plagued the community of cardiologists. As practitioners grew somewhat more comfortable writing prescriptions on a preventive basis, leaders in both fields looked to federally funded long-term, multisite, randomized placebo-controlled clinical trials to settle their respective clinical merits. However, whereas the VA study of asymptomatic hypertension became a model for generating clinical consensus and translating therapeutic experience into a broad-based public health program, its diabetic analogue—the NIH-funded University Group Diabetes Project Study—tore open the fabric of therapeutic consensus. It is to that act of disruption that we will next turn.

CHAPTER FOUR

Risk and the Symptom

The Trials of Orinase

> Everyone here is claiming to be in the public interest, and when everybody
> is riding the horse of the public interest in a different direction, there is
> great controversy.
>
> —ROBERT F. BRADLEY,
> Committee on the Care of the Diabetic, 1975

Early in the afternoon of May 20, 1970, a report was leaked over the Dow
Jones newswires that Orinase (tolbutamide), "a drug used to lower blood sugar
in diabetic patients," might be harmful.[1] Orinase had been Upjohn's showcase
success and sales leader for nearly a decade, and news of its possible toxicity
spelled disastrous things for the company and its investors. Before the New
York Stock Exchange closed that afternoon, Upjohn's stock had fallen in heavy
trading. The next morning, the *Washington Post* reported the preliminary findings from the federally funded University Group Diabetes Program (UGDP)—
the largest, longest, and most definitive study of diabetes therapy yet performed—with the implication that at least eight thousand patients a year may
have already died as a result of Orinase consumption. As the story was picked
up by the Associated Press, the Food and Drug Administration (FDA) hastily
issued a press release that provided only the briefest abstract of the study and
pronounced that the agency intended to revise the labeling of tolbutamide and
other oral hypoglycemic drugs.[2] In the meantime, all hell broke loose.

At that time, hundreds of thousands of Americans were taking Orinase
every day for mild (asymptomatic) diabetes, largely on the premise that the pill

reduced their long-term risk for diabetic complications and heart disease. Over the next few days, FDA commissioner Charles Edwards received hundreds of phone calls and letters from patients concerned to find that the drugs they were taking to reduce their health risks might actually be increasing them. "C.P.," a Virginia man, wrote: "I have a mild diabetes—the kind that shows up in blood tests only it does not show in normal urine tests. For the past two years have been taking 2 Pills daily Upjohn *Orinase* and using saccharin . . . Last week both were pointed out as dangerous to use as reported in the Wash Post. My Doctor thinks he should have more authoritative information before advising discontinuing or curtailment of these items. Could your office please advise on continued use of these items in view of frightening reports of Wash Post newspaper. I am 65 years old."[3] C.P. and the rest of the *Washington Post*'s readership were among the many Americans who learned of Orinase's putative toxicity before their physicians did.[4] This "premature announcement" of Orinase's toxicity, before the FDA had issued any warning to physicians and before the UGDP study was published in the clinical literature, unleashed a public debate over risk and asymptomatic disease that lasted more than a decade; created rancorous divides between advocates, researchers, and regulatory agencies; and left hundreds of thousands of diabetics, their families, and their physicians in a muddle of uncertain practice, contested information, and strained trust.[5]

If patients like C.P. were disturbed by the news of Orinase's toxicity, the news hit their physicians twice as hard. Those who learned of the controversy through the newspaper were relatively lucky compared to the thousands of physicians who first learned of the debacle from their agitated patients. Commissioner Edwards subsequently received the following letter from a "Poor Practitioner," who complained hotly of the difficulty he was thrown into due to the study's untimely publicity and the regulatory and epistemological uncertainty that followed:

Dear Dr. Edwards:

Now that my nurse, receptionist, and bookkeeper are no longer tying up the three telephone lines to discuss with patients who are extremely worried and apprehensive about the Orinase situation, I am able to obtain a free line to dictate this letter to your attention.

I sincerely believe that the "public leak" by the F.D.A. to the newspapers, and Walter Cronkite in particular, is not only a very stupid and indiscreet action on the part of your agency, but I firmly also believe this transgresses any and all medical ethic. This is a scare tactic to the general public who are using an ethical and adequate drug program and in so doing this, you are disrupting control balance of relationship of physician to patient and all other such relationships. You are further dictating by fiat medical practice and to make matters worse, you are using an unpublished study which has no logic, inadequate statistics and improper evaluation.

I deplore such action. I trust that it will not occur in the future over this or any drug. In the past, your direction has been to have the drug company release a news letter to physicians regarding dangerous or untoward side effects of drugs when they have been proven. In this case, you have done neither. I would hope very much that the F.D.A. would retract publicly its stand and correct this situation and future ones as they may occur.

Sincerely yours,
David L. Roberts, M.D.
Poor Practitioner[6]

The poor doctor Roberts and the hapless patient C.P. are but two of the thousands of minor actors in the public drama that became popularly known as the tolbutamide controversy. As it unfolded over the full course of the 1970s, this fight about Orinase and the UGDP trial proved to be one of the ugliest conflicts in the history of therapeutic investigation and came to involve a set of congressional hearings, an FBI investigation, and a court ruling challenged all the way up to the U.S. Supreme Court.[7] Eminent clinicians, typically reserved in public comments, took to calling each other "snake-oil salesmen," "unbridled sensationalists," and "drug-house whores."[8] Although the court proceedings had ended by 1984 and the dispute gradually disappeared from the pages of medical journals and popular newspapers, the debate never did reach a point of resolution.

I do not attempt in these pages to resolve the long-unsettled issue of whether Orinase reduced or increased the cardiovascular mortality of its consumers: now that Orinase has been replaced by newer generations of oral antidiabetic agents, the question has become largely irrelevant. Instead, I explore materials documenting the experience of how "street-level" actors like Dr. Roberts and C.P. came to terms with the diagnosis and treatment of asymptomatic diabetes at a time when its entire therapeutic rationale was under public scrutiny. The thousands of letters stored in the FDA dockets and administrative files during the tolbutamide controversy form a semi-ethnographic set of resources, documenting the voice of the patient as consumer and coming to terms with the pragmatic and moral issues that connect pharmaceuticals, risk, and asymptomatic disease. The letters preserve expressions of a changing ethos of patienthood and provide a perspective on the relationship of pharmaceuticals and disease in practice that is not widely available.[9] Historians, sociologists, policy analysts, journalists, and others have examined the tolbutamide controversy as a case study in clinical trial methodology, an exercise in failed public relations and communication breakdown, and a demonstration of the incommensurability of clinical and biostatistical logics.[10] These analyses, however, fail to convey the extent to which this controversy blew up around a disease in the process of shedding its symptoms and a drug that was instrumental in that transformation.

The displacement and attempted restoration of the symptom in the diagnosis of diabetes are central to this story. As illustrated in chapter 3, Orinase helped catalyze the transformation of diabetes from a symptom-bound disease into a numerical diagnosis treated on a preventive basis. One of the original goals of the UGDP study—proposed just one year after the 1957 launch of Orinase—was to interrogate whether the treatment of diabetes in terms of number rather than symptom provided any measurable benefit for these newly diagnosed "mild diabetics," "chemical diabetics," "latent diabetics," and others with laboratory-detected abnormalities of carbohydrate metabolism. Consequently, when the preliminary UGDP results suggested that tolbutamide *harmed* its consumers, the FDA's initial actions focused attention on the symptom as a vital site of regulation. As tolbutamide's identity shifted from risk-reducing agent to risk-augmenting agent, reexamination of the drug's efficacy crept backward from therapeutic agent to disease entity, casting the validity of asymptomatic diabetes itself into question.

The attempt to "roll back" diabetes from an asymptomatic condition to an exclusively symptomatic disease, however, did not go unchecked. Once the curtain of diagnosis had shifted outward to include the asymptomatic, after hundreds of thousands of symptomless patients had come to think of themselves as diabetics, there was no simple path back. The ensuing controversy offers a unique opportunity to understand how the regulation of pharmaceutical products—and the corresponding definition of patients as consumers—became entwined in a crisis over medical authority in late-twentieth-century America.

Origins of the University Group Diabetes Project Controversy

In spite of the rapid adoption of oral hypoglycemics in clinical practice during the 1960s, not all elements within the diabetic community were satisfied with the new logic of asymptomatic diagnosis and pharmaceutical prevention it supported. Criticisms of the oral drugs came from a surprising set of positions within the medical profession. On the one hand, many supporters of strict blood sugar control, including Elliot Joslin, were concerned that the ease of use of oral medications would lead diabetics to abandon the temperate discipline of right living that had long been the essence of good diabetic care.[11] They were joined, from the other end of the spectrum, by a "new school" of therapeutic reformers who saw rigid control of blood sugars in the absence of symptoms as a sort of physiological Puritanism, an unhelpful and potentially unhealthy by-product of precision measurement that had nothing to do with good clinical practice.[12] The two camps could not have been more different in their approach to diabetes control, but both agreed that the widespread use of tolbutamide was questionable. Even as late as the 1960s, the unsettled issue of the treatment of diabetes on the basis of blood sugar level was considered one of the great controversies in internal medicine, though the arguments had taken place largely within the polite context of the clinical literature.[13]

In the winter of 1958–59, a study section of the National Institute of Health's National Institute of Arthritis and Metabolic Diseases (NIAMD) began tentative discussions to support a long-term trial that might resolve this longstanding issue of glycemic control and also assess the long-term benefit of the recently released Orinase. A team of investigators met in the spring of 1959 in Atlantic City to begin discussions, and by early 1960 a grant application was submitted to the NIAMD.[14] The study would address three layered questions of import to the mild, asymptomatic diabetic:

(1) did tolbutamide have a favorable impact on vascular disease?
(2) did lowering blood sugar levels help decrease the risks of vascular disease?
(3) what methods were useful in clinical trials for diabetes?[15]

The young epidemiologist Christian Klimt—who argued strongly for the use of a multisite, randomized, double-blind, placebo-controlled design—helped to persuade the institute to support the proposed trial as much on grounds of innovative methodology as on the merits of the scientific questions it set out to resolve.[16] By 1961 the study protocol was approved with seven different re-search sites (Baltimore, Boston, Cincinnati, Minneapolis, New York, Cleveland, and Williamson, West Virginia) from which subjects newly diagnosed with diabetes on the basis of abnormal glucose tolerance tests (GTT) without regard to symptomatology, were recruited and randomly assorted into four treatment arms. A control arm (PLAC) would receive diet therapy and a placebo pill, and this arm would be compared with a tolbutamide arm (TOLB), which would receive diet plus 1,500 mg of Orinase daily; a "standard insulin" arm (ISTD), which would receive diet plus sufficient insulin for symptomatic control of blood sugars; and a "variable insulin" arm (IVAR), which would receive diet plus a more vigilant insulin regimen to ensure strict glycemic control. In 1962 a fifth treatment arm (PHEN) was added to evaluate phenformin (trade name *DBI*), a subsequently released oral hypoglycemic of a different therapeutic class from tolbutamide. By 1963 five more sites had been added (Birmingham; Chicago; St. Louis; San Juan, Puerto Rico; and Seattle), and by 1965 the full patient complement of 1,027 patients—roughly 200 in each arm—had been recruited and started on their respective interventions.[17]

The study was intended to determine whether aggressive treatment of largely asymptomatic patients offered long-term benefit for the prevention of diabetic complications and cardiovascular disease, and the desired study population represented the "new diabetics" for whom numerical diagnosis was often the only justification for long-term pharmacotherapy.[18] In essence, the four active arms of the trial compared two therapeutic strategies: a set of interventions producing strict glycemic control (TOLB, IVAR, and PHEN) versus more symptomatic management (ISTD), both compared with placebo (PLAC). In the ensuing debate, it became evident to all parties that the study was expressly *not* designed to assess tolbutamide for potential toxicity. The initial study documents made no mention of the possibility of detecting harm, only of measuring relative benefit. Principal end points of the study included cardiovascular, retinal, neurological, and renal complications of diabetes, evaluated on a regular schedule of quarterly exams. Mortality was not expected to be a significant end point. Indeed, one of the selection criteria of the study was to include only subjects whose disease was mild enough to guarantee a "minimum life-expectancy of at least five years," in other words, to guarantee no mortality from diabetes during the anticipated duration of funding for the study.[19]

Shielded from the eyes of all but Christian Klimt, the Coordinating Center of the UGDP in Minneapolis tallied results as the study progressed. In addition to the clinical end points, the Minneapolis office received a series of forms after the death of any subject, including the death certificate, autopsy report (if available), and clinical opinion regarding principal cause of death. Klimt's office then decided whether the cause of death could be considered cardiovascular or noncardiovascular. The Coordinating Center routinely ran analyses on the numbers for interim reports, and during one of these regular tallies, six years into the study, Klimt noticed a surprising and disturbing trend in the data. More deaths were appearing in the tolbutamide group than in any other group, including placebo. Furthermore, the majority of these deaths appeared to come from cardiovascular causes.

Nothing in the previous literature on Orinase had prepared the UGDP researchers for such a result, and Klimt was at first unsure how to proceed. As late as 1968, the longest-term studies of tolbutamide had at worst found no difference in cardiovascular mortality between patients treated with insulin and patients treated with sulfonylureas, and they had frequently found that drugs like Orinase produced significant reductions in cardiovascular mortality.[20] Klimt's first response, therefore, was to search for baseline differences in the two populations that might explain the difference in mortality.[21] In subsequent months, as week-by-week monitoring of trial data showed a steady increase in this higher mortality rate, Klimt became more anxious about the ethics of continuing the tolbutamide arm: at what point did the negative findings of a trial in progress make further conduct of the trial unethical?

The ethical question was compounded by a pragmatic one: how would it be possible to prematurely end a trial of this magnitude in such a way that the data could be salvaged and analyzed in a satisfactory manner? Klimt appears to have first alerted the other UGDP investigators of the early results in 1967, as the group was preparing a petition for renewed federal funding. By that time, Klimt was convinced that even if tolbutamide was found *not* to be toxic, there was no longer any possibility of demonstrating benefit; therefore the study arm could not ethically be continued. The other UGDP investigators, less confident in the data, were not convinced that the study arm should be stopped; they appealed to two outside statisticians for counsel: Jerome Cornfield of the National Institutes of Health (NIH) and Byron Brown of Stanford University.[22] The statisticians supported Klimt's initial analysis, and an executive meeting of the UGDP investigators was held in June of 1969 to decide the future course of the study. After two days of debate, the majority of investigators voted (21 to 5) to stop administering the Orinase and to immediately notify the FDA and the drug manufacturers of the findings.

As evidenced by the vote, a minority of UGDP investigators continued to question the strength of the evidence. Differences in medical management among the study's sites, some thought, might account for the differences in mortality.[23] This argument was compounded by the observation that the excess mortality in the tolbutamide arm was almost exclusively concentrated in four of the twelve clinics, suggesting that the finding might be due to confounding factors in the study populations or variations in clinical practice or protocol implementation. Nevertheless, the majority of investigators were convinced of the strength of the findings, and the tolbutamide arm of the study was formally discontinued on October 7, 1969.[24]

Much of the contemporary and subsequent discussion of the UGDP study has focused on the contested internal validity of the study design (whether the conclusions the investigators drew were valid), the contested external validity of the trial design (whether the conclusions of the UGDP study, valid or not, had any relevance beyond the universe of the study itself), and whether the decision to discontinue the tolbutamide arm was hasty or necessary at all. A great deal has already been written about these debates, and it is not my intent either to "get to the bottom of it" or to provide an epistemological account to explain the ruptures.[25] Suffice it to say that, even among the UGDP investigators in 1969, this internal debate was not resolved: a few of the investigators later resigned from the study in formal dissent and publicly attacked its conclusions. But the debate among the researchers was tepid compared to the violent external debate to come.

After the June 1969 meeting, the UGDP investigators met in closed-door sessions with the FDA and the Upjohn Company to determine—with an eye toward public and professional relations—how to best publicize the study results. It was agreed that the next annual meeting of the American Diabetes Association (ADA), scheduled for the following June (1970), would be the best possible moment for public statements. The intervening year would provide both Upjohn and the FDA time to fully evaluate the UGDP trial and its significance for product regulation. Well in advance of the ADA meetings, representatives from Upjohn, the UGDP, and the FDA made their way to Bethesda for a May 21 meeting to finalize their course for public communications and proposed revisions to the tolbutamide label. Little did they know, as they woke up that spring morning, that they were already too late.

The Study and Its Publics

The tolbutamide controversy achieved publicity at a pivotal moment in the relationship between therapeutic research and its multiple publics. The years through which this controversy smoldered were a period of crisis for the paternalistic model that had characterized relations between the American medical profession and its patient public for at least a century.[26] Letters archived by the FDA during the 1970s reflect the growing influence of the consumer advocacy and patient autonomy movements, which sought to replace medical paternalism with an open egalitarian approach to medical information.[27] For advocates of transparency, the *Washington Post*'s early release of trial data was no "leak" but rather was an important step toward openness in communication at a time when increasing media coverage was devoted to the topic of consumer health.[28] For many physicians, discomfort with the media's handling of the University Diabetes Group Project was directly linked to fears that an era of relative professional autonomy and uncontested authority had ended. As physicians tried to defend their traditional position as mediators of medical information, their efforts would be stymied by the multiple publics who claimed a right to information about Orinase.

The Financial Public

As we now know, on the afternoon of May 20, as the participants in the FDA Expert Advisory Committee were readying their presentations and making their way to the Washington area, news of the UGDP results broke over the Dow Jones ticker. That the first public news of tolbutamide's risks occurred not as a general press release or in an article in the science section of the newspaper, but as a report over the financial newswire, is highly significant. The first public for the tolbutamide controversy, then, was the broader financial community surrounding the pharmaceutical industry. Needless to say, this particular sort of publicity emphasized a concern for the welfare of the drug itself—as a product—that could be easily differentiated from a concern for the welfare of the diabetic patient. The Upjohn Company, a particularly interested public of the study, was able to insert itself swiftly into a counternarrative, mobilizing critiques from noted statisticians Alvan Feinstein and Stanley Schor that disputed the study's results.

In the early twenty-first century it has become common practice to see clinical trial results receive their first publicity in the business sections of newspapers, but in 1970 this was no common occurrence. At midcentury pharmaceutical companies had tended to produce a wide range of therapeutic agents that overlapped with other companies' offerings; the failure of any one product would not necessarily disturb the financial well-being of a company (see chapter 1). But as a smaller number of exclusive, branded, multi-million-dollar drugs came to dominate the interests of the industry, and as the industry grew to make up a larger portion of the national economy, pharmaceutical news became big business news, and the results of one clinical trial could affect the portfolios of thousands of investors. A direct consequence of this transformation was that by 1970 detailed information on ongoing clinical trials was eagerly sought by financial analysts, traders, and individual investors. Knowledge of the progress of a clinical trial itself became a valuable currency that circulated through a private-sector information economy.

As significant as the Dow Jones ticker was for the financial community, however, it was the next morning's *Washington Post* article—which representatives from the UGDP, FDA, and Upjohn read with their breakfasts before they went to their now preempted strategy meeting—that revealed the story to the broader clinician and consumer publics. The excess mortality in the tolbutamide arm of the UGDP study had alarmed investigators regarding the fate of the two hundred patients they had placed in tolbutamide treatment, but their concern paled in comparison to the dilemma now facing America's physicians and the eight hundred thousand patients estimated to be taking the drug.[29]

The Physician Public

The American medical profession can be viewed as a second specialized public with a set of interests in the study results that was distinct—if overlapping—from those of the financial sector. Physicians were bewildered to learn that a pharmaceutical agent and therapeutic rationale they had been recommending with confidence for more than a decade were deemed worthless and potentially injurious on the basis of a single study, but they were equally disturbed by the publicizing of the study through a medium that side-stepped the traditional role of the physician as broker of health information. Physicians were accustomed to receiving their news from the FDA in a more direct man-

ner—letters and bulletins—and understood themselves to be the vital link between the FDA and the consumer; a "first public" that would receive product
warnings and transmit them to the ultimate consumers.[30] The immediate contestation of the study's results by Upjohn-associated statisticians, coupled with
the fact that the study itself had not yet been published and therefore could not
be evaluated by physicians, further compromised the position of the physician
as knowledge broker.

As suggested by Dr. Roberts's letter to Charles Edwards, though, physicians
needed answers immediately, and some wrote letters in the earnest hope that
the FDA might supply guidance in time of crisis. Many, like Philadelphia physician Norman Knee, politely appealed to the FDA for more information and
offered their clinical reasoning for the continued usage of tolbutamide in
asymptomatic diabetics based on the primacy of clinical experience.[31] The
tone of many letters from physicians to the FDA at this time was still cordial,
information-seeking, hopeful that in spite of the "unfortunate publicity" of the
study, the FDA and the front-line physician might patch up this singular breach
and resume their typical relationship in distributing health information.

Physicians involved in state and national professional organizations, who
had been engaged in a longer contest against the authority of the FDA since the
1940s, tended to be more strident in their critique of the FDA's breach of
agency-physician privilege.[32] "I am disturbed," the president of the Kansas
Medical Society wrote, "by the recent publication of a directive released from
your office concerning the use of the drug 'Orinase'": "The release of such information precipitously and without prior notification of physicians causes excessive disruption of the care and welfare of patients. There is considerable
emotional disturbance of patients and undue excessive demands on the time
of the physicians to try to reassure the patient about his course of treatment . . .
When such items as the above are released by your department, it quite thoroughly disrupts the physician-patient relationship and further dilutes the time
of the physicians in caring for acutely ill patients."[33] A few days later, another
Kansas physician said in a separate letter to the secretary of Health, Education,
and Welfare (HEW) that the parties responsible for the news leak should be
"severely reprimanded" for their lack of concern for patient anxiety as well as
their "crucifixion of the honest practitioners of medicine and of the honest
producers of the medicinal substances that are required for good patient care
in this country."[34]

M. J. Ryan, the director of legislative services for the FDA, quickly replied to these physicians in a letter that lamented the leak of the UGDP summary over the Dow Jones News Service as an unfortunate "case in which the financial community and newspapers got reports of this medical research before physicians, who might find themselves beset by troubled patients." After the *Washington Post* article broke the story to the nation, the FDA needed to issue a press release as soon as possible and had not had the time to notify physicians first. Ryan apologized that the FDA had not developed a practical way of informing physicians in such instances before the information reached the patient via the lay press, and he assured physicians that the FDA would arrive at a better system in the future.[35] In a separate letter to physicians, Surgeon General Jesse Steinfeld apologized for the breach and reiterated that the government's hand had been forced by the early publicity.[36] By June 8 the FDA was taking steps to restore the primacy of FDA-physician relations and had prepared the first in a series of letters to the nation's doctors, suggesting that although the individual physician must ultimately decide the utility of oral hypoglycemics, "they could no longer be given simply on the ground that they might help and could do no harm."[37] This basic logic, in which the warning was tied tightly to the product and still allowed the physician freedom to evaluate its relative merit, seemed a step in repairing the breach between physicians and the FDA by reinstating the physician as the central mediator of individual health information.

The Consumer Public

As a result of the study's publicity, members of the public who were taking Orinase every day were trapped in a situation of limited information and vital consequences. Patients and family members inundated their physicians' offices with calls and filled the mailboxes of their senators, the Department of Health, Education, and Welfare, the FDA, and even President Richard M. Nixon. In the first few weeks of the affair, the tone of most letters was hopeful, if somewhat frayed with desperation, appealing to the FDA as a trusted authority that might help resolve a pressing and confusing issue. Many letter-writers stated that their own consumption of Orinase, like that of the bulk of the UGDP study subjects, took place in the treatment of mild, asymptomatic diabetes. As C.P.'s letter (at the beginning of this chapter) noted, they tended to have "a mild diabetes— the kind that shows up in blood tests only it does not show in normal urine tests."[38] A context of perceived urgency is conveyed by the physical appearance of many of these letters, handwritten on torn sheets of papers or hastily typed with numerous errors. One Brooklyn patient, "using *Oraniss* Tablets as prescribe by my Doctor," asked the FDA to "let me know what are the Harmful effects, and what are the available Treatment for these patients who are considered (slight) Diabetic. I understand the death rate among these patient are very High."[39]

This critical need for information was coupled with a sense of unease regarding the possibility of critiquing their own physicians' diagnostic and therapeutic decision-making. Patients felt particularly uncomfortable about discussing the propriety of the prescription regimen with their physicians and appealed to the FDA as a defender of the public interest. Even when they did speak to their physicians, patients' references to newspaper and radio programs were often dismissed offhand. "When we questioned our doctor about this news report," one patient complained, "he (1) knew nothing about it and (2) regarded our question as a personal offense. At best, he says he'll wait to see what the Food & Drug Administration says. Meanwhile, we don't know whether to allow continued use of the drug."[40] Another person, whose son had been on Orinase for a year, was told by her doctor that the drug was perfectly safe to use since the Joslin Clinic continued to recommend it: "Will you kindly let me know whether the drug orinase is still safe to take?" she asked the FDA, "now this article has come out according to which there is proof against it . . . why is it then still on the market? And why have not the doctors been advised not to prescribe it?"[41] Faced with a visible rift between public knowledge and professional practice, patients and their families appealed to the federal government to explain and mend the breach.

As much as these early letters suggest that patients felt they had recourse to federal protection of their consumer rights, their appeals to the FDA were frequently tinged with institutional mistrust. J.F. wrote to the FDA in late May of 1970, "Could you please get me some more down to earth *100% correct* info on this problem. Unfortunately I am one of those people with a Diabetic Problem, and I have been taking *Orinase* for 3½ years. After reading the enclosed article I am very much upset and afraid."[42] As he appealed to his senator to make his case to the proper parties, including the Food and Drug Administration, however, J.F. warned of a more general unraveling of medical authority that included the FDA in its critique: "If this article is true I would say that this drug killed more people than the Viet Nam War. I hate to be a conclusion Jumper But—It appears that the F.D.A., Upjohn Co., the Amer. Med. Assoc., and The Physicians Who Prescribed This Drug are to Blame . . . The American People deserve a better system of Protection From Harmful drugs than is at the Present Time in operation."[43] Although J.F. included the FDA in the parties he blamed for lack of oversight, he appealed to it as an entity capable of acting in the public interest, even if it had failed to perform its duties well in previous handling of tolbutamide. For other letter-writers, a taint of scandal indicted the FDA as possibly duplicitous in its responsibilities toward the welfare of the American people. Comparisons between a war on disease and the war in Vietnam spoke of distrust between citizens and state. As another diabetic patient wrote a few days later, "The F.D.A. has now recommended Orinaise be given only to a select group in case it is harmful, well, if it is, the harm is already done and they are 14 years to late. It is unbelievable a drug can be used that amount of years before possibley being declared unsafe, doesn't the Govt. care at all? . . . It does shake your faith in a country when you are told that such a thing could happen, it is not only a very expensive drug but could be like slow poisen to the Diabetic if not of any help . . . We can get to the moon and constantly be at war but our medical standards still rank lower than many other countries. Makes you wonder, doesn't it?"[44]

It fell to Marvin Seife, then director of the FDA's Office of Marketed Drugs, to respond to consumer queries. In a standardized letter, he replied that the FDA had convened an expert advisory committee to review the results of the UGDP study and stated that "despite a number of limitations in this study," both the FDA and the advisory committee agreed with the UGDP conclusions. "In the near future," Seife concluded, "we will inform your doctor, along with all other practicing physicians, of the findings and medical implications of this study. In the meantime, we suggest that diabetic patients now taking tolbutamide or chemically similar agents continue on their current regimen until advised otherwise by their physician."[45] Seife self-consciously hedged his direct communications with consumers by referring them to their own physicians for the individually tailored health information they often demanded. Although his agency was responsible for responding to consumer safety concerns, he was required to tread carefully lest he be accused of arrogating the role of the private physician in determining a course of therapy for an individual patient. All parties, it seems, were constrained by the cloud of uncertainty surrounding the relation of the UGDP trial to clinical practice.

Publicity and the Maintenance of Controversy

Once the study became public, however, it was no longer an entity to be contained and managed by the FDA alone. Instead, after evidence of harm had been publicly presented, the defenders of oral antidiabetic therapy recognized that the production and maintenance of controversy through news media was perhaps the only strategy that could help to prevent any stable consensus from forming around the study results. Generating a public controversy required continued publicity efforts and in return offered a sustainable space in which widespread usage of the oral diabetics could continue to be regarded as a legitimate therapeutic rationale. This process was initiated by Upjohn public relations personnel and several leading diabetologists within hours of the Dow Jones report and quickly found its way into public reporting of the UGDP results.[46] By mid-June of 1970 the issue—presented in the form of "balanced controversy"—had made the front page of the *New York Times,* with a headline reading "Pills for the Diabetic: Dilemma for Doctors."[47] In mid-November, a group of diabetologists wrote a letter to the *New York Times* describing the study as "worthless," and the Upjohn Company found multiple venues—in newspaper articles and journal advertisements—to make its case that the UGDP study simply flew in the face of all other trials and all clinical experience.[48] One of Upjohn's more direct strategies was an advertising campaign directly addressing clinicians' own personal judgment. "When you prescribed Orinase (tolbutamide, Upjohn) 14 years ago, you had to rely on our experience," Upjohn advertisements read. "Today you have your own . . . In short, Orinase is a drug you're familiar with, and probably have confidence in."[49]

In the fall of 1970, the Joslin Clinic—which eventually became the nerve center for organized resistance to the extension of the UGDP study results into clinical practice—held its first major publicity stunt.[50] The annual meetings for the American Medical Association (AMA) were to be held in Boston at the end of the month, providing an ideal time to capitalize on the distress many physicians felt toward the AMA's initial support of the FDA position on UGDP. Physicians of the Joslin Diabetes Clinic—most notably Robert F. Bradley, Holbrooke Seltzer, and Peter Forsham, recruited thirty-four leading diabetologists from around the country to sign a statement dissenting from the AMA-FDA decision. Along with Cornell's Henry Dolger, the three leaders then held a press conference, dubbed its "Boston Tea Party," to publicize their discontent with federal interference into diabetes clinical practice.[51] The group named itself the Committee for the Care of the Diabetic (CCD), seizing rhetorical high ground by characterizing its adversaries as a set of distant technocrats who meddled in clinical realms they did not properly understand.

Publicity was a weapon vital to the armamentarium of all stakeholders in this controversy.[52] The "Tea Party" paid off for its ringleaders: shortly after its initial press conference, the CCD was invited into negotiations with the FDA and UGDP investigators over labeling. Because of this move the FDA was delayed five years in implementing its proposed changes. The labeling negotiations also led to Senate hearings in the fall of 1974 and the spring of 1975, recursively generating further rounds of media publicity.

The tolbutamide controversy was not merely about diabetes care: had it been limited to therapeutic decision-making, it would never have reached such proportions. The publicity of the affair arose from the public nature of the therapeutic agent at the center of it. Three factors conspire to make a pharmaceutical like Orinase inherently more public than other therapeutics such as surgery, diet, or wound care. First, the pharmaceutical is a product and therefore belongs automatically to the public world of goods, services, and trade. It is no accident that the first news of this study came over the Dow Jones newswire. Second, because of its identity as a commodity, the late-twentieth-century pharmaceutical encoded a consumer-oriented approach to medicine. The pharmaceutical tablet is the perfect image of health care as commodity: a compact unit of therapy, portable, exchangeable across borders, universalized in shape to a nearly spherical form, its therapeutic value having nothing to do with the local circumstances of administration (unlike a surgical procedure or an injection) and having everything to do with a highly abstract network of data, research, and therapeutic information of which it is both product and emissary. Finally, as a highly regulated consumer product, the pharmaceutical represented an early site for federal intervention in medical practice, and actions of the FDA have tended to define patients according to a consumer model. In extending the public duty of consumer protection into the private realm of diagnosis and treatment, however, the FDA faced a difficult new task.

Regulating the Symptom

In spite of patient and physician concerns to the contrary, the FDA never seriously considered a categorical ban of Orinase. The FDA was particularly constrained in its ability to regulate Orinase, which was one of a special subset of drugs that, by 1970, had been doubly approved. The first new drug application (NDA) for Orinase was approved in 1957, before the passage of the 1962 Kefauver-Harris bill mandating proof of efficacy in addition to the proof of safety already required for FDA approval. Along with three thousand other drugs, Orinase was subjected to a retrospective efficacy review (termed DESI) by the National Academy of Sciences and the National Research Council between 1966 and 1969.[53] Upjohn's combination antibiotic Panalba had fared less well in the evaluation, but Orinase passed through the review process relatively easily.[54]

In addition to insisting on efficacy, the Kefauver-Harris bill expanded the FDA's authority from the safety of the drug per se to a more formal evaluation of the appropriateness of the drug for a particular therapeutic usage, termed a "therapeutic indication."[55] In a move with political significance for the later tolbutamide controversy, the FDA's attempts to set out a formal policy regarding clinical trials, therapeutic indications, and labeling change were stalled for eight years largely by a series of lawsuits from the Upjohn Company regarding Panalba.[56] The new process was not formalized until May of 1970, immediately before the UGDP controversy erupted. In the case of Orinase, the DESI review affirmed that Orinase was "an oral anti-diabetes agent which effectively restores blood sugar to normal ranges in selected diabetes patients."[57] In a proximate, short-term sense, Orinase had been proved both safe and effective. However, Upjohn's promotional claims for the drug had by 1968 extended to a second, long-term indication for the treatment of asymptomatic diabetes, and it was the safety and efficacy of these broader therapeutic claims that the UGDP results challenged.[58]

The FDA had never received a clear mandate on how to regulate physicians' use of drugs, and the scope of its authority over therapeutic indications had been murky ground since the passage of the Kefauver-Harris amendments.[59] Before 1962 the FDA had mostly restricted its regulatory responsibilities to the makers of pharmaceutical products. As the more formal therapeutic indication gained relevance in clinical practice, the FDA began to issue "Dear Doctor" letters giving warnings on new adverse effects that came to light after drugs were launched; nonetheless, the FDA's authority was still tightly limited to the product itself and particularly product labeling. Whereas the agency had an obligation to inform physicians and the general public regarding product claims, federal law gave the FDA no direct jurisdiction over the physician's actions as prescriber.[60]

The FDA related as much in its first public statements on the tolbutamide controversy, the day after the story broke in the *Washington Post*. Choosing its words carefully, the FDA agreed with the UGDP findings that, in the treatment of mild, adult-onset diabetes mellitus, the use of Orinase was "no more effective than diet alone, and as far as death from heart disease and related conditions is concerned, may be less effective than diet or diet and insulin." The statement was careful to point out that this warning only regarded mild, adult-onset diabetics, and that the drug might still be found to be useful in diabetics with more symptomatic disease. Pending results of further studies, the FDA's only specific recommendation was that Orinase and all other sulfonylureas should be used "only in patients with *symptomatic* adult onset diabetes mellitus who cannot be adequately controlled by diet alone and who are not insulin dependent"; a letter to that effect was sent out to physicians in June of 1970.[61]

In an attempt to avoid dictating medical practice, the FDA had turned instead toward the regulation of its consumers, delineating which populations of patients could be considered to have treatable disease. To this effort the FDA cautiously recruited allies from physician associations such as the American Diabetes Association and the American Medical Association. Whereas the ADA had noted in June that "the evidence presented does not appear to warrant abandoning the presently accepted methods of the treatment of diabetes," by late October, the FDA could announce that the AMA and the ADA agreed with its intention to insert a warning in all oral antidiabetic drug packages.[62] On October 30, 1970, the FDA further clarified its position on the symptom in a widely circulated bulletin, specifying that "the oral hypoglycemic agents are not recommended in the treatment of chemical or latent diabetes, in suspected diabetes, or in pre-diabetes."[63] A joint press release from the AMA and the FDA endorsed the bulletin, and the AMA's Council on Drugs agreed that the only legitimate use for the drugs was in the "symptomatic, maturity-onset diabetic" who could not be managed with insulin.[64] It seemed, at first, as though the medical profession would adopt the symptom as a site of risk differentiation and consumer protection.

As evidenced a few weeks later at the "Boston Tea Party" held by the Committee for the Care of the Diabetic, however, the medical profession as a whole did not agree with the AMA-FDA alliance on tolbutamide. At their initial press conference, the Boston group charged that the proposed labeling would unfairly "restrict treatment of patients with latent or asymptomatic diabetes who do not respond to diet alone."[65] For the federal government to dictate who was and who was not a valid patient, these clinicians argued, was an unprecedented "compromise of the physician's freedom to prescribe."[66] The CCD argued that the FDA's jurisdiction was limited to the *product* and not the definition of disease. From January to October of 1971, the two parties worked to find middle ground.[67] By June some headway had been made—the FDA appeared to be backing away from its strong claims on asymptomatic diabetes—but the CCD was concerned that the FDA label still "dictate[d] preferences of therapy of adult type diabetes." At a 1971 meeting in San Francisco, the CCD adopted a resolution stating that "it is the place of the FDA to give adequate warning documented by relevant reference but it should not give indications for therapy by mentioning preferences."[68]

Talks dragged on for more than a year and then broke down entirely.[69] Successive events, intended to resolve the controversy, instead provided new foci for publicity of dissent. For example, by late 1974 the FDA had decided to postpone labeling until a third-party report by a statisticians' group known as the Biometric Society was issued. Although the Biometric Society report was not published until February 10, 1975, its results were leaked to the press before publication in the medical literature. Newspaper reports across the country quoted the claims of the NIH's Thomas Chalmers that the persistent use of oral hypoglycemics was responsible for ten thousand to fifteen thousand unnecessary deaths each year in the United States. Chalmers's statements, publicized before any physician had been able to see the Biometric Society report, managed to both renew consumer anxiety and reactivate the hostility of a great number of practicing physicians. Harry Marks points to the publication of the Biometric Society report as a point of diminishing returns in the scientific debate over the UGDP, but it was perhaps at this moment that the public deliberation of risk in diabetes practice reached its clearest demarcation.[70]

Angry physicians wrote in to the *Journal of the American Medical Association* incensed about the early release of data and the fact that Chalmers—who had commissioned the study and whose career was materially interested in its results—had been allowed to write an accompanying editorial that many saw to be inaccurate, inflammatory, and unscientific. One physician asked the editors, "How is it possible that such a gross distortion of scientific data has been presented to the public rather than being quietly evaluated in scientific meetings?"[71] Among others, Robert Bradley of the Committee for the Care of the Diabetic saw the leaking of the Biometric Society report as more than coincidental. He claimed it was a deliberate strategy of the UGDP and the FDA to use the press to bypass learned debate: "History now repeats itself, in that furor has again been created by reckless extrapolation of a small and temporary cardiovascular mortality trend observed in the UGDP to the entire diabetic population of the United States, and by the unilateral publicizing of the controversy."[72]

After the Biometric Society report, the FDA moved to hold a "legislative-type public hearing" to openly discuss—and publicly resolve—its proposed labeling for oral hypoglycemics. In the official notice of hearing, the FDA published new labeling changes, which centered on four points: (1) the restriction of tolbutamide to symptomatic patients, (2) the notice that tolbutamide, when used, should be used only after diet and insulin had failed, (3) the publication of a special "warnings" section that advised physicians and consumers of the increased risk of cardiovascular disease, and (4) the requirement of informed consent from the patient before prescribing. The proposed regulation of the symptom was announced for public discussion, and all interested parties were invited to submit written comments or request time during the hearing on August 20, 1975. The overwhelming response from consumers, physicians, and the pharmaceutical industry ultimately filled twenty-four volumes.[73]

In a reversal of its earlier position, the AMA argued in a tersely written brief that—in spite of the FDA's claims that it had no intention to interfere with medical practice—the proposed labeling changes directly encroached on the physicians' prerogatives in the choice of medical therapy. The AMA's concerns were broader, suggesting that package inserts should all bear statements that they were merely limitations on the parameters for pharmaceutical advertising and "should not be considered a legal controlling influence over drug use by a given physician in the management of an individual patient." Otherwise, the AMA argued, the proposed labeling would impinge upon free exercise of medical judgment in the practice of medicine.[74] State and local medical societies supported the AMA and pulled together to defend the logic of professional sovereignty. The North Carolina Medical Society warned in a letter that "The Food and Drug Administration does not have the legal authority to establish a requirement that physicians obtain the informed consent of their patients before starting therapy with a particular drug." Such a requirement, they wrote, "would constitute an unacceptable Federal interference in the practice of medicine."[75] Similar statements were filed by the medical societies of New Jersey, Maryland, and Texas.[76]

Although a few physicians wrote letters in support of the UGDP results, the vast majority of physicians who wrote to the FDA also criticized the proposed labeling as an infringement of the physician's role.[77] Many simply denied the ability of the FDA's decision to influence their practice. As one internist noted in a letter to the agency, "Whether so labeled or not, many conscientious and careful physicians will continue to use these drugs in selected patients until such time as the issue is finally decided."[78] And use them they did. By 1975 it had become clear that—in spite of a slight dip in sales in the immediate aftermath of the UGDP, oral antidiabetic prescriptions had continued to climb at a remarkable rate.[79] "You must realize," another internist wrote, explaining the limits of the FDA's influence on therapeutic practice, "that I will not really be influenced by any warning that you eventually put on a label."[80] Amid this bravado and bluster, however, not all physicians were so sanguine about their insulation from the FDA's influence, particularly in the context of malpractice litigation.

In the decade of the 1970s, malpractice suits had become a tangible reality to practicing physicians on a scale heretofore unknown.[81] As Neil Chayet—the Committee for the Care of the Diabetic's most prominent attorney—had commented in the *New England Journal of Medicine* three years before the release of UGDP results, the pharmaceutical package insert was increasingly employed as a standard of practice to which physicians might be held legally accountable. In the late 1960s, the first malpractice conviction regarding a package insert was issued against a dentist for using an anesthetic in a dosage unapproved in its labeling, and the prospect of package-insert malpractice suits remained an open issue in the rest of the nation.[82] Although physicians were empowered to use therapeutics in any way they saw fit, off-label usage of drugs opened physicians up to possible liability for adverse outcomes.

The asymptomatic nature of mild diabetes made the liability regarding Orinase prescription more complex than for other agents. Since the prescription of Orinase was partially based on a logic of decreased cardiovascular risk, and since the warnings section on the proposed package insert was based on a logic of increased cardiovascular risk, the physician who prescribed Orinase could become a particularly broad target for litigation. One internist noted in his written submission to the 1975 hearing: "If this regulation passes, I have no doubt that a spate of new malpractice suits will arise. Whenever a diabetic on oral drugs has a CVA [cerebrovascular accident, or stroke] or an MI [myocardial infarction, or heart attack], his attorney will claim that the condition would not have occurred, had the patient not taken the medication. I suspect that the majority of such cases will be won by the defendant, but the stress of the case and the court costs and defense would remain. In view of the malpractice situation today, I am opposed to giving plaintiff's lawyers an additional tool."[83] Other physicians pointed out that these regulations would place an additional regulatory burden of malpractice risk on physicians. "If I am to follow the rules implied in the proposed labeling I will be required to discontinue the use of oral hypoglycemic agents entirely," a South Carolina physician sardonically observed, "or face the probable legal implications in a wrongful death action brought on by some enlightened family . . . During this day and time of *suing the doctor* I feel that your proposed labeling . . . put[s] the conscientious family level physician in an untenable situation. A damned if you do or damned if you don't position."[84]

The FDA did receive several letters from malpractice lawyers seeking to make cases. The following excerpt from an Alabama attorney's letter is typical: "I represent the estate of a deceased who died after being administered the drug Orinase. I am interested in gathering all information possible, concerning the effects of this drug, both good and bad, on humans. I would appreciate you forwarding me any information that your department has concerning Orinase."[85] Although it is unclear how many cases of this kind came to court, the concern over malpractice had an evident material basis. The FDA's response to such requests was to claim that the package insert was the only information on Orinase that it was legally allowed to share with the public due to confidentiality agreements.[86] At the same time, the looseness of causal association between asymptomatic diabetes and cardiovascular mortality, combined with the looseness of causal association between Orinase and cardiovascular mortality, meant that any strong wording on a package could be perceived as a legal trap.

To characterize the violent reaction with which physicians received the FDA recommendations solely as a concern over malpractice, however, is to miss the more significant gap perceived between the knowledge required to label a drug safe or unsafe and the knowledge required to judge a therapeutic practice as effective or not effective. Many physicians pointed out that the FDA recommendations themselves illustrated the regulators' ignorance of the logic of diabetes therapy in practice. Physicians particularly objected to the FDA's suggestion that Orinase should be thought of as a third-line agent, for consideration only after both diet and insulin had failed. As one physician pointed out, Orinase was only really useful as a first-line agent, because it was much more difficult to convince a patient with no symptoms that the needle-based practice of insulin therapy was worthwhile.[87] A Miami physician wrote, "it is a common observation that diet alone is sooner or later unsuccessful in treating diabetic patients with mild maturity onset diabetes . . . They are more easily controlled with the oral drugs, often refusing to consider self-administration of insulin."[88] Some insisted that many patients achieved better control of blood glucose with Orinase than with insulin or diet.[89] They argued that the FDA's evaluation of risk was singularly one-sided, focusing on sins of commission while neglecting the sins of omission if patients denied oral therapy went untreated: "Have they projected the statistical impact of such proposed labeling on oral agents as such directed labeling directions will apply to larger numbers of people who are denied the use of oral agents? I feel that in my own family practice there will be deaths *due* to your proposals."[90] Several physicians testified at the hearing that from a public health perspective the treatment of asymptomatic patients was of critical importance.[91] Although established forms of therapy for symptomatic diabetic patients offered no possibility of cure, early detection and preventive treatment offered an effective public health solution to the mounting incidence of adult-onset diabetes. By limiting the use of drugs in asymptomatic patients, he argued, the FDA labeling would halt preventive efforts:

> If we stop using these oral anti-diabetic agents then we may be losing an opportunity to prevent some of the problems of diabetes. Now, many diabetes patients become diabetic after they have had chemical diabetes. If we can do something about preventing the conversion of chemical diabetes to overt diabetes, this may be worthwhile. Now, chemical diabetes by itself does carry hazards, so it ought to be treated, and it is all very well to say to lose weight . . . [but] you know yourself that this is not an easily achieved goal. Therefore it seems to me that it is worthwhile to explore the potential usefulness of a variety of agents in preventing the progression of chemical diabetes to overt diabetes.[92]

UGDP investigators present at the hearing scoffed at such preventive claims. Paul Lavietes maintained there was "no scientific evidence that treatment of asymptomatic hyperglycemia by either insulin or oral agents improves the lot of the person with maturity-onset diabetes," criticizing what he called a "superstitious" climate among physicians resulting in a "compulsion to do something about the blood sugar."[93] When the Committee for the Care of the Diabetic and others claimed to speak for the populace's rights to access to effective preventive medicines, the UGDP investigators countered that this mass of deluded physicians were in fact placing an unwitting populace at significant risk. Both sides claimed an exclusive therapeutic rationalism, and both claimed to represent the best interest of the consumer, though this subject of representation proved elusive indeed.

Representing the Consumer

Several historians and sociologists have described the scientific debate over the UGDP trial as a case study of incommensurable dispute, in which each participant claimed to represent objectivity while deriding its adversaries as deluded by ideologies of self-interest.[94] Simultaneous and parallel to this scientific debate, however, an equally important pragmatic debate occurred regarding the continued day-to-day use of Orinase in a time of fractured medical opinion. Like the scientific dispute, the clinical contestation of tolbutamide in the decade of the 1970s was also marked by accusations of interest. In the latter debate, however, each side claimed not to represent objectivity, but rather to represent the best interests of the patient as consumer. Many physicians articulated a common paternalistic logic by which the medical profession was responsible for the consumption habits of patients; this was countered by an emerging radical consumerist lobby arguing that the state, and not the physician, had the responsibility for protecting consumers. But not all consumers felt themselves represented by radical consumerism, and several patients and physicians explicitly invoked the newer language of medical egalitarianism to argue for or against the FDA's actions in sometimes surprising ways. In this tangle of imputed self-interest and contested authenticity, representing the interests of the consumer became a crucial but elusive goal.

Perhaps the most visible accusations of self-interest were levied against the pharmaceutical industry. Proponents of the UGDP liked to explain the tenacity of their opposition in terms of the deep pockets of the Upjohn Company. Thaddeus Prout, a UGDP investigator, suggested that the UGDP was unfairly persecuted by skilled industry public relations specialists, complaining that "one of the things that this controversy has brought out in the last 5 years, I guess—it seems longer—is the incredible way in which a group of physicians teamed up with industry to attack the only scientific evidence there is on the use of these agents, at a time when we sorely need it."[95] Many clinicians, in turn, accused the UGDP investigators of operating within a self-congratulatory incentive structure of bureaucratic and academic promotion with little regard for "clinical reality." One clinician complained: "It is always interesting to read the glowing reports of headline physicians who are responsible in part for many such infringements on the working physician's relationship with his patient. They write and report from a biased Ivy-Tower situation—certainly not from the day to day relationships with the vast majority of people seen by the grass-roots physicians of this country!"[96] Direct allegations of financial interest were also made against the UGDP, especially after a former UGDP researcher suggested that Christian Klimt had received bribes from a rival pharmaceutical firm seeking to divert market share away from Orinase toward its own oral diabetes drug DBI (phenformin).[97]

Other critics noted that federal bureaucracies such as the NIH and the FDA became interested parties in any trial representing a significant expenditure of the taxpayer's money. "Why should we quote this study when other studies published before and since seem to indicate just the opposite?" one physician asked, adding, "I agree that we spent a lot of money on the study, but this does not make it a good one."[98] Another practicing internist complained: "It seems to me a classic case of conflict of interest in which the Federal Government is approving the study which the Federal Government has funded, in spite of the voluminous exterior criticism. This is an irresponsible act upon your part and can serve only to alienate the medical profession even further than some of your past half assed actions have already done."[99] All parties in this debate pointed out the self-interest of their opponents in order to present themselves as disinterested servants of patient welfare.

In their attempts to position themselves as the rightful representatives of the patient-consumer, stakeholders found themselves muddled by the peculiar nature of pharmaceutical consumption. As a result of a long history of professional and governmental regulations, every prescription drug can be seen to have at least two consumers: the physician who chooses which pharmaceutical to prescribe, and the patient who chooses whether or not to ultimately buy and consume the drug. For the greater part of the twentieth century, physicians had understood their own position as "mediate consumers" of pharmaceuticals to be a sort of natural type, a rightful interposition between pharmaceutical manufacturer and patient that was rooted in the paternalistic ethos of medical practice. To many physicians, the Orinase labeling proposal was part of a much broader trend of consumerism that was believed to be inimical to the patient's best interests. How could a federal agency understand the complexity of an individual patient's situation well enough to take on the responsibility of mediate consumer? Furthermore, in appealing directly to the patient as consumer, the FDA was making an egregious trespass into an area well defined by Hippocratic code. As one practitioner asked, "Why should the patient be forced to make these decisions rather than relying upon the best judgment and advice of his physician, when that same patient can go to grocery store—or even a gasoline station—and pick up over counter drugs which are far more hazardous to his health, such OTC drugs which the FDA does not seem able to control? Rather your agency seems bent on a course which implies that the medical community is less conscientious in its use of medications than is the non-medically educated to prescribe for themselves."[100] Another internist objected, in an acerbic tone, that the FDA's attempts to appeal directly to the consumer through labeling and remove the physician from the role of mediate consumer were simply absurd: "You have hit a new high in bureaucratic label-pollution! . . . Good Heavens! I would hope that diabetic pills would be sought by diabetics after consulting their physician who would indeed do all the thinking as to what's best for his particular patient's needs and weight the dangers of these medications against their benefit for the patient. What would you gain by putting such labels since the decision is the physician's and not the patient's? . . . The F.D.A. needs a label 'Warning: liable to waste tax dollars.'"[101]

Not all physicians saw consumerism as a corrupting influence on medical practice. In addition to a more general critique of paternalism within doctor-patient relations, a radical consumerism movement flourished in the early 1970s as an agent of progressive political change, particularly after Ralph Nader's *Unsafe at Any Speed* demonstrated that consumers could form a political base to effect change with broad public health implications.[102] Sidney M. Wolfe, one of many physicians inspired by the political possibilities of a "republic of consumers," joined Nader to form the Health Research Group within Public Citizen and came to represent the heart of the radical consumer movement in health care. Wolfe viewed the Committee for the Care of the Diabetic with conspicuous disdain. He saw the court-required delay of warning labels as a move that had caused patients to remain on "dangerous, ineffective, and expensive drugs" for five extra years: the moral equivalent of mass manslaughter. "During this interval of irresponsible delay by the FDA," Wolfe noted, "approximately 250 million dollars worth of these drugs have been consumed in this country alone, and according to experts, 20,000–30,000 unnecessary deaths due to these drugs have probably occurred."[103] Unlike the physicians of the CCD, who saw the FDA's efforts as excessively intrusive, radical consumer groups such as Public Citizen argued that the agency's backpedaling on the issue of asymptomatic treatment had already allowed too much. "The problem is you are granting the indication," Wolfe's colleague Anita Johnson objected during the August 20 hearing. "Here you are granting an asymptomatic indication. Then you are holding your breath a little bit after you grant it. Our position is that it should not be granted at all."[104]

Between October 20 and November 12, 1975, the FDA received more than two hundred letters from concerned patients, many of whom agreed with Wolfe and Johnson that the FDA should severely restrict the use of these "toxic agents." Family members of patients who had died of heart disease while on oral hypoglycemics wrote in to show support for a ban, as did other patients who had sustained reversible side effects and offered their individual testimonials as public evidence in support of regulation.[105] At one extreme, a Virginia woman detailed the decline and death of her husband from a pancreatic cancer that she insisted could only be the result of eight years of oral antidiabetic therapy. "The enclosed autopsy report," she added, "speaks for itself."[106]

Although these individuals—and the thousands who financially and materially supported Public Citizen—felt their identity as consumers was well represented by Wolfe's position, many other consumers disagreed and characterized Wolfe's protectionism as merely another form of paternalism that ultimately misrepresented the voice of the consumer. The majority of letters from consumers found in the hearing dockets appear to be concerned that the FDA might *overly restrict* consumer freedoms, not that the agency needed to increase its regulatory activities as Wolfe suggested. A.M., a Wisconsin man diagnosed with "chemical diabetes" and treated on an asymptomatic basis first with Orinase and then with DBI (phenformin), wrote to Senator Gaylord Nelson early in the UGDP controversy insisting that the consumer deserved continued access to risk-reducing treatments: "Since consulting Dr. Parks, I have checked in to his office about every two months or so for blood sugar tests, and have been told each time that my blood sugar is at satisfactory levels. I have assumed that I would be on this medication indefinitely, and I personally have never felt any unusual symptoms . . . The alternative to taking this oral drug would probably be taking insulin by needle. I have no faith in my ability to administer insulin to myself without causing serious harm to my blood circulation system."[107] Dozens of letters like A.M.'s insisted that proper defense of consumer rights should focus on the freedom to take on risks involved in consuming a given product, rather than on regulatory activities that restricted consumer choice.

Ironically, some physicians allied with the latter group of consumers and used the language of egalitarian patient-physician relations to criticize what they saw as Wolfe's overly paternalistic view of the patient-as-consumer. One physician remarked at the FDA hearings that overly protective regulation would rob the consumer of vital rights: "[The patient] has the right to decide if he should make that kind of change. If he has made that kind of decision that he wishes to continue on his food habits, and that is incompatible with his diabetic management, it then, according to the package insert as it is being positioned here, remains for the physician to choose insulin."[108] Restricting the use of these drugs, especially in the care of older patients, was portrayed as a tyrannical abuse of protective power that stifled the will of the patient. Another physician, specifically responding to Public Citizen's claims to speak for the consumer, wrote an angry note to the FDA that ended, "I hope that your final statement will not persuade successful patients to abandon the use of these drugs because of what they read in the newspaper. Who is responsible for them—you, me, or Dr. Wolfe?"[109]

Different factions could claim to represent the interests of consumers because the patient as consumer had no unitary voice. Was consumerism in medicine an authentic grassroots movement, or was it an intrusion of the marketplace into the sacred space of doctor and patient? To engage in the tolbutamide controversy was also to come to terms with the multiple roles that risk had come to play in the regulation of therapeutic agents and diagnostic categories. Who was best qualified to balance the risks and benefits of Orinase for an individual person: the physician, the government, or the consumer herself? Many of the consumers motivated to write to the FDA on this issue insisted that they should have the ability to perform their own risk analyses. T.H., a gentleman from Dallas, Texas, called the proposed labeling an "unnecessary and cruel ruling as it would submit us to the danger of insulin shock which could be much more dangerous than the possibility of heart involvement." T.W.H. argued that he was perfectly competent to perform his own risk-benefit calculations; he was "willing to take my chance on heart involvement in exchange for the comfort and convenience of being free from the danger of insulin shock or reaction."[110]

No party in the debate could properly claim to represent the interests of the consumer, but representing the consumer had become a political necessity in the changing health care climate of the 1970s. That the political value of consumer representation was evident to many patients themselves can be seen in the closing statement that A.M. added to a letter he sent to the FDA during the 1975 hearings: "I also wish to say that I have no interest in any drug manufacturing firm and that my comment is entirely based on my fear that any FDA rule which would lead to Dr. Parks cutting off my DBI-TD prescription would be a direct threat to my life expectancy."[111] The presence of this disclaimer suggests that the author's self-identification as a diabetic patient and hypoglycemic consumer was not sufficient, in his eyes, to elevate him from a situation of interestedness. The voice of the suffering patient had already been appropriated and mobilized by interested parties; and any authenticity of the patient's voice had become complicated. Of the two hundred letters the FDA received in the fall of 1975 in support of the drugs, many contained messages too similar in argument, tone, and metaphor for them to be merely independent occurrences. The following text was received verbatim by Edwin M. Ortiz (director of HEW's Division of Metabolism and Endocrine Drug Products) in three different letters from "concerned friends" in three different geographic regions:

> Dear Sir:
>
> A number of my friends have diabetes, and are controlling the disease with oral medication (DBI, Dymelor) and diet. Insulin is neither required nor suitable for them. I am concerned, on their behalf, to learn that you are seriously considering depriving them of their rights by removing the oral medication from the market. The risk of long term side effects seems minute compared to the psychological and physical shock of ingesting insulin.
>
> Please consider all the factors and do not remove these life-prolonging drugs from sale.
>
> Very truly yours[112]

Given that both of the drugs named were Lilly products, it is overwhelmingly likely that these letters originated in the marketing or publicity offices of Eli Lilly and Company. The tactic of the "concerned friends" letter was perfect: untraceable, with an air of authenticity and urgency. Who could venture to demand proof of whether this writer really had any friends with diabetes? Although these ghost letters present an extreme example, it is clear that the struggle to claim the interests of the consumer had deeply complicated any possible authentic voice of patient experience.[113]

Conclusion

Partly as a result of continuing contestation over which logic best represented the proper defense of the consumer, the 1975 hearings ended in stalemate and court proceedings continued much as before. Upjohn's lawyers and the Committee for the Care of the Diabetic put up a long legal fight, resisting every attempt of the FDA to take labeling action and demanding at every turn an unbiased evidentiary hearing and examination of the raw data of the trial. After a second public hearing in 1978 and a lengthy reanalysis of the data, these parties eventually yielded in 1984 to a labeling change for Orinase and all other drugs in its class.

To read this final verdict as a vindication of the UGDP and its proponents, however, would be a misinterpretation. Proponents of oral hypoglycemic drugs had not lost the debate by 1984; they merely lost interest in continuing it. Orinase was by then off patent, and the newer oral diabetic agents, like Upjohn's Micronase (glyburide), had cleverly obtained an independent therapeutic class designation. These new "second-generation sulfonylureas" were categorically differentiated from drugs of Orinase's class and less affected by labeling changes or package warnings. Second-generation sulfonylureas, along with newer classes of oral antidiabetics, remain a vital cornerstone in diabetes therapy today.[114]

Indeed, throughout most of the 1970s, while the tolbutamide controversy reappeared in newspaper headlines on a periodic basis, sales of oral hypoglycemics not only were maintained but continued to increase. To those who found in the UGDP irrevocable proof of the drug's toxicity, this information was deeply puzzling. Asymptomatic diabetes and its associated pharmacotherapy of prevention, they argued, were relatively recent phenomena in clinical medicine. Why did this preventive pharmacotherapy prove so resistant to attempts to "roll back" to the older symptomatic basis of diabetes treatment?

Some explained the "irrational state of affairs" in terms of physician pride. Frank Davidoff—a diabetologist at the University of Connecticut—testified before Senator Gaylord Nelson's oral hypoglycemic hearings that, while it was one thing for the FDA to challenge the safety of a drug, the message that physicians had been prescribing an ineffective and perhaps harmful drug to diabetics for twelve years "was, as I see it, a more serious blow to our professional pride . . . We submit that the pride of doctors is standing in the way of giving the best treatment to their patients and that this is irresponsible medicine if not malpractice."[115] Others saw the marketing of convenience surrounding oral hypoglycemics as evidence of a deeper lapse in medical ethics. "The drugs are probably being used excessively because the physician wants to do something when he makes a diagnosis of diabetes, and the patient wants something done," UGDP researcher John Davidson noted in 1975. "The easiest thing to do is to write a prescription for a pill, because it takes only a little time and it necessitates no significant change in the patient's life-style."[116] Public Citizen's Sidney Wolfe agreed that such laxity of convenience needed to be opposed with discipline and responsibility: "Unless a person knows why they are being taken from the easy path to what might be a harder path, they won't do it . . . I think the idea here is to motivate patients and their doctors toward treating, if necessary, the hyperglycemia in asymptomatic patients by diet and other means which have been shown to work as the primary mode of therapy, not again for convenience reasons."[117]

The UGDP supporters cited the example of hypertension trials, and specifically Edward Freis's landmark VA study, to ground their argument that the continued prescription of oral antidiabetics represented irrational behavior. Framingham investigator Jeremiah Stamler noted that, compared to the UGDP, "Just the very opposite consequence occurred from the VA hypertension studies, namely, the burden of proof was much heavier after those studies on those who say there's no benefit from antihypertensive medication. Now, I think, the burden of proof is much heavier on all those who say there's reason to believe that treating the blood glucose really is important, in terms of the control of the mild, mature-onset diabetic, in regard to vascular disease—particularly, treating the blood glucose with a drug."[118] Continuing the comparison, Thaddeus Prout and Thomas Chalmers noted, somewhat jealously, that Freis's study was "no better than the UGDP . . . Yet his study's results were immediately accepted and Freis won a Lasker Award." Their implication, as Gina Kolata captured in an article for *Science* that year, was that the UGDP study was only disputed because it was a negative study, the entire UGDP controversy being a case of publication bias writ large.[119] Chalmers and Prout raise a useful question, though perhaps they do not go far enough.[120] For while it is correct to note that the VA study achieved a positive result, the published results of the UGDP were far more threatening than negative (or absent) results. They were antegrade results. The University Group Diabetes Project did not merely document absence of proof; it claimed proof of harm.

These antegrade results were vitally important to their physician audience, for prescribing physicians were not mere spectators to the UGDP debate or narcissists with excessive pride in their own therapeutic powers. Physicians themselves were implicated parties. Inculcated over the past decade into believing and enacting a systematic program of early diabetic pharmaceutical prevention, they recognized their own agency and culpability if that system was overturned. In the crisis over Orinase, much more than a package label was at stake. The proposed FDA changes threatened to undermine the central principles on which the preventive pharmacotherapy of diabetes had been based. As they complained about malpractice implications of changed labeling, physicians were not only thinking about lawsuits based on their future actions but also grappling with the theoretically far broader culpability for their past decade of participation within a therapeutic system now being considered potentially harmful. Consequently, the burden of proof for practitioners to accept the UGDP results was much higher than that required to accept the results of the VA Study.

This observation is central to the argument of this book. The pharmaceutical mobilization and expansion of a disease category—whether hypertension or asymptomatic diabetes—is a complex process involving the coordination of many stakeholders; the structures of research and marketing must play multiple overlapping roles in overcoming resistance to therapeutic expansion. But as labor-intensive and plodding as the expansion of a disease category from symptomatic to asymptomatic might be, the process of *restricting* a disease entity once it has successfully expanded is an effort on an entirely different order of magnitude. It is one thing to convince doctors a condition is worth treating and to convince otherwise healthy individuals to identify themselves as having an unseen condition that demands treatment. But once those actors have been mobilized, once physicians have formed their practice around such labels and once pharmaceutical consumers have formed corresponding disease identities, any process of disease contraction must be contested by the embodied inertia of these newly diagnosed populations. After a decade of pharmaceutical therapy, it is difficult to tell a patient that he never really had a treatable disease without calling into question the entire edifice of medical knowledge and previous trust in the doctor-patient relationship.

We have seen how clinical trial evidence that threatened to overturn a logic of therapeutic risk reduction did not actually do so, even though the data was never proved to be false and there was not another trial conducted that directly controverted the study and supported the orthodox position.[121] The narrative of this crisis also reflects the sweeping challenges and changes in medical and governmental forms of authority that took place during the 1970s. Because pharmaceuticals represent a vital intersection of the federal government's ability to regulate consumer goods, the American medical profession's ability to determine medical practice, and the will of the citizen as consumer, drugs like

Orinase can bring conflicts over changing forms of medical authority into sharp relief. From the 1970s onward, pharmaceutical regulation has increasingly involved the public, and the assembled structures of publicity, in extraprofessional disease negotiations. As seen in the case of asymptomatic diabetes, this expansion of the arena in which disease is defined has broadened a conversation formerly limited to doctors and patients into a very large-scale conversation indeed. Included in the conversation now are many actors: NIH scientists, academic physicians, consumer groups, malpractice lawyers, federal regulators, and, increasingly, the pharmaceutical industry. Furthermore, as the pharmaceutical has become a principal site in the regulation of medical practice, it has become equally apparent that the patient's voice is most audible when it represents a consumer. As we have seen, this process has led to convoluted problems of co-optation, crises of authenticity, and a general confusion over what it means for a patient to be a consumer.

The trials of Orinase in the 1970s illustrate succinctly how difficult it is to undo a condition of risk and its pharmaceutical prevention after both drug and disease have been effectively marketed to a population of clinicians and patients. In chapter 5 we will explore the converse: what happens when a condition of risk, lacking a suitable therapeutic agent, fails to gain currency altogether.

Part III / Mevacor and Cholesterol, 1970–2000

CHAPTER FIVE

The Fall and Rise of a Risk Factor

Cholesterol and Its Remedies

> Those drugs thou hast, and their adoption tried; grapple them to thy soul with hoops of steel; But do not dull thy palm with entertainment of each new-hatch'd unfledged remedy.
>
> —*The Pharmacological Basis of Therapeutics*, 1954

Cholesterol is a familiar figure in contemporary American life. Even if the average consumer is not conversant with the chemical structure of this five-ringed sterol or its role in the biosynthesis of bile acids, sex hormones, and gallstones, chances are that he or she knows cholesterol to be an agent of progressive disease of the heart and blood vessels, to be avoided in one's diet and minimized in one's bloodstream to prevent illness and promote longevity. High cholesterol was one of the first identified risk factors for coronary heart disease. It shared top billing with high blood pressure as one of the two "prepathological" categories found, in the initial Framingham Study publications of 1957, to be firmly predictive of heart disease.[1] By 1974 these two had been joined by diabetes, gout, smoking, obesity, and a host of other behavioral and physiological states to form the Framingham risk factors for heart disease.[2] As we have seen, by the mid-1970s asymptomatic treatment of such categories on a basis of pharmaceutical prevention had already become standard practice in many medical arenas.

Not so with cholesterol. Over the course of the 1970s, while consensus developed around high blood pressure as a condition demanding preventive treatment, the status of high cholesterol as a treatable state had, if anything, deteriorated. It had not, like hypertension, attracted effective and palatable new medications in the 1950s and 1960s. Drugs devised to lower cholesterol either didn't work, weren't safe, or were found to have unpleasant side effects that made the calculus of preventive therapy untenable. The benefits of nonpharmaceutical interventions such as low-cholesterol diets were difficult to substantiate. Over the course of the 1970s, popular accounts of cholesterol and health began with broadly confident attempts to mobilize readers against high cholesterol and shifted to paranoid criticism of elevated cholesterol as a health-fad fabrication. In 1980 the National Academy of Sciences issued a report suggesting that widespread efforts to control cholesterol levels lacked justification in the clinical and scientific literature, sparking a new wave of controversy over the value of cholesterol to individual health.

How did cholesterol lose its relevance in the growing canon of physiological prevention? And how, in the period since 1980, did the perseverant compound find its way back into the sphere of mainstream clinical activity and mobilized public anxiety? Although pharmaceutical developments did not single-handedly determine this trajectory, both the fall and the rise of cholesterol as a risk factor were intimately related to the performance and promotion of discrete drug entities. Chapter 4 depicts a category of preventive pharmacotherapy—the treatment of asymptomatic diabetes with oral hypoglycemics—whose usage persisted even after the principal therapeutic agent was deemed ineffective and potentially harmful by a supposedly definitive clinical trial. This chapter offers a complementary contrapositive, for the story of cholesterol in the 1970s and 1980s recounts the failure of a category of risk reduction to survive in the absence of an appealing intervention.

Atherosclerosis and Its Germ Equivalent

By the mid-twentieth century, cholesterol had accumulated the most convincing causal claims of any putative etiological agent of coronary heart disease. Unlike the obliquely relevant physiology of blood pressure and blood sugar, or the behavioral fuzziness of weight gain and smoking habits, the role of cholesterol in the development of heart disease was borne out by an explicitly lesion-based model of disease. The molecule of cholesterol itself was present, and always present, in the fatty streaks, plaques, and clots that plagued the inner arteries of patients with atherosclerotic heart disease. It is deeply ironic that cholesterol—perhaps the closest thing to a germ that the search for causal agents of coronary heart disease could muster—would take so long to be legitimated in clinical practice.

To describe cholesterol as equivalent to a germ is not to claim that cholesterol was ever seen as an infectious agent. But the advent of germ theory in the late nineteenth century and its widespread medical and popular promotion by the early twentieth had ramifications that extended well beyond what we would today consider the scope of infectious disease. As the microbe worked its way into American popular culture as both explanation of illness and justification for preventive hygienic measures, enterprising investigators and marketers sought microbial bases for almost every variety of disease and often linked such pursuits to specific antimicrobial products.[3] At the same time, the theoretical emphasis on specificity implied by the germ theory—that all diseases were distinct species that could be linked to causal agents and understood on a mechanistic level—was gradually folded into many areas of medical research.[4] Although most chronic diseases could not be associated with specific microbes, some were explainable in terms of similarly minuscule agents. Pollen took on many of the roles of the germ in early hay fever research, studies of vitamin deficiency yielded clear and discrete causal agents for chronic diseases such as pellagra (niacin) and pernicious anemia (B12), and toxicological investigations suggested that exposure to nonliving toxic agents, such as silica, could produce specific chronic diseases like silicosis.[5]

None of these causal agents of chronic disease were discovered in microbiology laboratories, but the rational framework for their elucidation bears a strong kinship to the postulates set out by the German physician and bacteriologist Robert Koch in his widely circulated demonstration that the *Mycobacterium tuberculosis* bacteria was the cause of the disease tuberculosis. Koch elaborated a series of postulates as conditions that needed to be fulfilled to demonstrate causality between agent and disease, which would become central to the logic of specificity in early-twentieth-century American medicine: for every disease, there is ideally a single causal agent that can be understood in mechanistic terms.[6] Identification of the agent provided not only a more fundamental knowledge of disease but also the hope that its containment, prophylaxis, and eradication might follow.

Such was the context of optimism and activism in which cholesterol was advanced as a causal agent in the production of chronic ailments of the heart and the blood vessels. Early-twentieth-century texts had uniformly described *arte-*

riosclerosis—literally, the hardening of the arteries—as an inevitable degenerative process associated with aging. After it became apparent that most heart disease was associated with a particular type of hardening, called *atherosclerosis*, which involved the accumulation of fatty substances called atheromata on the inner lining of the arteries, some investigators began to search for microscopic mechanisms that might explain the fatty accumulation. In 1913, under the waning aegis of the last czars, Russian pathologist Nikolai Anitschkow demonstrated that lesions that looked remarkably similar to human atherosclerosis could be reproduced in experimental animal models supplied with a high-fat diet. The chief compound in these fatty streaks was a chemical known as cholesterin. Anitschkow subsequently showed that injection of cholesterin alone was sufficient to produce atherosclerotic lesions in previously healthy rabbits, and he argued that cholesterin would soon prove to be the causal agent of atherosclerotic heart disease.[7]

By the time the structure of the molecule, now known as cholesterol, was clarified in 1932, its role in the formation of steroid hormones, bile acid production, and fat digestion and transport had assured it a vital and vibrant place within medical research.[8] Nevertheless, Anitschkow's work did not immediately produce a sea change in medical practice relating to atherosclerosis. Critics objected that forcing an animal product into the diet of an exclusively herbivorous species like the rabbit was not a fair demonstration of pathological agency: these arguments gained sway when similar studies injecting cholesterol into dogs, a carnivorous species, failed to produce atheromatous lesions.[9] By the late 1930s, however, a series of studies of families with symptomatic cholesterol tumors showed that such populations experienced dramatically higher rates of angina pectoris and atherosclerosis than the rest of the population. And by the early 1940s, a set of researchers had been able to produce a more credible model of cholesterol atherosclerosis in the chick, which, like humans, was an omnivore.[10] This experimental vindication of cholesterol led Louis Katz—one of the leading cholesterol researchers of the mid-twentieth century—to proclaim, in 1952:

> It is upon the basis of the *cholesterol concept of atherogenesis* that fruitful research in this field is proceeding apace in a number of laboratories. The basic tenet of this concept may be simply stated: without an altered lipid-cholesterol metabolism little or no atherosclerosis will develop regardless of any other alterations in the arterial wall, including senescent changes. Obviously, if atherosclerotic le-

sions are the result—or even only part of the result—of altered lipid-cholesterol metabolism, then they are not inevitable. The whole foundation of the senescence theory itself is rendered untenable. The possibility, nay inevitability, presents itself that preventing or reversing the altered lipid-cholesterol metabolism will eliminate atherosclerosis. Thus, a hopeless situation is changed to one full of promise.[11]

Such optimism regarding cholesterol's causal role in the development of atherosclerotic heart disease reached its apogee in the 1961 edition of the classic *Cecil-Loeb Textbook of Medicine,* which argued that sufficient evidence had accumulated to demonstrate that cholesterol fulfilled Koch's postulates of agent-disease causality.

Koch's first postulate of sufficiency, the *Cecil-Loeb* entry argued, was "easily satisfied," since the cholesterol-containing low-density lipoproteins (also known as LDL, LDL cholesterol, or, more commonly in the early 1960s, beta-lipoproteins) were present in the plasma in every case of the disease and could be found within all atheromatous lesions. The second of Koch's postulates—that the agent must be "isolated in pure form," was also satisfied by both X-ray crystallography of the cholesterol molecule itself and the molecular characterization of the low-density lipoprotein. Koch's third postulate, that the "agent, in pure culture must, when inoculated into a susceptible animal, give rise to the disease," was considered to be satisfied by studies showing atherosclerotic development in several previously healthy experimental animals following injection of cholesterol or LDL cholesterol. Koch's fourth postulate, that "the agent must be observed in and isolated from the experimentally diseased animal," was more than evident in the discovery of LDL cholesterol in both the blood and the atheromatous lesions of these experimental animals. In comparison with microbial agents such as Koch's *M. tuberculosis* or B. anthrax, the textbook concluded, "the agent for atherosclerosis is unique only in the sense that it rises within the host and becomes a threat to the host as part of his internal instead of external environment."[12] With reference to Koch's postulates, the highest standard by which a laboratory could name an agent as the cause of a disease, this clinical textbook named cholesterol to be the causal agent of atherosclerosis.

In spite of such exultant formulations, many practicing physicians were less confident that the causal role of high blood cholesterol in producing atherosclerosis and coronary heart disease had been definitively established. The asymptomatic nature of the atherosclerotic process (prior to the symptomatic end-stage events of ischemic heart disease, myocardial infarction, or stroke) complicated the translation of cholesterol's significance from the laboratory to the clinic. Unlike a microbe, cholesterol wasn't something one had or didn't have; rather, all humans synthesized their own cholesterol and required it for survival, and cholesterol levels in the American population followed a roughly bell-shaped curve. Only some people with high cholesterol would ever show symptoms of heart disease; conversely, not everyone who experienced a heart attack, stroke, or angina had measurably high levels of plasma cholesterol. Sensing that Koch's experimental animal was insufficient to answer such fundamentally epidemiological questions, cholesterol researchers in the 1940s and 1950s increasingly turned to the field to demonstrate cholesterol's relevance.

The career of Jeremiah Stamler exemplifies this movement from the laboratory to the field. Stamler's early career as an experimental pathologist culminated with the 1953 publication of Experimental Atherosclerosis, a collaborative text Stamler authored with Louis Katz, the developer of the "omnivorous chick" cholesterol model. By the late 1940s, however, Stamler had become involved with a group of cardiologists and epidemiologists who were developing a population-based study to determine factors predicting the development of coronary heart disease in the entire population of a small industrial city an hour outside of Boston. Stamler oversaw the measurement of the cholesterol levels of thousands of residents for several years before his Framingham Study produced its first confident pronouncement of the statistical association between cholesterol and coronary heart disease in 1957.

By the time of the first Framingham publications, data linking cholesterol and heart disease had also emerged from other field studies, most notably a series of cross-cultural studies led by Ancel Keys of the University of Minnesota, which demonstrated a correlation between populational levels of dietary fat and cholesterol and incidence of coronary heart disease. One particularly influential study tracked the varying rates of coronary heart disease among Japanese-born men raised in Japan, Hawaii, and Los Angeles to show the significance of dietary practice over genetic substrate.[13] Routine autopsies on young men killed in action in the Korean War had revealed that a surprising amount of advanced atherosclerosis and coronary artery obstruction existed in soldiers under the age of fifty, with no reported symptoms.[14] A causal connection was assured, as the University of Pennsylvania's David Kritchevsky predicted in 1958: "In the popular mind, cholesterol and atherosclerosis are regarded almost as synonymous . . . a connection between the two seems definite."[15] As two prominent cardiologists declared in a review of cholesterol-lowering agents in 1961, "although the role of serum lipids in the genesis of human atherosclerosis is uncertain, the likelihood of an important cause-and-effect relationship is sufficiently great that . . . there is no question of the desirability of reducing serum lipid levels in hypercholesterolemic and 'hyperlipemic' states in which early and severe atherosclerosis are the rule."[16]

Indeed, by the early 1960s cholesterol was simultaneously promoted to the general public as a symbol of progress in the fight against heart disease and an increasing source of consumer anxiety.[17] Within the medical profession, discussions of the Framingham risk factors foregrounded the role of cholesterol:

> Recent studies of the epidemiology of coronary heart disease have led to the concept of the "coronary prone" individual. Such a person, as characterized especially by the United States Public Health Service study in Framingham, Massachusetts, is a male, is usually overweight, smokes cigarettes, is hypertensive and has a reduced vital capacity. Above all, however, he has an elevated serum cholesterol level . . . With this concept in mind, let us direct our attention to the regulation of cholesterol concentration in the plasma, for if plasma cholesterol is maintained at a low but healthy level throughout a patient's lifetime, any "atherosclerosis" of clinical consequence can largely be prevented.[18]

Based on the logic of molecular mechanism and proofs from the laboratory and the field, cholesterol had been indicted. In the early 1960s it was a highly visible target awaiting its magic bullet.

The Pharmacopoeia of Failure

As evidence for the association between cholesterol and heart disease mounted, cardiologists and pharmaceutical industry executives saw great opportunities for this specific and broadly applicable therapeutic target. Early pharmaceutical attempts at cholesterol reduction in the 1930s and 1940s had relied on nonspecific biologic remedies using intermediates in lipid metabolism and endocrinologic pathways, including bile acids (choline and inositol), emulsifiers such as lecithin, thyroid preparations, estrogens, and a variety of plant and animal products—artichoke extract, garlic, seaweed, extracts of animal brain and pancreas, and others.[19] After the growing popularization of cholesterol-as-nemesis in the late 1950s, elements within the pharmaceutical industry began a concerted effort to develop a specific chemotherapeutic agent for lowering cholesterol.[20] Unlike the cases of hypertension and mild diabetes, however, a series of pharmaceutical setbacks would greatly complicate the project of turning elevated cholesterol into a treatable condition.

The first drug to be explicitly marketed as a specific cholesterol-reducing agent emerged from the same promising wave of chemotherapeutic developments that produced Merck's blood-pressure-lowering Diuril and Upjohn's blood-sugar-lowering Orinase. Approved by the FDA in 1960, MER/29 (triparanol) was launched by a well-oiled marketing machine with the careful integration of respected opinion leaders in medical research and practice. Unlike Diuril and Orinase, however, a few years after its launch MER/29 was found to be materially harmful to its consumers, was removed from the market in scandal, and led to a series of criminal convictions for its manufacturers that carried one of the largest corporate payouts yet recorded in settlements for injured consumers.

MER/29 was a product of the Richardson-Merrell Company, owner of the Vicks brand of over-the-counter remedies; it was a consistent Fortune 500 company throughout the 1950s. Merrell invested in basic science research on small molecules that could interfere with cholesterol and steroid synthetic pathways; in 1958 an R&D team, under the direction of Merrell chemist Frank Palopoli, announced the discovery of a promising compound and issued practicing physicians a confidential brochure to recruit patients for early trials. By spring of 1959, clinical trials in humans reported that MER/29 significantly reduced serum cholesterol and was "well-tolerated clinically" with no reported side effects. The compound appeared to function by blocking the final step in cholesterol synthesis, in which desmosterol—the final precursor in the biosynthesis of cholesterol—was turned into cholesterol.[21] In the presence of an apparently safe, effective pill that reliably reduced blood cholesterol with no unpleasant effects, Merrell saw significant sales potential and set out to bring the agent to its broadest possible market.

In July of 1959 Merrell president Frank Getman issued a memorandum to his chief of sales titled "Let's Start Selling," declaring, "This is the year when we have every reason to believe Merrell should break into the truly 'big time.' Let's take a close, critical look at the way we are stimulating the field force on MER/29."[22] The marketing of MER/29 presented unique opportunities and challenges for Merrell. The company stood a chance to become an industry leader in the field of preventive cardiology, with no significant competitors among cholesterol-reducing agents. As a breakthrough agent, though, MER/29 would face a population of relatively naive physicians with no clinical consensus on the pharmacological treatment of cholesterol. Marketing analysts noted in 1961 that "this was an entirely new product—to this day it still has no direct competitors—and this necessitated a broad scale informational and educational program to brief the medical profession on what MER/29 is, what it is supposed to accomplish, and how it should be used."[23]

Clinical trial data would be vital to both the regulatory approval of MER/29 and its subsequent marketing, and the Merrell network of clinical researchers soon included many of the same institutions and individuals involved in the launch of Diuril. By April of 1959, a press release announced that research teams in Philadelphia and Boston had simultaneously demonstrated the efficacy of MER/29 in reducing cholesterol up to 35 percent with "few side effects . . . taken orally in doses as low as 250 milligrams a day."[24] The marketing value of these clinical trials was known currency, as one marketing analyst noted: "For the advertising copywriter, this evidence was like money from home—ready ammunition."[25] The kick-off event for the MER/29 promotional campaign was a symposium held in Princeton, New Jersey, in December of 1959, featuring luminaries of cardiology such as Irvine Page, John Moyer, and Robert Wilkins. Their presentations were subsequently published as a supplement to the journal *Progress in Cardiovascular Disease;* Merrell's president later called the supplement "the most terrific selling tool Merrell had ever had."[26]

Following FDA approval of MER/29 in June of 1960, Merrell accelerated its promotional efforts. A former Merrell employee who had since become an advertising executive at a leading pharmaceutical marketing firm was named director of advertising and provided with an ample budget. A year later, MER/29's marketing campaign was written up on the pages of *Drug and Cosmetic Industry* as a case study in the effective deployment of "the varied marketing tools used today in bringing a new pharmaceutical to market."[27] The article enumerated the steps of MER/29 promotion as a primer for future drug launches:

1   On June 1, a first-class letter signed by the president of Merrell went out to some 160,000 practicing doctors announcing the availability of MER/29.

2   This was followed immediately by the delivery via Western Union of hard-cover binders to some 100,000 doctors. This loose-leaf binder contained basic data on MER/29, including a précis of the Princeton

conference. Doctors were advised that further reports of MER/29 activity would be issued and they were urged to add this data to their binders . . .

3   On July 1, journal advertising began with the placement of eight-page inserts in a basic list of 15 publications. Advertising was restricted to mass medical magazines such as the Journal of the American Medical Association, MD, Modern Medicine, and Medical Economics, and to several specialty journals such as the American Journal of Cardiology, Annals of Internal Medicine, and Circulation.

4   The same insert was repeated in August issues and Merrell then reduced the size of insertions to double pages and full pages, running them through the rest of the year at the rate of one a month.

5   The direct mail program, calling for three contacts a month with doctors, was maintained. The basic vehicle consisted of "Facts About MER/29" folders which were designed to be added to the binders already distributed to doctors. The tenth folder in this series was an impressive annotated bibliography of "Cholesterol and Atherosclerosis," covering the years 1958 through 1960.

6   Supplementing the advertising program was a concentrated detail effort, organized at a national sales conference of all Merrell detailers at French Lick, Ind. Merrell has about 300 detail men and this was the first national conference for them in a number of years. The detail effort was considered quite important, since doctors, unfamiliar with therapy of this kind, would tend to turn to detailers for more information.[28]

Merrell's marketing sought to encompass the entire universe of information sources by which a practicing physician might learn of the drug. The first year of MER/29 promotion involved an advertising outlay of eight hundred thousand dollars, as much as the entire Merrell ad budget for all products combined the year before. The campaign was beautifully synchronized and thematically linked: all advertisements were coordinated in form and style to the surrounding promotional effort—down to the color-matching of the ink to the exact pearl-grey hue of the MER/29 capsule itself. By the end of the year, MER/29 had reached three hundred thousand consumers and had generated a sales volume of $5 million, prompting *Drug and Cosmetic Industry*'s Milton Moskowitz to proclaim that next to the best-selling tranquilizer, Librium, "MER/29 was probably the outstanding new product of 1960."[29]

Amid the glowing praise of market analysts, however, some physicians expressed concern that a novel and unknown compound would be intended for widespread and long-term consumption by such a large number of people. Although MER/29's blockade of the final step in cholesterol synthesis did indeed reduce cholesterol levels, it also led to a buildup of the precursor molecule, desmosterol, which bore a significant chemical resemblance to cholesterol. As some voices in the clinical literature asked whether high levels of desmosterol in the blood could be just as bad as high cholesterol, reports began to document that desmosterol might actually be worse. Excess desmosterol in the blood tended to accumulate within cholesterol-rich structures such as the lens of the eye, the hair follicles, and the skin—producing cataracts, unusual changes in body hair, and occasionally ichthyosis, a dermatological disease in which the skin becomes hard and covered with reptilian scales.[30]

As reports of cataracts and hair loss began to flow in to Merrell's public relations offices, the company continued to reassure physicians that such "idiosyncrasies" amounted to an "insignificant incidence." Merrell actively resisted posting warnings to physicians until the FDA forced its hand in December of 1961. By the spring of 1962, publications documenting the side effects of the drug, coupled with a reexamination of Merrell's own original clinical trials by FDA statisticians, which suggested that patients on the drug had a threefold increase in cataracts, prompted the FDA's deputy commissioner to note that "in retrospect, it is apparent that the drug should not have gone on the market in the first place." In April of that year, Merrell removed its flagship product from the market at precisely the moment when its promotional efforts were beginning to bear fruit.[31]

It was a difficult year for Merrell. As MER/29 was being removed from the market in scandal, an FDA officer named Frances Kelsey—in a decision that would launch her into the public spotlight and ensure the passage of the flagging Kefauver-Harris legislation in Congress—rejected Merrell's application for its next anticipated blockbuster medication, the now infamous anti-emetic known as thalidomide.[32] MER/29's failures, however, ultimately proved more injurious to the firm than the negative publicity surrounding thalidomide. Merrell, it appeared, had known of MER/29's side effects for several years before launch but had chosen to hide them from investigators, regulators, physicians, and patients. Documents surfaced that suggested that Merrell had actively encouraged its salesmen to shift blame for any possible side effects, as evidenced in this 1960 memorandum advising MER/29 salesmen: "When a doctor says your drug causes a side effect, the immediate reply is: 'Doctor, what other drug is the patient taking?' Even if you know your drug can cause the side effect mentioned, chances are equally good the same side effect is being caused by a second drug! You let your drug take the blame when you counter with a defensive answer."[33]

A disgruntled Merrell employee in Cincinnati, Beulah Jordan, provided laboratory notebooks to FDA inspectors indicating that her laboratory had fabricated large portions of the safety and efficacy data it originally submitted to the agency. In a first for the twenty-four-year old agency, the FDA moved that sufficient evidence had been obtained to support prosecution of the corporation and the individuals involved. Shortly afterward, a federal grand jury issued a twelve-count indictment against the William S. Merrell Company, its parent company, Richardson-Merrell, two laboratory supervisors, and a Merrell vice president on charges of lying, fraud, and intentional misleading of government agencies and the American public. Seeking to avoid a public trial, the corporations and individuals pleaded no contest to eight counts; sentencing in 1964 resulted in nominal fines and six months of probation in lieu of five-year prison terms.[34]

Merrell had become the first drug company to be criminally convicted for failing to provide complete information about a drug to the FDA, and the one-two punch of MER/29 and thalidomide had seared Merrell in the public's mind as a potent symbol of the pharmaceutical industry's corruption and abuse of public trust.[35] Although the fines were minor, the reputation of the firm was shattered, and Merrell soon faced a series of costly civil trials from injured patients that continued for the rest of the decade. A loose affiliation of several hundred lawyers and injured claimants that called itself the "MER/29 Group" assembled a mounting body of material regarding Merrell's actions. Settlements in these cases—typically resolved out of court to avoid precedent—averaged from $25,000 to $125,000, though a few did surpass $1 million; by 1972, Merrell had paid an estimated $45–55 million. In its short career, MER/29 had witnessed the full promotional slingshot of a state-of-the-art drug-marketing machine. It had been, briefly, the drug industry's shining avatar of research, development, and promotion, but its hasty launch had cost Merrell more dearly than any American drug company had ever been fined.

This enormous precedent of criminal charges and civil payouts levied against a reputable pharmaceutical firm left a deep impression upon prescribing physicians and leaders of the pharmaceutical industry. Chief among their concerns was the increasing scrutiny of risk and efficacy in arguments of responsible and irresponsible drug promotion and prescription. Whereas a side effect like a cataract might be deemed a worthwhile risk in the treatment of cancer or a devastating systemic infection, FDA investigators considered MER/29 usage to be particularly problematic precisely because it was used for a symptomless condition that had not been decisively linked to the successful prevention of heart disease. It was evident, after MER/29, that the scrutiny of any cholesterol-lowering agent would require not only rigorous long-term safety examinations but also a stronger demonstration that lowering cholesterol levels actually decreased mortality enough to justify long-term exposure to a pharmacological agent. For more than a decade, the development of cholesterol-lowering agents was tainted by the memory of MER/29.

During the remainder of the 1960s, the pharmaceutical industry took a more cautious approach to the development of cholesterol-lowering medications. In 1960, the year MER/29's cholesterol-lowering effects were initially published, Upjohn's antibiotic neomycin was found to have cholesterol-lowering properties equivalent in magnitude to triparanol's.[36] Although neomycin had some role in cholesterol reduction in ensuing decades, Upjohn's marketing teams largely downplayed this aspect of their product, and neomycin was promoted chiefly as an antibiotic. At roughly the same time, the intravenous blood-thinner heparin sodium was found to reduce cholesterol levels in many patients. The dosing of heparin for cholesterol reduction required subcutaneous injections twice daily; in addition to side effects, hemorrhage, and bruising at the site of injection, heparin's role in producing occult or excessive bleeding severely curtailed its usage for cholesterol lowering.[37] Niacin (nicotinic acid), the B-complex vitamin at the end of Joseph Goldberger's search for a specific cause of pellagra, was also found to reduce blood cholesterol levels.[38] To be effective as a cholesterol-lowering agent, however, niacin had to be taken in a dose much higher than when it was used as a metabolic vitamin. Three to six grams daily was a typical dose, and a highly predictable niacin toxicity accompanied it: generalized flushing in almost all patients, coupled with extensive and unremitting bodily itching and severe gastrointestinal complaints. Due to these regular occurrences, it was extremely difficult to keep a patient on niacin for more than a few years, making the collection of outcome data difficult and niacin's use in long-term prophylaxis "largely experimental."[39]

Largely because of the unpleasantness of these agents, it was difficult to determine whether they actually worked beyond the short-term end point of immediately lowering cholesterol. The drugs were chemically effective, commentators noted, but was there any proof of long-term clinical efficacy? Did lowering people's cholesterol really help them to live longer? Did it actually prevent heart disease? Clofibrate, a derivative of fibric acid marketed by Ayerst as Atromid-S, was perhaps the best-tolerated cholesterol-reducing agent of the 1960s and 1970s, but optimism surrounding the drug was mixed with significant therapeutic nihilism, and clinical reviews suggested that, until the long-term toxic and therapeutic effects for the drug were clarified, the use of clofibrate should be restricted to high-risk patients with symptomatic disease.[40] Shortly after clofibrate's release, a randomized, double-blind clinical trial of more than eight thousand men, the Coronary Drug Project, compared nicotinic acid and clofibrate along with estrogens. When the study concluded in 1974, its results indicated no benefit for niacin or fibrate usage.[41] To make the chance of a meaningful result as likely as possible, the group of subjects enrolled in the Coronary Drug Project were "high-risk" individuals who had already suffered a cardiac event; even so, the trial indicated that "there is no evidence of significant efficacy in the drug[s] with regard to total mortality or cause-specific mortality."[42] As the *New York Times* reported in 1975, these two drugs widely used to prolong the lives of men who have recovered from heart attacks had "proved useless for that purpose."[43]

Whatever optimism Atromid-S had inspired in the treatment of high cholesterol was effectively gutted by these 1975 negative trial results. Clofibrate was subsequently found to cause gallstones and various liver abnormalities, and, although Ayerst was never criminally implicated in the same sense as Merrell, vocal critics within the medical profession began to call for the drug's withdrawal. Opponents of clofibrate argued that the known short-term adverse effects of the drug mitigated any hypothetical long-term improvement in cardiovascular mortality. A lead article in the "Science Times" section of the *New York Times* summed up the expansion and bursting of the clofibrate bubble with the headline "'Miracle' Drug Discredited; Health System is Faulted."[44]

Cholesterol Diets and the Collapse of Consensus

In the absence of conclusive evidence that lowering cholesterol produced any health benefits that would justify the potential hazards of long-term drug therapy, less-threatening dietary interventions became the central focus for those who saw cholesterol as the key to cardiovascular risk reduction. By the late 1960s, many physicians agreed with this clinical reviewer that "because of uncertainty regarding the toxic effects of many drugs, some agents remain experimental and have only special applications. Probably the safest, most healthful, and least expensive treatment measure to use . . . is dietary modification, involving restrictions on foods rich in cholesterol and fat."[45] The low-fat, low-cholesterol diet had been a favorite for health boosters long before Framingham, having its roots in turn-of-the-century critiques of new "diseases of civilization" and in more diffusely Puritan critiques of excess.[46] Epidemiological studies that demonstrated lower coronary mortality associated with the lower dietary cholesterol levels of Japanese populations were mobilized into a form of cardiovascular Orientalism, which identified the Asian diet as inherently more healthful for the cardiovascular system than its American counterpart. Ancel Keys's cross-cultural accounts of diet and coronary heart disease were soon joined by other condemnations of the "high-fat, high-cholesterol American diet," a critique that spread quickly through the popular press.[47]

Although a handful of small-scale diet intervention studies had, by the 1950s, indicated some benefit of low-fat diets in cardiac patients, their widespread extrapolation had become controversial because of small size and lack of methodological rigor.[48] Moreover, studies in the early 1960s showed that the body synthesized most of its own cholesterol de novo rather than absorbing it from dietary fat, suggesting that reducing dietary cholesterol might not have any significant effect on one's blood cholesterol levels.[49] Nutritionist Frederick Stare and prominent preventive cardiologist Irvine Page (we remember him from chapter 2 as an early supporter of antihypertensive medications) spoke with influence from within the American Heart Association (AHA), insisting that dietary faddism was not necessarily healthful for the American population. As coauthors of the first AHA policy on diet and heart disease, they warned against the "flood of diet fads and quackery" that could easily surround low-fat, low-cholesterol diets.[50] Amid a swell of low-cholesterol food products presenting cardiovascular health claims in the late 1950s, such as Mazola Corn Oil and Emdee polyunsaturated margarine, the American Medical Association described dietary interventions for heart disease as a "near hysteria," and the FDA formally reprimanded food product manufacturers in 1959 that "a causal relationship between blood cholesterol levels and this disease has not been proved."[51] Elements of the beef and dairy industry also mobilized in the early

1960s to defend "the traditional American breakfast" in general and their industry's cholesterol-rich products in particular, against what they considered unsupported claims of unhealthiness.[52]

By 1960 diet critics Stare and Page, along with diet boosters Ancel Keys and Jeremiah Stamler, had been elected to the committee responsible for the American Heart Association's dietary policy. A compromise was reached in an AHA policy statement admitting that dietary intervention could be useful in certain "high risk" cardiac patients. But the fundamental question of whether cholesterol-related dietary intervention actually prevented heart disease was still unanswered, and all members of the committee felt the question could be answered only by a long-term double-blind prospective clinical trial.[53] From opposite sides of the issue, Page and Stamler independently petitioned the National Heart Institute (NHI) of the National Institutes of Health (NIH) for funds to perform a "national cooperative study" that would study the effects of diet on the incidence of heart disease, and they persuaded the NHI to grant them a provisional budget.[54] The planning group for the National Diet-Heart Study, convened in the spring of 1960, included Keys, Stamler, Page, and Stare, as well as a handful of cardiologists, nutritionists, and the NIH statistician Jerome Cornfield. The proposed study, a "large-scale public health field trial," would test whether dietary intervention—if possible at all—would have any effect on the incidence and mortality of heart disease.

The study was never conducted, but not for want of effort. Randomized clinical trials of this scale had never been performed with dietary interventions, so the planning group conducted a one-year feasibility study to see whether the model of the trial could even function. In the philosophy of double-blind, randomized clinical trial design, the placebo should ideally be identical to the intervention in all respects except for the presence of the active agent being tested, and that agent should be standardized across all participants. This was difficult enough to conduct for a pill-based intervention; with diet, however, such standardization seemed nearly impossible to attain. To circumvent the problem, the Diet-Heart researchers required participants in the pilot studies to purchase all of their meals in randomized, standardized units provided only at local study centers, dispensing diet through the equivalent of a pharmacy. To the surprise of many involved, the study design proved both feasible and reasonably consistent in pilot studies, and participants on experimental intervention diets were found to experience an average reduction of 11 percent in serum cholesterol levels. However, at such a modest level of change, the investigators projected that even if reducing cholesterol did have a robust effect on the incidence and mortality from heart disease, it would take as many as 115,000 people over seven to ten years to demonstrate it, at a cost of over $1 billion in 1968 dollars. By 1971 the study was deemed unfeasible and relegated, unfunded, to posterity.[55]

In its place, the National Heart Institute (soon to be renamed the National Heart, Lung, and Blood Institute, or NHLBI) moved to fund a much different trial. Instead of following a population at average risk over many years, the Multiple Risk Factor Intervention Trial (MRFIT) would select a subjectively healthy population with membership in several high-risk categories; in these people the likelihood of demonstrating the value of risk-reduction therapy would be most evident. In this federally coordinated research effort, one group of high-risk men would be coached to quit smoking, eat more healthily, and reduce their blood pressure, while another would be followed in routine primary care as a control.[56] At $112 million, MRFIT represented one of the largest and most expensive clinical trials the federal government had ever supported. Unfortunately, its primary results, far from proving the value of risk reduction, were almost entirely negligible. Although the intervention arm managed to quit smoking, change their diets, and lower their blood pressures to an impressive degree, there was no measurable difference in incidence of cardiovascular disease or mortality caused by it between the two groups. From the perspective of justifying the long-term value of cholesterol reduction, MRFIT was a failure.[57]

The release of the disappointing MRFIT results in 1980 was accompanied by a report in the same year from the National Academy of Sciences—the nation's most prestigious scientific body—whose Food and Nutrition Board (FNB) had recently reviewed the status of cholesterol in the prevention of heart disease.[58] The 1980 FNB report cited high blood pressure as an example of an epidemiologically labeled risk factor that had been demonstrated through intervention trials to be a condition worth treating. However, the report pointed out that the association between cholesterol and heart disease was still merely an association. Without the validation of a successful program of intervention, high cholesterol could not properly be considered a treatable condition.

Manufacturing Consensus: The National Cholesterol Education Program

In January of 1984, just a week before Ronald Reagan was sworn in for his second term in the White House, the director of the National Heart, Lung, and Blood Institute held a press conference to unveil the results of a trial that claimed to definitively prove the merit of cholesterol reduction in the prevention of heart disease: the Lipid Research Clinics Coronary Primary Prevention Trial (LRC-CPPT).[59] Following this widely publicized trial, the NHLBI swiftly set up the Consensus Conference on Lowering Blood Cholesterol to Prevent Heart Disease, which then moved to enact the National Cholesterol Education Program (NCEP), a public-private hybrid institution that has since worked to educate the medical profession and the general public on the value of lowering cholesterol levels by all possible means.

Pharmaceutical agents were central to this process. Like the Multiple Risk Factor Intervention Trial, this new study trial arose from the ashes of the Diet-Heart Study as the governing board of the NHLBI worked to allocate clinical research funds into long-term prevention trials that promised greater likelihood of demonstrating clinical benefit. A crucial difference between the two trials was that the Lipid Research Clinics trial used pharmaceutical prevention rather than behavioral change as its central modality of intervention. Proponents of cholesterol control, who were increasingly coming to populate the NHLBI, realized that if they wished to establish any value of cholesterol lowering, they would need to find a way to demonstrate a stronger "signal" of efficacy, to find an intervention that would cause a more marked reduction in cholesterol level than diet alone. That clearer signal could be provided only by a pharmaceutical intervention. As the planners of the LRC-CPPT noted, the failure of the Diet-Heart Study suggested the need for an "alternative test" more pragmatically suited to clinical trial than diet: "The use of the drug cholestyramine resin permitted a double-blind design. This drug . . . was selected on account of its known effectiveness in reducing total cholesterol and LDL-C levels, the availability of a suitable placebo, its nonabsorbability from the gastrointestinal (GI) tract, its few systemic effects, and its low level of significant toxicity."[60] The drug would enable the clinical trial. Given the limitations of niacin and clofibrate, cholestyramine in 1971 was indeed the most attractive drug available for the treatment of cholesterol. But hidden in the phrases "few systemic effects," and "low level of significant toxicity," was an uglier side of cholestyramine that explains why the investigators decided to use it as an experimental proxy for diet rather than as an intervention in itself.

Cholestyramine was a Merck-developed cholesterol-reducing agent that the company found so unpalatable that it refused to market it as such.[61] It was an effective drug, even a safe drug, but in terms of palatability, cholestyramine was about as ugly a pharmaceutical as they come. Alfred Alberts, who for many years managed the Merck Sharp & Dohme cholesterol project, described the compound in 1987: "The dose is very large, and patient compliance is low, because they don't want to take all of this stuff. The side effects: it has an odor of rotten fish, which we had some trouble getting rid of. It was a granular material, which sandpapered a part of your anatomy on the way out, it also caused fecal impaction in old people . . . but still, nevertheless, it's still a drug that actually works. It's safe, it's not absorbed, it's safe."[62]

Unlike Diuril or Orinase, cholestyramine was not a pill. Rather, it was presented to its consumer as a small paper sachet of gravel to be swallowed at a dose of two to eight grams per day. Cholestyramine was a bile-acid-binding resin known to be locally active in the small intestine, where it latched onto cholesterol-rich bile acids and blocked their reabsorption into the body. Because it was not an internal agent (though swallowed, it was never absorbed into the body) cholestyramine technically had no internal adverse effects. That is why Alberts could confidently declare the drug safe. Nonetheless, the presence of gravel in the intestine—particularly a gravel that interfered with the process of fat absorption—reliably produced constipation and a particularly unpleasant form of diarrhea. The marketing team at Merck felt that the drug's discomfiting taste and administration would make it a hard sell as a preventive measure. By the mid-1970s, cholestyramine had been licensed out to another company, Mead Johnson, for marketing as a cholesterol-lowering agent and was given the brand name Questran.

Patients were loath to follow their Questran prescriptions, sales were low, and refills of prescriptions were infrequent. The low level of patient compliance with cholestyramine helps to explain why NHLBI investigators saw the drug as a proxy for successful dietary reduction of serum cholesterol rather than as an intervention in itself. Although cholestyramine was crucial to the study's design, once the results were tallied, the drug was effectively written out of the study, its results being interpreted as a vindication of dietary intervention, not of pharmacological intervention. Rather, the cholestyramine study was announced as a double-blind, placebo-controlled clinical trial that tested, broadly, the "efficacy of lowering cholesterol levels for primary prevention of CHD."[63] Cholestyramine was useful as a research tool but not as a public health measure in itself.

As a primary prevention study, the LRC-CPPT enrolled only individuals whose sole risk for heart disease was their high cholesterol. Any subject with symptoms or prior diagnosis of heart disease was excluded from the study; those with other known cardiovascular risk factors, such as hypertension and diabetes, were excluded as well. Maintaining motivation among the volunteer subjects was no easy task: all four thousand of them, whether randomized to the cholestyramine arm or the placebo arm, were expected to swallow six packets of the odiferous gravel each day for ten years with no expectation of benefit.[64] Much effort was needed to convince people to stay in the trial; in one promotional strategy, the NHLBI issued a set of celebrity posters including tennis star Arthur Ashe to encourage study subjects to continue consuming their cholestyramine or placebo packets.[65]

At the end of the study, a statistical difference in outcomes was reported between the placebo and the cholestyramine groups. The difference was quite slim, however. The primary end points measured by the study were death from coronary heart disease (CHD) and nonfatal heart attacks. As the study's results announced, deaths and/or nonfatal heart attacks were 19 percent lower in the treatment group than in the placebo group. In absolute numbers, in groups of equal size, the treatment arm suffered 155 heart disease deaths or heart attacks compared to 187 such events on the placebo arm. In absolute terms, this difference was hardly world-shattering. It did not even meet the study's own initial criteria set for statistical proof, and the finding of proof at a lower threshold was achieved only by a questionable technique: applying a one-tailed test after a two-tailed test showed equivocal results.[66]

It is deeply ironic that such a slim margin of difference—amid several layers of troubling methodological critique—would lead not only to strong proclamations of "definitive proof" but also to policy recommendations supporting cholesterol-lowering efforts in all segments of the population, extending far beyond the bounds of the study's population (middle-aged white men) and intervention (cholestyramine).[67] The original publication of the LRC-CPPT acknowledged, "These results could be narrowly interpreted to apply only to the use of bile-acid sequestrants in middle-aged men with cholesterol levels above 265 mg/dL (perhaps 1 to 2 million Americans)."[68]

Nonetheless, in the next sentence the authors of the article shifted their tone from "is" to "ought": "The trial's implications, however, could and should be extended to other age groups and women and, since cholesterol levels and CHD risk are continuous variables, to others with more modest elevations of cholesterol levels."[69] Before the year was out, the NHLBI had assembled a consensus conference in Bethesda to swiftly translate this data into action and firmly "resolve some of these questions" surrounding the cause-and-effect relationship between cholesterol reduction and lowering of risk of heart disease.[70] The framers of the two-day conference explicitly asked its group of bench researchers, cardiologists, primary care physicians, epidemiologists, biostatisticians, and experts in preventive medicine to consider evidence and provide conclusive answers regarding the causal nature of blood cholesterol and heart disease. Particular emphasis was given to the value of intervention and the need for a broad-based public health effort around cholesterol.[71]

The conference was intended to diminish the controversy surrounding the study's results and present a unified statement of the trial's relevance. Prominent physicians, statisticians, and policymakers had cited the LRC-CPPT as an example of a poor study with low generalizability, dubious post hoc statistical wrangling, clinically insignificant demonstration of preventive power, and wildly unsupported extrapolation in generalizing the results of a study on middle-aged white men to the entire population over the age of two.[72] Amid such controversy, the promotion of a "Consensus Conference" with panelists cherry-picked from the NHLBI's own ranks struck many as a peculiar form of doublespeak. Michael Oliver, one of the most prominent critics of the NHLBI's actions, dryly suggested that the Consensus Conference would be better named a "Nonsensus Conference": "Clearly, the aims of . . . the consensus development conferences were to try to develop a consensus view and, not surprisingly, the final statements prepared at the end of each 2½ day meeting were biased. How could they have been otherwise? Those who initiated the idea were either naïve or determined to use the forum for special pleading, or both. The panel of jurists for each of the conferences was selected to include experts who would, predictably, say . . . that all levels of blood cholesterol in the United States are too high and should be lowered. And, of course, this is exactly what was said."[73]

As Oliver observed, the consensus panel participants ratified and extended the LRC-CPPT conclusions as a broad mandate. Deeming the study to be the solid kernel of a growing scientific basis for large-scale intervention in individual and populational cholesterol-lowering, they supported the founding of the National Cholesterol Education Program (NCEP), whose ranks they quickly filled. The philosophy and goals of the NCEP were unveiled alongside the original publication of the consensus report in the Journal of the American Medical Association: "Buttressed with these positive results from [LRC-CPPT] and acknowledging the overwhelming biologic and epidemiologic associations between serum cholesterol and coronary heart disease, it seems rational, if not imperative, to mount a major educational effort to lower plasma cholesterol levels . . . This sequence of logic is formidable and leaves unresolved only the implementation strategy and the degree of diligence that should be applied."[74] By the time this call to action was published in April of 1985, steps had already been taken to enact the National Cholesterol Education Program. That March, the leadership of the NCEP had explicitly adopted the organizational structure of the National High Blood Pressure Education Program (NHBPEP), described in chapter 2. Like the blood pressure education program, the cholesterol program began its efforts by organizing planning conferences and identifying numerous participants in public, private, and professional sectors to share financial, organizational, and promotional responsibilities; the actions of all stakeholders would be coordinated by a central Coordinating Committee. NCEP organizers sought to portray the LRC-CPPT trial as a mobilizing event on par with Edward Freis's VA study of asymptomatic hypertension.[75]

The first NCEP meetings, in March 1985, focused on strategies for overcoming barriers to "cholesterol awareness" in professional and patient populations; these were followed by a set of meetings in April and May that focused on promotional techniques of mobilizing the general public.[76] "Cholesterol awareness" was a specially defined term, meant to imply something more than commonsensical awareness of the existence of cholesterol or of the arguments connecting elevated cholesterol with atherosclerosis and heart disease. The planners of the cholesterol education program defined "awareness" as a proclivity toward active intervention to reduce cholesterol levels on the part of professional, patient, and public audiences.

Among physicians, barriers to awareness included poor knowledge and attitudes toward cholesterol, existing "issues related to the misunderstanding/interpretation of the science," and the problem that cholesterol was an asymptomatic disease—"that elevated blood cholesterol doesn't 'hurt' (e.g., has no symptoms) and its treatment requires a complex lifestyle change, so physician motivation to intervene is often low."[77] Among the general public, according to recent surveys, over 80 percent had heard of the condition of "high blood cholesterol," but only 3 percent actually knew their cholesterol levels, and widespread confusion surrounding the significance of cholesterol continued even after the Consensus Conference.[78] Building "cholesterol awareness" thus became a key goal for early NCEP work.

The panels settled upon a series of tactics distributed across member institutions to build awareness of cholesterol as an actionable problem on an individual and a population level. One committee had the task of standardizing measurements, and another was to draft a set of detection and treatment guidelines; one was asked to come up with workplace interventions, and another with educational interventions. Considerable effort, however, was devoted to determining a central message and its modes of delivery.[79] The foremost candidate on the table was a "know your level" campaign, which would urge the general public to identify themselves with their cholesterol count.

> Participants felt that a "know your cholesterol level" message was essential since 25% of the population has undetected high cholesterol. Some participants urged that the message should include a specific number which represents a "desirable" cholesterol level. A number of participants urged that a single number be publicized (e.g., 200 mg/dL). They pointed out that the use of a single number would simplify the message, would get people's attention quickly focused on the issue, and was found to be effective in the hypertension campaign. Other participants were concerned that a message which specified a "target" blood cholesterol level would undercut the importance of all Americans reducing their blood cholesterol level. They felt that if we tell people that their blood cholesterol is normal they won't hear the rest of the prevention message which is to "change dietary behaviors now to avoid future problems."[80]

The function of the "know your level" message was to make cholesterol a part of every American's experience of self, and in this project the medium was at least as important as the message. In devising tactics to bring the message of cholesterol awareness to physicians, the NCEP borrowed most of the techniques used for pharmaceutical promotion, with the exception of hiring a sales force. Incorporating promotional techniques from the private sector into public health education was emphasized as a benefit of the NCEP's public-private collaboration. For example, the following list of recommended "communication techniques" for physicians reflected in form as well as content a set of precepts well known to pharmaceutical marketers by 1985:

— Articles in *JAMA, Annals of Internal Medicine,* etc.
— CME skill-building courses on diet/drug counseling/compliance techniques
— Self-teaching education modules
— Focus on changing practice habits of student role models (preceptors)
— Professional practice change through locally organized projects (e.g., American Cancer Society "check" programs)—through building public expectation about cholesterol intervention by professionals
— Work on lab standardization issue
— Don't reinvent the wheel—make use of National High Blood Pressure Education Program (NHBPEP) approaches that work.[81]

To reach the general public, an alternate set of communications resources were mobilized. The NCEP developed its own posters and brochures and worked with public relations brokers to develop articles on nutrition and blood cholesterol for women's magazines.[82] In December of 1986, the Coordinating Committee released its *Communications Strategy for Public Education,* which highlighted the function of mass media as a vehicle for public health promotion and also reflected the influence of a marketing and publicity perspective: "Role of the Mass Media Campaign: One important component of the NCEP's public and patient education program is the mass media campaign. The reason: surveys show that, after the physician, television and radio public service announcements constitute the most frequently reported sources of health information. In the terms of public education, the mass media campaign can be expected to help."[83]

The resultant "Know Your Number" campaign made the personal responsibility for cholesterol management a visible topic in the mid to late 1980s. However, the NCEP understood that mass media campaigns did not unilaterally effect large-scale changes in behavior without other supporting components of public and patient education.[84] An effort was therefore coordinated among the different NCEP stakeholders to link public and patient education with professional education.

Responsibility for the cholesterol education program was spread across a diverse group that included many private and public organizations, but no pharmaceutical companies. A partial list of stakeholders includes the American Heart Association, the American Medical Association, and the American College of Cardiologists, as well as state, local, and municipal government bodies and some wings of the federal government. This technique was borrowed directly from the structure of the National High Blood Pressure Education Program.

The cholesterol education effort had also learned from the experience of the blood pressure effort that the presence of a nominally consumer-driven lobbying body could be helpful in requesting federal appropriations and buffering interactions between the constituent parts of the organization. Mike Gorman and the Lasker foundation had set up a nominally populist lobbying body—Citizens for the Treatment of High Blood Pressure—that provided crucial support for the activities of the NHBPEP in the 1970s (see chapter 2). With the creation of the NCEP, the same group saw an opportunity to extend its efforts into cholesterol, creating first a joint body named Citizens for the Treatment of High Blood Pressure and Cholesterol and then splitting off another group—still headed by Mike Gorman—named Citizens for Public Action on Cholesterol.

The brief history of this organization, cut short by Gorman's retirement, demonstrates the increased role of health lobbying groups in brokering disease definitions and their promotion in late-twentieth-century health politics. In early 1985, while serving on the Coordinating Committee of the NHBPEP, Mike Gorman got word of the formation of the new cholesterol education program and dashed off a memo to LRC-CPPT researcher Antonio Gotto; a draft remains in his papers:

> Gotto Meeting week of july22 or 23 best
> Need at least an hour of your time—Bring my state man down
> Send you a March 4 game plan plus most recent one—more detail
> Idea of a separate Cholesterol letterhead—You be willing
> To serve as chairman and invite ten lipid experts to serve with you—
> Mike DeBakey, Chairman—Mary Lasker, Honorary Chairman
> You Chairman, Advisory Board=
> Possible members—
> Dr. Daniel Steinberg
> Michael Brown
> Joseph L. Goldstein[85]

Attached to the memo was a document outlining the role of Citizens for Public Action on Cholesterol in the new campaign against cholesterol. Gorman described a plan to translate the tactics learned in the promotion of hypertension to the promotion of cholesterol awareness. Citizens was prepared to "encompass this new cholesterol offensive," Gorman concluded, adding, "The very fact that a National Cholesterol Education Program mandates a revolutionary change in the life style of millions of Americans is a challenge that Citizens feels uniquely capable of handling . . . Our first few years were devoted to putting in place the various networks which eventually changed a part of the life style of the American people and their physicians to the point where not knowing your blood pressure was considered neither stylish nor smart. The fact that hypertension is now the most common reason for office visits to primary care physicians is the end point of 13 years of educational experimentation."[86] Gorman then detailed for Gotto a series of strategies "specifically tailored to the unique requirements of a mass-oriented cholesterol education program," including a newsletter sent to all federal, state, and local public health officials; a grassroots detailing effort to reach doctors, nurses, mass media, and legislators; a legislative action network to procure funds and state support; and large-scale screening efforts across the country.[87]

Most of the points on Gorman's memo had been actualized by December of 1987. Citizens for Public Action on Cholesterol had become an independent lobbying group with a Washington, DC, address, and Gotto had been named chairman of the organization's medical advisory panel, a body that included almost all the other prominent cholesterol researchers that Gorman had named. This panel read as a roster of key personnel involved in the LRC-CPPT and NCEP, including Nobel laureates Michael Brown and Joseph Goldstein, as well as Robert Levy, Jeremiah Stamler, Daniel Steinberg, Scott Grundy, DeWitt Goodman, and others.[88] The first issue of the newsletter Gorman had proposed, *Cholesterol Update,* was sent out that month. An accompanying memo noted that "*Cholesterol Update* will, of course, place priority on reporting budgetary, legislative, regulatory, and policy news and issues . . . Our goal is to revitalize a broad based constituency, once quite active on behalf of high blood pressure."[89]

With the help of partners like Citizens and similar private groups devoted to promoting the platform of cholesterol awareness, the cholesterol education program could claim a modest degree of success by 1987. Measuring and lowering cholesterol as a means of exerting control over one's cardiovascular health had become a popular news item once again, featured on the evening news and magazine covers. New surveys published by the NCEP that year suggested that "cholesterol awareness" had risen in the two years of the program's existence.[90]

But in spite of these efforts, the early years of the program were marked by frequent criticism: many physicians still forcefully disagreed with the authority, basis, or need for a National Cholesterol Education Program and contested the validity of the LRC-CPPT study that had been its key catalyst. The cardiologist and antihypertensive booster Irvine Page, for example, had remained an active critic of "cholesterol faddism": "Despite a determined effort on the part of the Lipid Research Clinics Coronary Primary Prevention Trial to convince the public that at long last we now had the answer for the cholesterol problem, the results were disappointingly unconvincing. They tried, as other world-wide studies had done, but only a partial answer emerged."[91] Michael Oliver was also still critical, suggesting that the NCEP had extended its recommendations far beyond what could legitimately be called a scientific basis.[92] In 1987 a still more rigorous critique came from Harvard internist and epidemiologist William C. Taylor, who published a model demonstrating that even if the LRC-CPPT results were valid, the NCEP recommendations would add only eighteen days to the average American's life, at significant cost to comfort and pocketbook.[93]

Laymen writing about cholesterol were not all willing to take the NCEP at face value, either. Thomas J. Moore, an investigative journalist researching a book project on preventive cardiology in the early 1980s, published an influential denunciation of the National Cholesterol Education Program in the *Atlantic Monthly* entitled "The Cholesterol Myth." The article was effective muckraking: it was widely read and helped lead to a congressional hearing on the NCEP in late 1989.[94] The congressional investigation resulted in no charges or changes, but it was clear to the NCEP that widespread professional and public criticism limited their effectiveness at making cholesterol a national health priority. As Mike Gorman had noted, the NCEP was "deeply aware of the fact that a widespread educational and treatment offensive against cholesterol is a much more complex and demanding task than the one we encountered in high blood pressure." In this comparison, Gorman maintained, one of the NCEP's biggest detriments was the lack of an attractive pharmaceutical intervention.[95]

Enter Mevacor

The year 1987 would prove to be pivotal for the National Cholesterol Education Program; it included two fortuitously linked deployments. In October the program released the first national guidelines for the detection and treatment of high blood cholesterol, known as the Adult Treatment Panel (ATP) guidelines. Roughly equivalent to the NHBPEP's Joint National Committee guidelines, the ATP rules circulated through all medical institutions, ran through several sets of revisions, and became a central reference point in the practice of American primary care medicine.[96] The launch of these guidelines, however, was both preceded and made possible by the launch, one month before, of Merck's Mevacor (lovastatin), the first of a class of drugs now known as the statins. If the birth of the NCEP was intimately tangled with the life history of cholestyramine, the expansion and growth of cholesterol treatment guidelines in the late 1980s and 1990s had everything to do the with the rise of the statins.

From its celebrated launch until its quiet patent expiration in 2001, the promotion of Mevacor introduced the word *statin* and the concept of *cholesterol pill* into the lives of millions of Americans. Mevacor's success helped Merck become, for many years, the largest and most profitable pharmaceutical company in the world and launched a public image of Merck's CEO, Roy Vagelos, as the personification of an industry committed to a mutually advantageous union of science, humanitarianism, and profit. The development of Mevacor involved multiple Nobel Prize laureates and became a textbook example of drug design. As with Diuril, the long development time and substantial R&D costs of developing Mevacor were repeatedly publicized to argue that American pharmaceuticals remained a bargain in spite of rising prices. This position has subsequently been formalized in the shape of the yearly rising number of dollars spent per new drug released, now claimed to be somewhere in the vicinity of $1 billion.[97]

Merck did not spend $1 billion to develop Mevacor, but it is impressive that a line of research into cholesterol-lowering medications that had not produced any viable products since its origins in the early 1950s was allowed to continue its operations long enough to produce lovastatin in 1987. Merck's anticholesterol drug project began in the postwar Merck laboratories of Karl Folkers and Carl Hoffman, whose search for a putative growth factor (hopefully named vitamin B-13) turned up a tropic factor in bacterial growth that they called mevalonic acid. Jesse Huff, a Merck biochemist whose laboratory had been investigating cholesterol synthesis since the early part of the decade, ran radioactively labeled mevalonic acid through a cholesterol assay and found it was swiftly incorporated into newly formed cholesterol.[98] Subsequent research suggested that of all the steps involved in the biological production of cholesterol, the enzyme that mediated the production of mevalonic acid—identified in 1958 with the ponderous name 3-hydroxy-3-methylglutaryl-coenzyme A reductase, or HMG-CoA reductase for short—was the major rate-limiting process in the molecular assembly line. The enzyme quickly became a therapeutic target of considerable interest, but efforts to chemically synthesize an agent to block its activity remained unproductive for decades.

Where American pharmaceutical chemists had failed, however, a team of Japanese microbiologists succeeded. Akira Endo, a researcher with the Japanese pharmaceutical firm Sankyo, began to adapt microbial screening methods— which had proved highly successful in the development of antibiotics—to the problem of cholesterol synthesis.[99] Beginning in 1971, Endo's team ran a series of microbial strains through an in vitro assay that tested for the ability to block HMG-CoA reductase activity. By November 1973, after more than six thousand microbial strains had been screened, a substance extracted from the *Penicillium* mold—the same class of bread mold that had famously yielded penicillin—showed promising ability to block the target enzyme. This substance, initially designated ML-236B, was analyzed by spectroscopy and X-ray crystallography and named both mevastatin (after its ability to block mevalonic acid synthesis) and compactin. By 1980 testing in rats, egg-laying hens, dogs, monkeys, and finally in humans showed mevastatin's significant promise in reducing cholesterol levels with few apparent side effects.[100] Subsequent trials revealed that mevastatin was indeed a far more potent reducer of cholesterol levels in humans than any other intervention yet known, lowering cholesterol levels by 29 percent alone and up to 60 percent in combination with cholestyramine.[101]

Endo's work quickly received international attention in the clinical literature and the pharmaceutical trade press, and Sankyo signed a disclosure agreement with Merck, agreeing to share data and samples in the hope that Merck in return would use its larger resources to assess the drug's potential for development by Sankyo. Their relationship was terminated in 1978, when it became evident that Merck had, in the interim, developed its own HMG-CoA reductase inhibitor—alternately called mevinolin, lovastatin, and, ultimately, Mevacor.[102] This compound, extracted from an *Aspergillus* mold instead of *Penicillium*, became the subject of a patent dispute between Merck and Sankyo that was geographically resolved largely in Merck's favor by 1980.[103]

Immediately after Merck received patent rights on lovastatin, however, it also received rumors that Sankyo's agent had been found to cause tumors in dogs.[104] Lovastatin was jerked from its leading position in the Merck pipeline and sent back to the medical division for additional safety studies. In 1982 the company cautiously made the drug available to clinicians only on an experimental basis for severe cases; in 1984, after a four-year hiatus in the development process, the drug was again cautiously released for clinical trials. These early clinical trials were conducted with patients diagnosed with severe hypercholesterolemia at Oregon Health Sciences University and the University of Texas, and the trials were directed by Scott Grundy—a prominent member of the NIH Consensus Conference and active in the National Cholesterol Education Program. With promising initial results in severe patients and a clean bill from extended toxicology studies in October 1986, Merck finally filed a 160-volume new drug application covering twelve hundred experimental subjects in November 1986 and received FDA approval to market the drug for severe hypercholesterolemics in August of 1987.

The research project for Mevacor intersected closely with the research program of Michael Brown and Joseph Goldstein. Significant revisions of the definitions of pathological lipid disorders had hinged on the duo's Nobel Prize–winning research in the late 1970s and early 1980s, which provided a new mechanistic explanation for the treatment of high blood cholesterol as a metabolic disorder (see chapter 6). Previous cholesterol-blocking efforts had been based on an assembly-line model of body-as-factory, in which intervention focused on the input of raw materials (via low-fat, low-cholesterol diets), the synthesis of cholesterol (via the fibrates), and the excretion of cholesterol (via bile-acid-binding resins like cholestyramine). Brown and Goldstein replaced the linear model with a homeostatic model based on receptors and feedback loops, in which the decisive factor in bodily cholesterol regulation—and hence the most effective site of intervention—was not cholesterol itself but its cellular signaling patterns, which are governed largely by a cell-surface protein called the LDL-receptor.

In the mid-1980s, Brown and Goldstein had theorized this system but were not fully able to demonstrate their model using cholestyramine. In their model, what made "good cholesterol" (HDL) good and "bad cholesterol" (LDL) bad was where each species deposited itself. The "good" HDL cholesterol was whisked toward the liver and out of the body, whereas the "bad" LDL cholesterol had a proclivity for sticking itself onto artery walls and working its way into tangled plaques and thrombi. Unless, that is, the LDL cholesterol was plucked out of the bloodstream by cell-surface LDL-receptors and put to more constructive use elsewhere. When issued samples of lovastatin, the two Texan researchers were able to produce convincing evidence that lovastatin's dramatic cholesterol-lowering effect came not from its blockade of the cholesterol-production assembly line, but from a series of intricate homeostatic interactions involving the LDL-receptors. Blocking the HMG-CoA reductase step in the liver appeared to produce a cascading effect through biofeedback pathways, resulting in an up-regulation of LDL-receptors. Put simply, lowering the cholesterol level in the liver created a local demand for cholesterol in the liver and caused the liver to produce more receptors to remove LDL cholesterol from the bloodstream for its own purposes. Mevacor, then, became an important proof of Brown and Goldstein's new vision of cholesterol metabolism as an modifiable homeostatic system. As Alfred Alberts, the head of Merck's cholesterol research program, recalled: "Without the receptor concept, lovastatin wouldn't be the drug it is . . . With this compound they substantiated their own thesis, and with their thesis they proved the mechanism of action of our compound, so the two go hand in hand."[105]

Along with Grundy, Brown and Goldstein linked Mevacor's research project with the National Cholesterol Education Program, in which all three were deeply involved, and with Citizens for Public Action on Cholesterol, on whose medical advisory panel all three sat. Without any direct participation of Merck in NCEP, Mevacor's promotion was subsequently tightly bound up with NCEP promotion. Even though cholestyramine had been essential to the founding of the MCEP, Mevacor was the drug most publicly associated with the National Cholesterol Education Program at the time the NCEP guidelines were launched. By 1987 Merck's advertising budget was the highest in the industry; in 1988 Mevacor and the new antibiotic Cipro (ciprofloxacin) were tied as the most heavily advertised pharmaceuticals in the country.[106] Grey F. Warner, Merck's senior director for MSD marketing planning, told shareholders that the marketing of Mevacor would rely heavily on the public education campaigns of the NCEP and Citizens for Public Action on Cholesterol, as mediating bodies that could help foster "awareness" of cholesterol risks among physicians and the lay public.[107]

Merck's relationship with Citizens helps to illustrate how swiftly the field of disease-specific lobbying had grown since the 1970s. Citizens for the Treatment of High Blood Pressure had initially been funded solely by the Lasker Foundation, but by the late 1970s, representatives from Merck Sharp & Dohme contacted Mike Gorman to inquire into the possibility of establishing a parallel lobbying group modeled on the earlier Citizens, one that devoted itself to securing public funding and promotion for glaucoma detection and treatment efforts; Merck's interest in the topic was linked to its successful glaucoma agent Timoptic (timolol).[108] After an extended correspondence with Gorman and Lasker, Merck's director of public relations sent a proposed sketch for the creation of a glaucoma effort modeled after Citizens for the Treatment of High Blood Pressure. Like hypertension, glaucoma was a progressively degenerative condition that could be slowed or reversed if treated early but typically remained asymptomatic until it had advanced beyond the realm of effective treatment. The National Initiative for Glaucoma Control, which opened the following year, funded by a one-hundred-thousand-dollar grant from Merck, was ostensibly left to run its own campaigns independent of Merck's public relations office. However, Gorman's regular communications with Merck public liaisons Anthony Fiskett, Grey Warner, and Russell Durbin—detailing the strategies and successes of the organization and the barriers it encountered in its campaigns to popularize glaucoma detection and treatment and mobilize state and federal funds for the project—helped ensure the steady flow of funds from its founding donor.[109]

By the time Citizens for Public Action on Cholesterol was added to the Gorman-Lasker family of disease-specific lobbying groups, Gorman was well acquainted with the Merck publicity office and Merck was a reliable six-figure annual donor. After the founding of the NCEP, Gorman's records indicate that Merck donated similar levels of financial support to directly assist the operations of Gorman's cholesterol campaign.[110] In addition to its general activity to mobilize public health and legislative action on cholesterol detection and treatment, Citizens for Public Action on Cholesterol also served Merck-specific promotional and market-research projects. For example, in 1988 the organization distributed a Merck-prepared booklet entitled *Cholesterol and You* and promoted it through an advertising and press-release network of local newspapers. Hundreds of copies were requested in letters addressed to "Merck Sharp, and Dohme, The Citizens for Public Action on Cholesterol."[111] Placement of articles in local newspapers also gave a specifically geographic sense to cholesterol awareness as a part of the Mevacor promotional effort, which was printed up on individualized awareness maps (see fig. 5.1) of use to Merck and the NCEP alike.[112]

Moreover, in perhaps its most significant capacity, the nonprofit Citizens for Public Action on Cholesterol was allowed a presence inside the National Cholesterol Education Program that was denied to for-profit pharmaceutical cor-

Fig. 5.1. Cholesterol promotional map showing geographic penetration of cholesterol-awareness news clippings. *Source:* "Known Placements to Date of Your News Release," 1987, box 2, folder 4, Mike Gorman Papers, National Library of Medicine, Bethesda, MD. Courtesy of the National Library of Medicine.

porations. In the figure of Mike Gorman and the mediating body of Citizens, Merck could have at least an eye and an ear on the NCEP Coordinating Committee meetings, and occasionally even a mouth. Unfortunately for Merck, the entire Gorman-Lasker family of lobbying projects became entangled in a financial dispute in 1988—with Mevacor only one year on the market—and Gorman consequently retired and dissolved all three organizations.[113]

Nevertheless, the connections that Citizens provided between Merck and the NCEP and other associations continued to be valuable to the company. In a *Washington Post* article accompanying Mevacor's release, for example, Antonio Gotto, who was president of the American Heart Association and chairman of Citizens for Public Action on Cholesterol, was quoted as a Merck representative—"speaking for the drug's maker, Merck, Sharpe & Dohme." In this capacity, Gotto suggested that he would prescribe Mevacor for all adults over forty with cholesterol levels higher than 260 and for all ages twenty to forty with cholesterol levels above 240, as long as diet therapy had been ineffective after three months.[114]

In that first year of its release, Mevacor brought in sales of $260 million, the highest first-year sales figure yet recorded for any prescription medicine.[115] In the media landscape of the late 1980s, the promotion of the NCEP cholesterol detection and treatment guidelines and the emergence of Mevacor as a cholesterol-lowering "wonder drug" were quickly fused.

Agents of Consensus

Although the introduction of Mevacor provided a sense of optimism and progressive enthusiasm about the treatment of high cholesterol and also generated a powerful promotional framework encouraging physicians to detect and treat it, the launch of Mevacor was not by itself sufficient to dispel the broad disagreement still extant within the medical community regarding the treatment of high cholesterol. The coincident emergence of Mevacor and the NCEP treatment guidelines seems to have itself prompted a good deal of skepticism and conspiracy theorizing regarding the program, even though the NCEP guidelines explicitly referred to Mevacor only as an "experimental" therapeutic and not a first-line agent. Thomas J. Moore's widely read critique of the NCEP in the *Atlantic Monthly* noted, wryly, that the launch of Mevacor relied heavily on a small group of expert investigators—including Daniel Steinberg, Antonio Gotto, and Scott Grundy—who were both Mevacor researchers and architects of the NCEP guidelines.[116]

The Consensus Conference had not yet produced the desired consensus. Although, as mentioned above, the congressional hearings inspired by Moore's account did not lead to any rebukes or fiscal crisis for the NCEP, the inquiry demonstrated to stakeholders that the NCEP and its recommendations were far from bulletproof.[117] The weaknesses of the cholestyramine trial were well known, and although the leadership of the National Cholesterol Education Program continued to assert that its recommendations were based on firm scientific footing, the treatment of high cholesterol would remain a controversial issue until incontrovertible evidence of the benefit of reducing elevated cholesterol could be produced.

Unlike diet, which had proved simply too difficult to test properly, or cholestyramine, which had worked as a proxy for dietary change but in itself was an unappealing tool for mass intervention, Mevacor was an ideal nucleus around which to gather clinical trial evidence for the treatment of asymptomatic elevations of blood cholesterol. Whereas the LRC-CPPT had faced enormous difficulties in maintaining a population of four thousand subjects on their daily sachets of resin, tens of thousands of clinical trial subjects were swiftly integrated into long-term clinical prevention trials for Mevacor and the subsequent statins Zocor and Pravachol by 1990. The statins helped validate asymptomatic hypercholesterolemia in the 1980s and 1990s in the same way that thiazide diuretics like Diuril had helped to validate the treatment of asymptomatic hypertension in the 1960s and 1970s. As the Mevacor product director at Merck Sharp & Dohme predicted in 1987: "Based on the experience with lovastatin, inhibitors of HMG-CoA reductase are likely to prove a major advance in the treatment of hypercholesterolemia. They may well usher in a new era in the management of this disorder, playing a role comparable to that of the thiazides in hypertension a quarter of a century ago."[118] Indeed, the statin prevention trials would be found by the mid-1990s to have an effect rivaling that of the VA study for hypertension. As explained in chapter 6, compelling evidence published in 1994 linked Merck's second-entrant Zocor (simvastatin) with a significant decrease in cardiovascular mortality in patients with known heart disease and high blood cholesterol (secondary prevention); by 1996 two other major large-scale trials had demonstrated that Mevacor and Bristol-Meyers Squibb's Pravachol (pravastatin) significantly decreased cardiovascular events in asymptomatic patients with no other risk factor besides high blood cholesterol (primary prevention).[119] Retracing the ground where clofibrate's Coronary Drug Project had failed and where cholestyramine's LRC-CPPT trial had produced only the slimmest margin of evidence, these prevention trials provided a more tightly defensible argument for the value of cholesterol detection and treatment.

The history of the statin prevention trials and their relationship to the shifting definition of treatable elevations in cholesterol are explored in more detail in chapter 6. For the present discussion, it is enough to observe the effect these trials—and their widespread publicity and popularization—had in restricting the intellectual space in which one could dispute the rationale for detecting and treating high blood cholesterol as a pathological condition. The career of Michael Oliver, perhaps the most prominent and well-respected critic of the LRC-CPPT and NCEP, illustrates particularly well the narrowing option for dissent. In 1988 Oliver could argue that there was no convincing evidence that reducing cholesterol did, in fact, reduce mortality.[120] The emergence of Mevacor and the NCEP did not in and of themselves silence Oliver's critical voice; in the early 1990s, Oliver penned several influential reviews concerning the possible dangers of cholesterol-lowering interventions.[121] But by the end of the 1990s such claims were no longer visible. Although occasional critiques of "cholesterol dogma" continued to appear, by the end of the twentieth century they had been effectively marginalized, no longer appearing in refereed journals or the popular media but now reduced to a conspiratorial feature of the far left press.[122] As one commentator observed in a retrospective on the influence of statins on the treatability of cholesterol: "The first studies using lipid-lowering drugs, such as the early fibrates or the bile acid sequestrant resins, provided, at best, equivocal results, which fueled the arguments for inaction or procrastination among physicians. When the statins were introduced into clinical practice and, more importantly, when they were tested in the fire of the randomized controlled clinical trial these arguments disappeared."[123]

Conclusion

The career of cholesterol in the late twentieth century demonstrates that epidemiological data was not in itself adequate to bring about the widespread adoption of risk factors as clinically relevant entities. The best proofs of the laboratory and the field study were insufficient to provide the key pragmatic justification required to mobilize attention around cholesterol. To accomplish this last step, discrete interventions were necessary, and these interventions met with variable fates. In contrast to the cases of hypertension and mild diabetes, individual pharmaceuticals were as likely to hurt (MER/29) as to help (Mevacor) the status of elevated cholesterol as a target for preventive medicine. In both extremes, the fall and rise of cholesterol as a condition worthy of widespread detection and treatment were highly dependent on the availability of therapeutics.

For asymptomatic diabetes, the successful pharmaceutical mobilization of a condition of risk was able to persist even after the relevance and safety of treatment had been questioned. In the case of asymptomatic elevated cholesterol, it was difficult to build widespread consensus around the treatment of risk without an appealing pharmaceutical intervention. Taken together, the narratives of diabetes in the 1960s and 1970s and of cholesterol in the 1980s and 1990s form parallel testimonies to the central role of pharmaceuticals in the widespread promotion of conditions of risk in late-twentieth-century American medicine. As large-scale, long-term pharmaceutical trials became crucial tools for the validation of preventive medicine, events such as the UGDP or the LRC-CPPT developed vast public audiences, with multiple parties working to closely manage their influence on the public.

The importance of managing the public reception of clinical trial information is particularly visible in the functioning of the National Cholesterol Education Program, which explicitly incorporated private-sector strategies of promotion and publicity into its function as a public health body. As an early model of the sort of public-private collaboration that became increasingly important in the health policy environment of the 1980s and 1990s, the NCEP worked to carefully draw boundaries so that it would not be perceived as merely a front for pharmaceutical promotion. It did so by including a diverse set of stakeholders, by holding public meetings, by insisting on a rigorous scientific background for its arguments, and by including in its ranks expert cholesterol researchers such as Scott Grundy and the Nobel laureates Michael Brown and Joseph Goldstein. Nonetheless, as figures such as Grundy, Brown, and Goldstein illustrated, most experts in cholesterol research had, by the 1980s, already developed extensive ties to pharmaceutical companies. Moreover, as the figure of Mike Gordon also demonstrates, there were many other third-party entities through which a company like Merck could involve itself with the NCEP without explicitly violating the independence of the program or its guidelines.

Supporters of cholesterol as a public health concern turned to pharmaceutical agents after dietary and behavioral means of demonstrating the value of intervention had failed. In the LRC-CPPT trial, public health advocates saw pharmaceutical agents such as cholestyramine as research tools calculated to show the benefits of cholesterol reduction. After the trial, the pharmaceutical itself was erased from the public face of the results, which were intended to legitimate a broad policy of dietary and lifestyle change. To the extent that the LRC-CPPT enabled the construction of the NCEP and its widespread efforts to build public and professional awareness of cholesterol, this move was highly successful. However, once incorporated into the structure of cholesterol activism, the pharmaceutical clinical trial did not go away. Rather, it went private. Having helped to build the prevention trial as an engine of public consensus, the architects of the NCEP would watch in later years as such trials became a powerful marketing tool used to promote broader and broader use of Mevacor and other cholesterol-lowering drugs.

This process is further explored in chapter 6, along with the question of whether high cholesterol was seen as a clear pathological category or merely as a manipulable physiological variable. In the consolidation of elevated cholesterol as a diagnosis that mandated action, generated a functional system of physiological surveillance, and simultaneously identified the aberrant as patients and as markets for pharmaceuticals, the interventions themselves were crucial actors. After Mevacor, the realized dream of the cholesterol pill would transform cholesterol into a disorder of highly mobile thresholds.

CHAPTER SIX

Know Your Number

Cholesterol and the Threshold of Pathology

> Under ideal conditions a company would be able to control all aspects of marketing, thereby developing a new pharmaceutical product effortlessly, bringing it to the marketplace smoothly, and capturing universal awareness, total acceptance, and maximum sales volume immediately. Unfortunately, ideal conditions do not now exist and are unlikely to come into existence . . . The manager in a pharmaceutical marketing company, therefore, must simply adapt to whatever conditions are encountered and try to control what can be controlled.
> —*Principles of Pharmaceutical Marketing*, 1983

What is your cholesterol? Odds are about even that you can answer this question with a number, or at least a value. Your cholesterol is 198. Your cholesterol is normal. Your cholesterol is 250. Your low-density lipoprotein (LDL) cholesterol fraction is above 130 milligrams per deciliter, or you have too much "bad" cholesterol, or, perhaps, you just "have cholesterol." Regardless of how the results are defined, surveys indicate that between 50 and 75 percent of Americans over the age of twenty have had their cholesterol checked in the last five years.[1] Knowledge of one's cholesterol levels has become for many adult Americans an essential act of self-surveillance, a window into one's inner health.

Over the last two decades of the twentieth century, the market presence of cholesterol-lowering agents ballooned outward from a minor therapeutic category to the leading class of prescription drugs sold in the world. This widespread enthusiasm for the detection and treatment of elevated blood cholesterol did not arise merely from the passive diffusion of a body of scientific knowledge through the general population. Rather, it was the product of a concerted public-private effort to make awareness of blood cholesterol a priority for American physicians and consumers, involving, to a significant extent, the material and commercial attributes of the class of pharmaceuticals called the statins. This chapter narrates the life and times of Merck Sharp & Dohme's Mevacor (lovastatin), the first statin in the American market, in relation to changing national guidelines on the treatment of high cholesterol.[2]

As I was writing one of the drafts of this chapter, a wave of newspaper and magazine articles announced another proposal to lower the definition of normal cholesterol levels, based on another large-scale privately funded prevention trial comparing Lipitor and Pravachol, two of the leading statins on the market today.[3] Statin trials are probably the most visible example of the hydraulics by which commercial clinical trials now drive the production of clinical guidelines and the standardization of clinical practice. I want to make it clear in this chapter that the relationship between pharmaceutical companies and expert committees is not merely a question of conflict of interest, bribery, scandal, or bad science. Rather, this relationship is encoded in the very practice of "good science" that is central to the circulation of medical knowledge. As a result, there is no organized opposition to the demonstration of benefit at more and more subtle levels of risk; there is no visible barrier to the continued expansion of the statin market.

Now that Mevacor is off patent, any study of its branding and marketing has become a historical project, and an interested historian can now pursue a Freedom of Information Act request for documents surrounding the drug's launch. Such materials, read in conjunction with the clinical literature, the industry trade literature, and the public records of the National Cholesterol Education Program, provide a sketch of how Mevacor's market expanded through a series of interconnected clinical trials and evidence-based guidelines. In the interests of dividing historical narrative from contemporary speculation, I begin with the end of Mevacor's life cycle and work backward.

The controversy surrounding the last days of Mevacor and the struggle for extended brand life illustrates both how much popular conceptions of cholesterol had changed over the course of the drug's career and, paradoxically, how constant several of the arguments surrounding the definition of normal and pathological have remained even as the threshold of pathology has shifted. As a prominent "blockbuster drug" in a period when the production of blockbusters was becoming increasingly important to the marketing priorities of drug firms, Mevacor captures the zeitgeist of the pharmaceutical industry of the 1980s and 1990s much as the sulfa drugs did in the 1930s and penicillin did in the 1940s.[4] Mevacor's clinical trials and product promotion walked a fine line between emphasizing elevated cholesterol as a legitimate disease with discrete pathology and de-emphasizing the severity of that condition so that it would be understood as a common and familiar condition and reach the broadest possible market. Consequently, as definitions of frank pathology came to approximate common numerical deviations of blood cholesterol, more and more Americans came to think of statin consumption as a relevant, nonstigmatized, and even desirable aspect of healthy living. The relationship between Mevacor and the evolving classification of high cholesterol it was indicated to treat illustrates the malleability of the numerical threshold dividing health and disease in late-twentieth-century medical practice.

The Twilight of a Drug

Mevacor's life as an active brand was cut short in the summer of 2000 by a federal tribunal in a suite of the massive Parklawn building in Rockville, Maryland, where most of the Food and Drug Administration (FDA) resides. Mevacor still exists today, in a ghostly sense: Merck plants continue to produce blue pills with "MSD" stamped into the side, and the *Physician's Desk Reference* still carries an entry for the drug alongside other generic versions of lovastatin. But Mevacor inspires no large-scale clinical trials or advertisements in medical journals or popular magazines; its name graces neither pen nor desk pendant in the pharmaceutical sales representative's bag of gifts. Its patent lost, it has receded gracefully into the ranks of discount medicines and medicines deemed essential for care in developing nations but not essential to Merck's own portfolio.

Perhaps the events in the Parklawn building that essentially ended Mevacor's life would best be described as a form of negative euthanasia, as a health care system refusing to provide life support for an ailing organism. Mevacor's situation by June of 2000 was already critical: the twenty-year patent on lovastatin was due to expire the following year; Merck's second-entry statin, Zocor, had sucked away most of Mevacor's promotional budget and was already more popular among consumers and physicians; and generic manufacturers were developing plans to produce their own lovastatin and eat into the remaining market.[5] Mevacor's last chance for survival as a brand was to follow the lead of other faded blockbuster drugs and weather the switch from prescription (Rx) to over-the-counter (OTC) status.[6] Like the anti-ulcer agent Tagamet—the first prescription drug to be called a blockbuster and one of the first to successfully switch to OTC—Mevacor had developed a well-known brand name and widespread confidence in its safety and desirability as a consumer product.[7] The firm announced plans for the Mevacor OTC project in the spring of 1999 and submitted a formal petition for nonprescription status to the FDA in early June of 2000.[8]

However, as many analysts pointed out, unlike the ulcer-blocking Tagamet, Zantac, and Pepcid and the pain-relieving Tylenol, Advil, and Aleve, the cholesterol-lowering Mevacor OTC would treat a condition unique for its lack of recognizable symptoms.[9] All the other drugs that had successfully switched to OTC status treated conditions that patients could easily self-medicate. If someone had a headache and then took two Advil tablets, she herself could judge when the headache went away. If the symptom wasn't relieved by the Advil, the consumer would know to seek more formal health care. The same argument could be made for persistent stomach pain that didn't respond to Tums or Zantac, or, indeed, for any other symptom unrelieved by available nonprescription drugs. As recently as 1997, the FDA had explicitly pronounced that OTC drugs were to be used only for "self-recognizable conditions that are symptomatic, require treatment of short duration, and can be treated without the oversight and intervention of a health-care practitioner."[10] Mevacor, it was argued, could not possibly relieve symptoms if high cholesterol did not present any symptoms to relieve.

Or did it? By 2000, some advocates argued, cholesterol was so widely "felt" by the consumer populace that it could almost be considered a symptom. Consumers had been educated to feel ill if they had high cholesterol and to feel healthy if their cholesterol was low enough, and many studies began to document the subjective illness felt once an individual received a diagnosis of high cholesterol.[11] Reliable finger-prick cholesterol monitors were available in most pharmacies by 1999. Many consumers with high cholesterol, concerned about their numbers, were already purchasing "nutriceuticals" and other alternative medical products that claimed to lower cholesterol levels. Some of these products, such as red yeast rice supplements, actually contained naturally fermented lovastatin in a quantity comparable to that of the proposed Mevacor OTC but in an unregulated, nonstandardized form.[12] If consumers were already keeping tabs on their own cholesterol levels, Merck's representatives noted, and spending large sums of money on treating these numbers to their own satisfaction, then perhaps cholesterol had, in a sense, effectively become symptomatic to many Americans.

The FDA had set July 13, 2000, as the date to hear Merck's arguments for Mevacor OTC, but given the breadth of interest, they also scheduled a public hearing two few weeks earlier to revisit the fundamental issues at stake in switching preventive medications from prescription to OTC status.[13] For the public hearing, held in a Holiday Inn in Gaithersburg, Maryland, an odd collection of groups gathered to protest the OTC-switch. Included were radical consumer groups such as the Ralph Nader–founded Public Citizen, physician groups such as the American College of Cardiologists, and Merck's chief competitor, Pfizer, which by 2000 had captured more than 25 percent of the market with its popular Lipitor and did not want to lose leverage as the newest prescription drug on the market to a heavily marketed and familiar brand available on an OTC basis. "We don't see how patients will be able to monitor their levels and treat to the right goal," the head of Pfizer's cardiovascular division was quoted in the newspapers. He added, "I think you really need a physician to check your levels and what your goals should be."[14]

It is not surprising that mainstream physicians' groups would oppose a switch that would remove their central role in the adjudication of risk. Ed Frohlich, a spokesman for the American College of Cardiologists (ACC) noted, "The ACC believes that the relief of symptoms should be an important requirement for OTC products . . . If relief requires a laboratory test, the consumer does not know whether he or she, in fact, [is] relieved. This is especially important for cardiovascular drugs which often can treat conditions which have no associated symptoms with which a consumer can assess the drug's efficacy."[15] Frohlich's argument that the consumer could not gauge cholesterol as a symptom, however, was immediately critiqued on cross-examination by an FDA panelist, who pointed out that such a distinction was overly simplistic: "Ed, you draw a sort of bright line between treating symptoms and treating signs, I guess you could say, and one of the reasons is that a patient can't assess whether his sign has improved without some external help. However, in two conspicuous areas, cholesterol and blood pressure, you can go to your Giant Supermarket and get your latest blood pressure. I don't know how accurate those are, but you can do it, and there are or will be simple tests of cholesterol available. So a person who was taking an over-the-counter drug in order to modify those signs would, if they were interested in the first place, be able to see how they were doing, if they bothered."[16] Merck advocates and several minority physician advocacy groups argued that it was incidental whether the patient perceived a symptom bodily or through a mediating consumer technology. As the Association of Black Cardiologists, and the Interamerican College of Physicians and Surgeons added, a more significant error lay in making the paternalistic and insulting assumption that only physicians could make sense of a number that the entire populace had been extensively educated to internalize.[17]

Merck lost its bid to extend Mevacor's branded life into the realm of consumer products.[18] On July 13 the FDA advisory panel voted eleven to one against Mevacor OTC, and the company, after a few last-minute attempts to gain further patent extensions, lost any hope of continued brand exclusivity.[19] But even if elevated cholesterol was not deemed a condition symptomatic enough for consumers to self-medicate, the debate itself was symptomatic of how much an abnormal cholesterol number had come to be considered pathological, on the part of both physicians and the public.

The Abnormal and the Pathological

One form of high cholesterol had been clearly delineated as a disease state even before the word *cholesterol* entered the medical dictionary. The pathological condition known as xanthomatosis, first described in 1851 and well known by the time the structure of cholesterol was identified in the early twentieth century, was characterized by cholesterol levels so high that small fatty tumors called xanthoma would become evident on the skin and in other regions of the body.[20] This was high cholesterol at its most obviously pathological, a lesion-based model of disease. Its status as a pathological condition was further grounded by an observed clustering in families that suggested xanthomatosis was an inherited metabolic disorder moving according to single-gene Mendelian principles, as an entry in the 1955 Cecil-Loeb *Textbook of Medicine* described: "There is a deposit of lipid, and in the hypercholesterolemic families chiefly of cholesterol, in subcutaneous and cutaneous tissues, tendons, and aponeuroses. Cholesterol crystals may be evident . . . The lesions are most easily recognized in the skin and tendons, but those which develop in the structure of internal organs such as the heart and arteries do not differ materially from the superficial lesions."[21] Clinical manifestations of xanthomata included xanthelasma, an accumulation of fat and cholesterol in the eyelids; arcus senilis, or a cholesterol streak in the cornea; xanthoma planum, a flat lesion commonly found in the creases of the palms, the folds of elbows, or the skin fold beneath the breast; and xanthoma tuberosum and xanthoma tendinosum, the more common hard nodular deposits found on the skin and the tendons, respectively.

By the early 1960s, the genetic basis of metabolic error had become more central to the diagnosis of xanthomatosis than the original symptom of the xanthoma itself. In the 1963 edition of the popular Cecil-Loeb textbook, xanthomatosis—by this time also referred to as "essential hypercholesterolemia"—could be diagnosed by a combination of laboratory measurement and genealogy. The presence of fatty tumors was "typical" but no longer necessary for diagnosis: "Diagnosis of essential hypercholesterolemia is made by the finding of an elevated serum cholesterol in patients with tendon xanthomas. *The disease can be presumed in a person who has an elevated serum cholesterol with or without xanthomas and a blood relative with hypercholesterolemia and xanthomatosis.* An elevated serum cholesterol level in a subject without additional information about the family is not sufficient basis for the diagnosis."[22] As the last sentence pronounces, a clear distinction existed between the *disease* of essential hypercholesterolemia and the mere *chemical marker* of elevated serum cholesterol. That the former should be treated with all available pharmaceutical agents was never in question. The latter category—the mere detection of blood cholesterol elevation—found no therapeutic consensus at that time.

Subsequent attempts to quantify the condition of high blood cholesterol reproduced the gap between the frankly pathological and the merely abnormal. Observations of the serum cholesterol levels of overtly xanthomatous patients allowed for the description of a numerical threshold of pathology. Because no such lesions existed in patients with serum cholesterol below 400 mg/dL, any cholesterol level above 400 could be a potentially pathological finding even in a symptomless individual.[23] This pathological threshold could be joined, from the other side of the spectrum, by a normal threshold. Using data from life insurance examinations and hospital laboratories, physicians, epidemiologists, and actuaries represented the distribution of serum cholesterol in the general population as a bell-shaped curve, its center around 195 mg/dL, with known variance and standard deviation. According to the logic of standard deviation, "normal" could be bounded as the set of values within two standard deviations of the mean, a cutoff that defined the middle 95 percent of the population as normal and bound the upper and lower 5 percent as abnormal extremes. By such a calculation, medical textbooks in the 1950s and 1960s listed 130–260 mg/ dL as a normal range for cholesterol and considered values above 300 mg per 100 ml to be abnormally high.[24] Abnormally high cholesterol was not the same thing as an overtly pathological lipid disorder; rather, it constituted a shadowy third space between health and disease.

Calculating hypercholesterolemia on the basis of deviation from the mean reflected an interpretation of the meaning of normal explicitly in terms of statistical norms. In this rather democratic regime, the populace itself became the reference point for health; disease could be defined by demarcating the statistical deviant, the small group of people quantitatively distant from the central majority of the population.[25] This scheme was dominant throughout the 1960s and 1970s and into the early 1980s.[26] That the threshold itself was arbitrary was understood by all; that it related to some real underlying boundary between pathology and physiology, however, was also universally implied. Just as the elevation of blood sugar over a certain level would "spill over" into the urine to produce the symptoms of diabetes, elevation of blood cholesterol past a certain level would accumulate in tissues to produce xanthomatosis.[27]

However, for many advocates who believed that cholesterol was a central culprit in producing heart disease, the normal American way of life—particularly the cholesterol-rich "American diet"—was implicated as a cause of cardiovascular pathology. Noting that the twentieth-century "American epidemic" of coronary heart disease was correlated with the rise of the well-fed and underexercised American body, and claiming that heart disease represented a true "disease of civilization," these cardiologists, nutritionists, and other public health activists argued that the bell curve of the American population should *not* be seen as the repository of healthy values.[28] Epidemiological field studies and cross-cultural studies drew attention to other social groups—particularly the so-called "preindustrial societies" of non-American populations, with lower mean cholesterol values—and pointed out that these people experienced substantially lower cardiovascular mortality.[29]

This argument was formally incorporated into the report of the National Institutes of Health Consensus Conference on Cholesterol and Atherosclerosis in 1985, a collective pronouncement of enthusiasts for cholesterol treatment that subtly but fundamentally redefined high blood cholesterol, at the same time announcing the creation of a federally funded program to fight it. The report of the Consensus Conference shifted the boundary between normal and abnormal in a step that retained normative statistical techniques while refusing to acknowledge that the mean of the American population represented the state of health: "Often, an abnormally high level of a biologic substance is considered to be that level above which is found the upper 5% of the population (the 95th percentile). However, the use of this criterion in defining 'normal' values for blood cholesterol levels in the United States is unreasonable; because, *in part at least, a large fraction of our population probably has too high a blood cholesterol level.* A review of available data suggests that levels above 200 to 230 mg/dL are associated with an increased risk of developing premature coronary heart disease. It is staggering to realize that this represents about 50% of the adult population of the United States."[30]

The Consensus Conference report detached the distribution of the normal population from the distribution of the American population and relocated the desirable mean leftward toward the distribution of an idealized "preindustrial" population. In this value-laden shift, the U.S. population was neatly transformed from arbiter of normality to locus of pathology. In place of a threshold defining the upper 2.5 percent as abnormal (300 mg/dL), the committee inserted a new threshold, 240 mg/dL, that intentionally defined the upper 25 percent of the adult population as abnormal, while labeling those with levels over 300 mg/dL as pathologically severe hypercholesterolemics.[31]

As quantitative definitions of normal cholesterol evolved, the qualitatively defined pathology of the disorders of lipid metabolism also shifted. What had been listed in textbooks as a single disease—"xanthomatosis" in the 1950s and "essential hypercholesterolemia" in the 1960s—had by 1970 bloomed into a verdant nosology of disorders of lipid metabolism. These were enumerated as Types I through V (based on the lipid subfractions involved) and subdivided into subtypes *a* and *b* depending on whether they represented genetic or acquired disorders. The presence of the symptom—xanthoma—was now a morphological term incidental to diagnosis.[32] What counted instead, diagnostically, was the analysis of lipoprotein profiles—the *molecular* rather than the morphological, histological, or genealogical presentation of metabolic error.

Type II lipid disorders became a particularly important site of negotiation between the manifestly pathological and the merely abnormal, for Type II could be translated roughly to mean "pathologically high LDL-cholesterol," with no other distinguishing abnormalities. Type II(a), genetically inherited high cholesterol, was also termed familial hypercholesterolemia, or "FH," a highly penetrant mutation that clustered in families and fulfilled the requirements for a single-gene Mendelian defect. Type II(b), the acquired form, however, was considered a "pattern," rather than a disease: a diagnosis of exclusion.[33]

The taxonomy of lipid disorders underwent a further contortion that significantly narrowed the intellectual space between abnormal and pathological. The 1980 *Harrison's* entry on lipid disorders was authored by Michael Brown and Joseph Goldstein, who in addition to describing the fate of cholesterol in the body as a homeostatic process with multiple feedback loops, had been able to characterize the gene that caused FH and the mutated molecular structure that produced the abnormality: the LDL-receptor. When the pair rewrote the lipid disorder section of *Harrison's*, they replaced the Type I through V system with a new taxonomy based on molecular genetics.[34] Familial hypercholesterolemia (FH), formerly known as Type II(a), was now reclassified as a LDL-receptor deficiency. Lipid disorders for which single-gene mutations had not yet been found were shunted to the end of the chapter. At the bottom of the list, the condition formerly known as Type II(b) had been relocated to the group of primary hyperlipoproteinemias of unknown etiology, with the new designation of polygenic hypercholesterolemia.

This new term, "polygenic hypercholesterolemia," can be roughly glossed as a molecular geneticist's shorthand for "We don't know why it's high." Polygenic hypercholesterolemia was a dummy variable, a placeholder that allowed statistical abnormality to become more easily commensurate with taxonomies of pathology.[35] Brown and Goldstein noted as much in their *Harrison's* entry, in a passage that came far closer to equating the abnormal with the pathological:

> By definition, 5 percent of individuals in the general population have LDL-cholesterol levels that exceed the 95th percentile and therefore have hypercholesterolemia . . . On the average, among every 20 such hypercholesterolemic persons, one person has the heterozygous form of familial hypercholesterolemia, and two have multiple lipoprotein-type hyperlipidemia. The remaining 17 have a form of hypercholesterolemia, designated *polygenic hypercholesterolemia*, that owes its origin not to a single mutant gene but rather to a complex interaction of multiple genetic and environmental factors. Most of the factors that place an individual in the upper part of the bell-shaped curve for cholesterol levels are not known.[36]

If these changing taxonomies of lipid disorders confused the average practitioner, in the public eye the overlapping terms of Type II hyperlipidemia, familial hypercholesterolemia, and the diffuse polygenic hypercholesterolemia were easily blurred. In practice, the individual clinician had a wide latitude within which to explain what a patient's high cholesterol meant in terms of disease. If your brother and sister were tested and found to have abnormally high cholesterol, or hypercholesterolemia, and you had it too, didn't that sound like "familial hypercholesterolemia"? It is reasonable to believe such confusion extended beyond patients to practitioners. Ultimately, this slippage between the abnormal ("having high cholesterol") and the pathological ("having a disorder of lipid metabolism") played into the broader goals of the National Cholesterol Education Program (NCEP) and other parties interested in mobilizing high cholesterol as a treatable condition or, ideally, a *disease*.

Market Expansion and Disease Expansion

This slippage between high cholesterol as distinct pathology and high cholesterol as quantitative variation was also ideal from a marketing perspective, and Mevacor's marketing team—well under way with product development by 1985—paid close attention to both the shifting textbook definitions of lipid disorders and the NCEP's activism regarding the lowering of the numerical threshold of normality.[37] Although Merck Sharp & Dohme marketing predicted that in effort and in return, launching Mevacor would be "its largest effort to date," Mevacor's marketers had to be very careful in their initial promotion of the drug.[38] To be approved by the FDA, every drug requires an indication, and ideally the indication should reflect an identifiably pathological state. To that end, the initial trials and new drug application submitted for Mevacor carefully limited their claims to the disease of familial hypercholesterolemia, not elevated blood cholesterol levels in general.[39] This tactic represented the first stage—legitimate market penetration—of what Merck's senior director of marketing planning, Grey Warner, explained to Merck employees and shareholders as a two-pronged marketing strategy:

> Our primary effort will be devoted to gaining physician awareness, trial and acceptance of Mevacor as a major breakthrough in the treatment of elevated cholesterol. This will be accomplished primarily through the efforts of our Professional Representatives. At the same time, we will work with various organizations including the American Heart Association, the National Cholesterol Education Program of the National Institutes of Health and Citizens for Public Awareness of Cholesterol to foster awareness and knowledge of the risks associated with el-

evated cholesterol among physicians, other health care professionals and, where appropriate, among consumers.[40]

To assure the first goal—establishing that Mevacor was an effective drug that treated an unambiguously pathological condition—Merck conducted all of its initial clinical trials in severely hypercholesterolemic patients with the single-gene diagnosis of familial hypercholesterolemia (FH).[41] Subsequent trials focused on other discrete disorders of lipid metabolism such as familial dysbetalipoproteinemia, diabetic dyslipidemia, and nephritic hyperlipidemia. Mevacor's product manager, Dr. Jonathan Tobert, emphasized the severity of familial hypercholesterolemia as justification for the use of a novel experimental therapy, citing the example of lipid-disorder specialists who worked with gravely hypercholesterolemic FH patients. "These clinicians," he noted, "said Mevacor might prove the only chance these patients had to lower life-threatening cholesterol levels."[42] Merck's director of cholesterol research, Alfred Alberts, recalled a conversation with a Merck-affiliated clinician in early 1987: "He was very anxious and he said all the cardiologists and all the primary care physicians down there want the drug, and I said, 'Well, that's a little different than what our marketing people tell us that these are the toughest people to convince,' and he had a very interesting retort to this, he . . . likened the disease—the severe form of hypercholesterolemia—to AIDS, in this regard: that people . . . were going to die in three, four years, unless you could do something drastic for them."[43] It was important that Mevacor first be given only to subjects who were clearly sick, since it was a potentially risky experimental agent. The comparison of the AIDS patient and the Mevacor patient was, in 1987, neither an accidental nor a trifling statement.

Attention to this small population of severely affected hypercholesterolemics (about four hundred thousand in the United States) was carefully balanced with the second goal of expanding Mevacor's marketing toward the one out of every four American adults estimated to have cholesterol values above the NCEP threshold of 240 mg/dL. Because early results of Mevacor in FH patients showed promising levels of cholesterol reduction with a good safety profile, Merck was able to organize a few trials testing the drug in patients who also had the cholesterol levels greater than 300 mg/dL that characterized FH but lacked the other affected family members—or monogenetic mechanisms—required for diagnosis of the familial disorder.[44] The new drug application that Merck eventually submitted in November 1986 was based on 750 of these mono- and polygenetically severe hypercholesterolemic patients. Consequently, when FDA approval of Mevacor was announced in September of 1987, both indications were included.[45]

The two indications allowed for a convenient blurriness in the terms available to promote the drug at the time of its launch. Merck's press release announcing the launch of Mevacor also listed names and phone numbers of Merck-funded researchers who were available to comment on the significance of the drug. These researchers are striking examples of the academic-industrial dual citizenship that had increasingly come to define a successful career in academic medicine by 1987. The *New York Times,* for example, used the press release to contact Antonio Gotto, who was simultaneously a professor at Baylor University, active as a Mevacor clinical trial researcher and listed as a Merck spokesman, the head of the NIH-funded Lipid Research Center, a planner within the National Cholesterol Education Program, and the president of the American Heart Association.[46] Even as he announced that Mevacor had been fully tested and FDA approved only for severe lipid disorders with cholesterol levels over 300 mg/dL, Gotto was able to simultaneously suggest that himself would enthusiastically prescribe the drug for all adults over forty whose cholesterol level was over 260 mg/dL—the 10 percent mark on the normal curve—"if they could not reduce their cholesterol by other means."[47] The overlap between the drug's formal indication for severe patients and Gotto's personal recommendation for more widespread "moderate" usage was subtle but definite. Moreover, because Gotto could claim that he was merely stating his own clinical opinion, neither Merck nor the NCEP could be held responsible for his comments, regardless of his strong ties to both organizations. Through such powerful intermediary figures as the president of the American Heart Association, then, the broad off-label usage of Mevacor could be advocated without negative consequence.

But even though off-label promotion could be a successful technique for expanding the market of a drug—and prescription volume as early as 1988 quickly indicated that the off-label use of Mevacor was indeed widespread—Merck understood that large-scale clinical trial data in patients would be necessary to achieve the larger potential market treatable by NCEP guidelines. Even before approval was announced, postmarketing plans were already under way to seek broader FDA indications that applied more generally to the adult population. A few months before Mevacor's launch date, product manager Jonathan Tobert announced to the rest of Merck that physician comfort and familiarity with the drug would soon be bolstered by "a large-scale Phase V study just getting underway that will eventually involve over 7,000 patients."[48]

The "Phase V study" was a Merck-specific term for a type of postmarketing clinical trial that was gaining prominence in the therapeutic landscape of the late 1980s. The clinical trial sequence codified in the wake of the 1962 Kefauver-Harris Act delineated four phases of clinical trial research: small Phase I trials emphasized tolerability in healthy subjects, larger Phase II and III trials emphasized efficacy and dose-response in patients, and Phase IV research was intended to emphasize the long-term safety and efficacy of a compound after FDA approval—a form of monitoring for adverse effects. What Tobert called Phase V studies (and other firms referred to as specialized Phase IV trials) were expensive, large-scale, long-term trials conducted with the aim of developing additional therapeutic indications for an already-approved drug or broadening the terms of an existing indication. In other words, these were trials of market expansion. To get a sense of the role of such trials in the changing fiscal priorities of a late-1980s pharmaceutical firm, one need only compare the sum total of Mevacor subjects whom Merck studied to obtain FDA approval—750 in all—to the size of just one of the many Phase V Mevacor trials Merck supported—listed at 7,000 subjects and growing when Tobert announced it in 1987.

Unlike Mevacor's earlier clinical trials in clearly pathological conditions such as familial hypercholesterolemia, these postmarketing studies explicitly staked out a role for Mevacor in the treatment of abnormally elevated cholesterol, using the population thresholds set out within the NCEP guidelines. Merck's first Phase V trial of Mevacor was called EXCEL, shorthand for "Extended Clinical Evaluation of Lovastatin," which enrolled more than 8,000 subjects with "moderately elevated fasting plasma total cholesterol" by the time the trial was finished. In addition to being the first study to use the national treatment guidelines as a guide to enrolling research subjects, the EXCEL trial was also unique in applying the guidelines' goals of achieving a total cholesterol of less than 200 mg/dL and/or an LDL cholesterol of less than 160 mg/dL as its target end points. In 1991 the study proclaimed success when the overwhelming majority of subjects taking Mevacor were able to achieve the LDL cholesterol levels set by the NCEP.[49]

Publication of the EXCEL trial validated Mevacor, and it also helped to validate the NIH Consensus Conference statement and the NCEP's program of guidelines that many clinicians had critiqued as a false consensus of arbitrary thresholds.[50] By generating data using the NCEP's numerical thresholds, the EXCEL trial helped to buttress them with a concrete empirical basis. After the trial, a direct flow linked the NCEP guidelines to Mevacor prescription in patients with abnormally high cholesterol.[51] The flow from therapeutic guideline to Mevacor prescriptions became even more pronounced two years later, when revised national guidelines reclassified Mevacor as a first-line agent.[52]

However, by the time the revised guidelines were released, another important shift had altered the dynamics between trial and guideline even more drastically: Mevacor was no longer the only statin on the market. The 1991 launch of Bristol-Myers Squibb's Pravachol (pravastatin), the second statin available in the United States, and the near-simultaneous release of Merck's own second-entry Zocor (simvastatin), transformed the landscape of cholesterol clinical trials from an ambitiously expanding monopoly into a fiercely competitive arena with a billion-dollar market already at stake.[53] The resulting outpouring of competitive trials—designed not just to expand the market but also to wrest it away from direct competitors—gave the commercial clinical trial a more pivotal role in marketing and development and an increasing scale of funding and influence.

Trials, Indications, and Guidelines in the Competitive Marketplace

EXCEL rapidly became an example of how a commercial clinical trial could validate guidelines and concretize them into more substantive forms of clinical knowledge, but the second generation of large-scale statin trials increasingly came to exert a formative influence on the guidelines themselves. Indeed, the statin trials of the 1990s provide a uniquely dramatic illustration of the central role that industry-funded trials have now assumed in the economy of clinical knowledge production. Although EXCEL was sponsored by Merck and conducted in academic medical centers such as the University of Kansas, Louisiana State University, and the Baylor College of Medicine, the "responsibility for the execution of the study" was contracted to a small company in Research Triangle Park called Clinical Research International.[54] The company—an early contract research organization, or CRO—was part of a nascent industry growing in this North Carolina academic-industrial suburb that marketed clinical trial services to pharmaceutical companies.

The CRO industry, which began in the late 1970s with a series of small statistical and regulatory consulting groups, had experienced a wave of growth and consolidation by the 1990s, due to the growth in size and number of industry-sponsored clinical trials.[55] The result was an industry of transnational corporations that could, for a price, organize all aspects of a clinical trial for a drug company: from trial manufacturing of pills, to producing the human subjects who would take them, to tabulating the results and easing passage through regulatory bodies.[56] By the end of the 1990s, 25 percent of drug development budgets were outsourced to CROs, and the industry had doubled to a $6.2 billion market worldwide with the top twenty companies making revenues of $50 million or more.[57]

The rapid rise of the CRO documents the increased demand for large post-marketing clinical trials and the augmented scale such trials required to show the benefits of pharmaceutical agents in subtle or asymptomatic conditions like elevated blood cholesterol. Whereas in the 1970s most truly large-scale post-marketing trials had been funded by large bureaucracies such as the NHLBI and the World Health Organization, by the 1990s every company with a statin interested in expanding market share needed its own version of EXCEL. At the beginning of the decade, 80 percent of pharmaceutical industry funds for clinical trials were channeled through academic medical centers; by 1998 this figure had been cut in half.[58] As the founder of one of the first CROs noted, the growing difference between "academic science" and "FDA science" made academic medicine an increasingly inefficient partner for a pharmaceutical development.[59] Over the course of the 1990s, the clinical trial had become an industry unto itself.

In this context, it is not surprising that when Bristol-Myers Squibb launched its own statin, Pravachol, in 1991, with the intent of redirecting as much of Mevacor's $1 billion annual market as it could manage, the drug's developers concluded that the obligatory FDA-level safety-and-efficacy demonstration would not suffice to obtain the prescription and sales figures it desired. Instead, Pravachol entered the market fully equipped with a variety of large-scale, long-term clinical trials actively investigating broader secondary and primary prevention outcomes that the firm hoped to tout exclusively.

Marketers at Merck suspected that they could deflect Pravachol's trajectory and keep it from wounding Mevacor's market share by demonstrating that Mevacor was the more potent lipid-lowerer. Merck swiftly initiated a series of head-to-head trials culminating in the Merck-funded Lovastatin-Pravastatin Study Group, which documented Mevacor's superior efficacy at lowering LDL

cholesterol and reaching the LDL cholesterol goal set by the National Cholesterol Education Program.[60] The bodies of clinical trial subjects became the bat-
tleground of a brand warfare among blockbuster cholesterol medications, a
field that soon expanded to include Lescol, Baychol, Lipitor, and several other
entrants. By 2000 at least thirty-five separate competitive trials comparing the
cholesterol-lowering abilities of two or more statins had been published, and
industry sponsorship had been central to all but three.[61]

Nevertheless, head-to-head trials were potentially damaging for either agent
involved and could easily backfire. Drug developers preferred instead to conduct trials in a way that delivered particular advantage to their own products
without risk of accidentally proving their own product inferior to the competing brand. By the late 1980s, the most promising avenue for expanded statin usage was the field of secondary prevention. Whereas *primary prevention* involved the difficult task of motivating a healthy and symptom-free population
to take a pill, *secondary prevention* addressed a population that had already suffered a heart attack or an ischemic event diagnostic of coronary heart disease
(CHD)—in other words, a population that already saw itself as diseased and
uniquely motivated to prevent further heart disease. Since patients with known
heart disease were at a much higher risk for further cardiac events, they also
represented a population more poised to accept and benefit from cholesterol-
lowering drug therapy.[62] In the terminology of commercial clinical trials, the
population of CHD offered a higher "signal-to-noise ratio" and a higher probability of positive results.[63]

Secondary Prevention Trials

Secondary prevention had also become a focus among national policymakers in cardiovascular health when the second NCEP treatment guidelines (ATP-
II) were published in 1993. As we saw with the national blood pressure treatment guidelines (JNC-I through JNC-VII) in chapter 2, each successive set of
treatment guidelines tended to expand the total treatable population. Unlike
the JNC series, however, which broadened the total pool of treatable hypertensives by successively lowering the numerical thresholds separating normal
from high blood pressure, the core threshold numbers dividing ideal, borderline, and high cholesterol levels remained constant between the 1987 and 1993
NCEP guidelines.[64]

Instead of changing the threshold values for normal and high cholesterol,
ATP-II carved out a separate population of coronary heart disease patients for

whom measurement of total blood cholesterol was simply irrelevant. Instead, cardiac patients were to be treated on the basis of their lipoprotein fraction, a measurement recommended only for the highest-risk group in primary prevention. And whereas drug therapy for other patients was considered only if their LDL fraction was higher than 190 mg/dL, drug therapy for cardiac patients was indicated at 130 mg/dL.[65] By the time the revised guidelines were published, Merck's and Bristol-Myers Squibb's dueling large-scale, long-term studies were already well under way in a race to market the first and therefore, for a crucial marketing window, the only drug with demonstrated benefit in the secondary prevention of recurrent heart disease.[66] Whichever drug could produce results first would likely, according to the teachings of brand psychology, produce the most significant brand association with prevention in the minds of physicians and patients.

The advantage this time was Merck's, but the drug was not Mevacor. The first published secondary prevention trial of a statin to be published featured Merck's newer product Zocor (simvastatin), which, though more potent than Mevacor, was less well known and not yet selling as well.[67] This trial, the Scandinavian Simvastatin Survival Study (popularly known as "4S"), was initiated four years before Zocor's FDA approval and involved nearly 4,500 patients and 500 clinician-researchers in five different countries.[68] When the trial's results were made public in November 1994, the significant price tag of the research was evidently justified. The group of subjects receiving Zocor had experienced roughly one-third less cardiovascular mortality than that seen in the control group. The results were promoted as international news, the headline "Cholesterol Drugs Found to Save Lives" made the front page of the *New York Times,* and Nobel laureate Michael Brown was quoted in national newspapers describing the results as "pivotal" and "absolutely astonishing." It was the first prevention trial that had satisfactorily documented that lowering cholesterol could actually reduce mortality.[69]

By July of 1995, the FDA had approved a new indication for Zocor's label, making it the only lipid-lowering drug allowed to claim the ability to reduce mortality from heart attacks and prevent recurrent heart disease. Previously, all cholesterol-lowering drugs had been required to include in their labels and advertisements a disclaimer that there was "no definite link" between lowering cholesterol levels and lowering the rate of developing a heart attack. This new promotional possibility gave Zocor room to grow its market, and surveys indicated that as much as 75 percent of secondary-prevention-eligible populations were not yet being treated.[70] Pravachol's own major secondary prevention study, entitled Cholesterol and Related Events (CARE) was not ready for publication until October of 1996—nearly two years later—and consequently did not afford Bristol-Myers Squibb the same level of publicity or market opportunity that the 4S had provided for Merck.[71] Nonetheless, Pravachol was rewarded for the $42 million it had spent on the CARE trial with an expanded indication that went beyond Zocor's: it could claim the preventive benefit of cholesterol-lowering pharmacotherapy among CHD patients with relatively normal cholesterol levels.[72]

The reception of Zocor and Pravachol also marked a turning point in the relationship between commercial clinical trials and clinical practice guidelines. Whereas EXCEL had used existing NCEP categories to support a role for Mevacor in the broader treatment of high cholesterol, these two secondary prevention trials themselves exerted a determining influence over the content of the ensuing guidelines. The public importance of the 4S trial data—the first data to show that cholesterol-lowering in any form could favorably affect the "hard end point" of mortality outcomes—necessitated a further revision to the NCEP guidelines regarding the role of statins in secondary prevention. In 1997 a supplement entitled *Cholesterol Lowering in the Patient with Coronary Heart Disease* was published to incorporate the 4S and CARE trials into treatment guidelines. The major alteration in treatment recommendations was to lower the LDL cholesterol threshold for pharmacological therapy in CHD patients from 130 mg/dL to 100 mg/dL. Overnight, this leftward sliding of the threshold created several million additional candidates for statin therapy. The new guidelines explicitly defined populations of patients in terms of the clinical trials—and hence, the pharmaceutical agents at the center of such trials—that had rendered them candidates for pharmaceutical therapy. As the guidelines noted, there were a projected 3.5 million "4S-like patients" to whom, in the aftermath of the trial, it was now unethical *not* to offer Zocor. "Extending the criteria for aggressive therapy to those used in CARE" they noted, "will produce an even greater impact."[73]

Primary Prevention Trials

The success of secondary prevention trials in broadening the potential market for statins, through expanded therapeutic indications and broader national guidelines, only enhanced the stakes for similar trials in the much larger market of primary prevention among otherwise healthy individuals. Once again,

Merck and Bristol-Myers Squibb challenged each other in evenly matched, $50 million dollar, five-year-long trials, each enrolling thousands of clinical trial subjects over several years. That the desired data would, in fact, emerge from these trials was to many industry analysts a tacit assumption; the question they focused upon was rather which drug, and which pharmaceutical firm, would be favored with the trial data and the new promotional license that would come with them. As a trade journal noted in the early fall of 1995, "Once the results are published (and assuming they are positive) the fight for market share will begin in earnest. With statins influencing both primary and secondary prevention, the total base will be huge. Jockeying for position will be fluvastatin (Sandoz) simvastatin and lovastatin (MSD) and pravastatin (Bristol-Myers Squibb), and poised in the wings atorvastatin (Parke-Davis)."[74]

Bristol-Myers was ultimately the fastest in this race for primary prevention results; in a public relations coup, the results of Pravachol's primary prevention trial were simultaneously published in the *New England Journal of Medicine,* announced at the American Heart Association meetings, and reported in a *Wall Street Journal* article declaring Pravachol's first significant victory over Mevacor.[75] The front-page splash in the *New York Times* was even larger than the analogous headline dedicated to Zocor's 4S study the year before, and it declared that the publication of Pravachol's West of Scotland Coronary Prevention Study (WOSCOPS) had shown "for the first time that one of the potent new cholesterol-lowering drugs can prevent heart attacks and coronary deaths in apparently healthy men."[76]

The West of Scotland study had enrolled a total of 6,595 "ostensibly healthy" adult male subjects with high cholesterol and LDL cholesterol levels and no previous history of heart disease, randomized into either drug or placebo arms for five years.[77] By May of 1995, a statistically significant difference had distinguished the two arms: the group of men taking pravastatin had experienced 31 percent fewer nonfatal heart attacks and 28 percent fewer deaths overall than the placebo group. The study's principal investigator, James Shepherd, noted that one could "now say with confidence that pravastatin reduces the risk of heart attack and death in a broad range of people, not just those with established heart disease, but also among those who are at risk for their first heart attack."[78] In a special editorial accompanying the WOSCOPS study, Torje Pedersen—the principal investigator of the Zocor 4S study—noted that the West of Scotland study had completed the final link in the logic of cholesterol lowering, offering hard outcome data demonstrating the benefits of lowering high cholesterol levels in an otherwise healthy population. "The benefits of reducing cholesterol," Pedersen concluded, "are now established beyond any reasonable doubt."[79]

The benefits of Pravachol's Scottish study extended well beyond this immediate publicity. Within eight months, the firm's new labeling submission had been approved by the FDA for a primary prevention indication.[80] Under the new provisions, Pravachol alone was licensed to claim the ability to prevent heart attacks in people with elevated cholesterol levels but no other risks of heart disease. This led to a massive direct-to-consumer (DTC) print advertising campaign—well in advance of the 1997 decision that allowed the expansion of broadcast DTC advertising.[81] Bristol Myers Squibb took out two-page spreads in the front sfection of the *New York Times* and other newspapers and magazines, proclaiming, in large print:

PRAVACHOL HELPS PREVENT FIRST HEART ATTACKS
If you have high cholesterol, there's something you should know. You may be at risk of having a first heart attack, even if you have no signs of heart problems. And the grim fact of the matter is, up to 33% of people do not survive their first heart attack. If improving your diet and exercise is not enough, you should ask your doctor about PRAVACHOL. The first and only cholesterol-lowering drug of its kind proven to help prevent first heart attacks. It may be able to help you live a longer, healthier life.[82]

WOSCOPS had caught Merck unprepared, as its own primary prevention study had required more time to develop a finding (eight years as opposed to five) and was not published until 1998. The gap gave Pravachol two years on the market as the only statin that could directly promote therapeutic claims for primary prevention, and when Merck's own study was eventually published, it received far less publicity. The editorial in the *Journal of the American Medical Association* accompanying Merck's study initially referred to it as "yet another randomized, placebo-controlled, clinical trial of a 3-hydroxy-3-methylglutaryl coenzyme A (HMG-CoA) reductase inhibitor, in this case lovastatin, as a means to prevent atherosclerotic coronary artery disease."[83] Nonetheless, Mevacor's major primary prevention study, the Air Force / Texas Coronary Atherosclerosis Prevention Study (AFCAPS/TexCAPS) proved to be a swan song of sorts for the aging drug. After eight years, Mevacor-consuming subjects experienced more than one-third fewer cardiac events than their placebo-consuming counterparts. What made the study results particularly impressive, however, was that the better part of the 6,500 overtly healthy men and women enrolled in the trial had cholesterol measurements within the contemporary boundaries of normal.[84] Less than one out of five of these subjects would have previously qualified for statin therapy under existing treatment guidelines.[85] The following year Mevacor received an FDA indication that industry analysts heralded as "the first approval to market a statin in a generally healthy population."[86]

As we have seen, from 4S and CARE onward, commercial postmarketing trials exerted an increasing effect on clinical practice and published guidelines. By the time the third revision of the NCEP guidelines—ATP III—was published in 2001, nearly thirty preventive statin trials were under way, all but three of them privately funded. As with the first and second guidelines, the numerical thresholds distinguishing normal from borderline and borderline from high total serum cholesterol remained constant in the third guidelines. At the same time, ironically, the number of total serum cholesterol had become largely irrelevant as a determinant of cholesterol-lowering therapy. New threshold numbers now enabled patients with normal total cholesterol to be classified as abnormal. Between the second and third guidelines, LDL cholesterol ("bad cholesterol") thresholds multiplied from a single line to a graded spectrum, including ideal, above optimal, borderline high, high, and very high. Following the Mevacor primary prevention trials, the line separating normal from abnormal HDL cholesterol ("good cholesterol") levels shifted upward from 35 mg/dL to 40 mg/dL, further increasing the ranks of the treatable. The new guidelines also denoted several qualifying conditions, such as diabetes, and a calculated risk score from a separate worksheet, which placed a person *without* coronary heart disease into a "coronary heart disease-equivalent" category that merited the aggressive lipid-lowering therapy of secondary prevention.[87]

Overall, the 2001 guidelines nearly tripled the proportion of the U.S. population that was eligible for lipid-lowering therapy to a market of 36 million people, and commercial clinical trials had played a key role in driving that expansion. When confronted with the observation that most of the guideline committee members had financial ties to the companies that produced statin drugs, the committee chair, Scott Grundy, responded that it was impossible by the end of the twentieth century to find any medical expert who did not have strong industry ties. "You can have the experts involved," he noted, "or you could have people who are purists and impartial judges, but you don't have the expertise."[88] By focusing on the possibility of illicit influence of the pharmaceutical industry, however, Grundy's critic missed the potentially more significant role of licit influence by which pharmaceutical manufacturers could use clinical research and clinical researchers to influence the NCEP guideline process. Even without the implication of any undue corporate influence in the guideline-setting process, it is evident that the commercial clinical trial was tremendously successful at expanding the role of statins to first fill the categories of NCEP guidelines and then exert an outward pressure on those categories, expanding the population of the treatable as thresholds for drug therapy decreased.

Defining the Bottom Line

The success of the statin trials in lowering treatment thresholds led many to wonder openly if there was indeed any limit to how far those boundaries could be pushed. As one Mevacor investigator noted at the June 2000 over-the-counter hearings, the curve of benefit from prevention trials at successively lower and lower degrees of risk implied a potentially indefinite extension of Mevacor's utility:

> I was an investigator in EXCEL [in which] the excellent tolerance, safety, and cholesterol lowering ability of lovastatin were impressive, but what remained to be shown was whether this reduction could, in fact, translate into a reduction in adverse events, heart attacks, and improved survival. This . . . has now been well demonstrated in a series of singularly successful and self-reinforcing studies published in just the last six years. These began with populations at highest secondary risk and then proceeded and concluded with those at average to slightly elevated to primary risk. In each of these studies the benefit of statins was shown . . . the Scandinavian Simvastatin Survival Study . . . the CARE and LIPID trials . . . the West of Scotland study . . . and most recently, in 1998, the Air Force, Texas Coronary Atherosclerosis Prevention Study, extended the demonstration of benefit in primary prevention to those with average cholesterol levels and no evident heart disease . . . It also indicated beneficial potential and safety in subjects resembling those who would be candidates for OTC statin therapy. Well, given that background, what then is the next step in primary risk reduction through cholesterol lowering? I believe the next logical step is to review and, if appropriate, then approve the statins for appropriate OTC use.[89]

At stake in the Mevacor OTC hearings was not just the principle of unmediated access to the drugs but also the designation of a new population of patients not currently understood to fall within the NCEP guidelines for pharmacotherapy. Using evidence from the Mevacor primary prevention study, Merck sought to market Mevacor OTC to a population of protopatients "with mild to moderately elevated cholesterol whose conditions are not severe enough to warrant prescription medicine." Based on its own primary prevention trials in patients showing benefit in patients with "borderline" levels of cholesterol (200–240 mg/dL), and borderline levels of LDL cholesterol (130–190 mg/dL), Merck argued that the data on Mevacor justified not only the over-the-counter marketing of the drug but also a further redefinition of who had "treatable" cholesterol.[90] Jerome Cohen of St. Louis University, speaking for Merck, clarified this position with a discussion of the arbitrary line dividing normal from pathological when risk follows a graded continuum:

> Let us examine the risk of [a total serum cholesterol of] 300. It's four times higher. When I began medical school, in fact, 300 was often called normal, and we've seen it drop to 280 and 250. Two-fifty offers twice the risk as 200, but let us look at 200, which is now considered so-called desirable. Do you want a level of 200? The answer, I would hope, when you know the data, is no, an ideal level which I would define as optimal levels of cholesterol is shown there at 150 milligrams per deciliter, which minimizes your risk for death from vascular disease, and we're moving in that direction . . . What you can see is the preponderance of cholesterol levels from which coronary disease eventually arises is in this so-called mild elevation of cholesterol range. That's where the action is. That's where the majority of people are. That's the group that's often dismissed by physicians and say, "Well, our cholesterol is a little high, 210, 220." It's almost normal; it's almost average. Well, the average person in this country dies from coronary heart disease, and so you don't want to have an average level. You want to have an optimal level. Remember that if nothing more.[91]

The suggested treatment of a borderline population with Mevacor OTC would bring an additional 30 percent of the U.S. population into the potential market for statins, and if Cohen's ideal value of 150 mg/dL were ratified, approximately 90 percent of the U.S. population would be defined to have higher than ideal cholesterol.[92] Compared to the logic prevalent in the early 1960s, which defined normal precisely in terms of the cholesterol level represented by 90 percent of the population, this represented only the latest movement of a historically mobile threshold.

Physicians knew that if a patient's blood pressure was lowered below a certain level, one risked entering a clearly pathological state of hypotension that was even more acutely dangerous than hypertension. The treatment of hyperglycemia, as any patient on insulin or Orinase knew, was similarly bounded by the potentially lethal dangers of lowering the blood sugar too rapidly. Although some rare metabolic disorders such as Gaucher's disease and Niemann-Pick syndrome were associated with lowered total and LDL cholesterol, however, the level at which excessively low cholesterol clearly caused harm to the organism was much harder to define. A literature describing epidemiological linkages between low serum blood cholesterol under 160 mg/dL and mortality from cancer and other noncardiovascular causes dwindled as the evidence of mortality benefit from the series of statin prevention trials continued to mount.[93] And LDL cholesterol seemed to have no bottom value at all; in 2003 the widely reported PROVE-IT trial comparing Lipitor and Pravachol greatly bolstered clinical consensus that there is little to risk, and much to gain, from lowering LDL cholesterol levels to smaller and yet smaller numbers.[94]

By the end of the 1990s, it had become a commonplace occurrence for cardiologists to suggest, only half jokingly, that statins should be included along with fluoride as a general additive to the nation's drinking water supply. From the pharmaceutical industry's perspective, lowering the threshold for treatment represented a "win-win" arrangement between private industry and public health. The lower the threshold, the larger the market, the healthier the pharmaceutical economy. The lower the threshold, the less the mortal and economic costs to the nation from heart disease and stroke. Who could argue against such a convergence of benefit? Critics of broader pharmaceutical treatment of cholesterol had found their space for argumentation within the medical literature steadily diminished. With no absolute physiological grounds for opposing the progressive lowering of the threshold of treatability, those opposed to widespread statin consumption turned instead to economic and moral arguments. The critique of widespread statin usage remains today split on the classical divide of ethical argumentation, between utilitarian calculations of efficient use of resources and deontological arguments based on absolute principles. The resulting field is fragmented and unable to mount a unified opposition to the recursively empirical rolling back of treatment thresholds.

Utilitarian arguments against widespread statin use tend to founder on questions of metrics. The shift in argument from a risk-benefit perspective to a cost-benefit or cost-effective approach can be traced back to a 1989 Canadian-

American health conference evaluating the role of policy on asymptomatic detection and treatment of high cholesterol.[95] In the same year the U.S. Office of Technology Assessment (OTA) worked to quantify the costs of the first NCEP guidelines, generating a report that suggested that in the Medicare population alone, applying NCEP guidelines would cost anywhere from $3 billion to $14 billion dollars a year, with unclear benefit.[96] These early cost-effectiveness analyses of cholesterol treatment were initially deployed as a critical methodology used by central planners to evaluate whether costly interventions served the best interest of the population as a whole. Both the OTA and the Toronto groups proposed the cost per life-year saved and the number of patients who needed to be treated in order to prevent one death as fungible metrics that could be used to compare interventions in a more pragmatic sense than the condition-specific concepts of efficacy and safety. As the Toronto group noted, "to permit meaningful comparison, it is useful to report such analyses in a common currency."[97]

Cost-effectiveness studies of cholesterol guidelines in the late 1980s calculated the costs of the NCEP guidelines to range from $32,000 to $606,000 per life-year saved, depending on age and risk profile of the population studied, with an average figure around $150,000 per life-year saved through drug therapy.[98] Based on such a high outlay per positive outcome, these critical analyses argued that screening the entire population was far from cost-effective. In a zero-sum economy, screening and cholesterol-lowering pharmacotherapy distracted funds that could be more efficiently directed toward other cardiovascular preventive efforts with lower price tags, such as drug therapy for mild-to-moderate hypertension (gauged at around $40,000 per life-year saved) or even the cost of educating ten-year-olds to reduce cholesterol (gauged at $7,000 to $50,000).[99] A swell of critical cost-effectiveness studies continued to critique cholesterol screening and pharmacotherapy in the early 1990s.[100]

Planners of the national cholesterol guidelines also recognized the importance of the new logic of cost-effectiveness, and they incorporated cost-effectiveness of cholesterol-lowering therapy in the system of the revised (1993) guidelines.[101] The same year that saw the publication of ATP-II also witnessed widespread debate over the failed Clinton national health care proposal and the publication of the World Bank Report *Investing in Health,* which introduced to the international community the concept of the disability-adjusted life-year—or DALY—as an idealized currency of comparison for various health plans.[102] The disability-adjusted life-year was considered a superior metric to the life-year saved, because it encompassed the social costs of morbidity as well as mortality. The quality-adjusted life-year, or QUALY, which the NCEP adopted for its 1993 report, represented a further extension of the same reasoning.

Echoing a prevailing trend in health economics, the NCEP declared that spending less than the per capita output of the U.S. economy (at that time, less than $20,000), per QUALY saved was "highly cost-effective" and that interventions ranging from $20,000 to $50,000 per QUALY—such as cervical cancer screening and hemodialysis—represented an "acceptable" range; a cost of $50,000 to $100,000, however, "raised questions" of cost-inefficiency, and any intervention costing more than $100,000 per QUALY represented an "excessive expense and inappropriate utilization of resources."[103] By the NCEP's calculations, cholesterol-lowering in secondary prevention was clearly worthwhile— with a QUALY price tag well below $20,000. Primary prevention in high-risk patients, with a QUALY price of $17,000 to $42,000, was also well within the acceptable range of cost-effectiveness. But general primary prevention with drug therapy in all categories of hypercholesterolemia, the NCEP noted, ran from $90,000 to well over $100,000 per QUALY, suggesting that the widespread drug therapy of elevated cholesterol could not be considered a cost-effective intervention.[104]

In a health care environment increasingly influenced by managed-care formulary decisions, the introduction of cost-effectiveness brackets into national treatment guidelines complicated the process by which pharmaceutical companies promoted cholesterol-lowering medications. Safety and efficacy might satisfy the FDA, but to gain access to the broadest possible markets, the pharmaceutical industry recognized that it would need to conduct its own cost-effectiveness research on a much larger scale.[105]

Pharmaceutical manufacturers, facing a potentially disarming critique, worked to co-opt the new field of pharmacoeconomics by incorporating cost-effectiveness outcomes as end points for major postmarketing clinical trials. In 1991 Antonio Gotto published a study in the *American Journal of Cardiology* that bounded a small group of "high risk" patients in whom the cost per life-year saved was calculated to be far more cost-effective, at $6,000 to $53,000. This fraction of the total treatable population, which Gotto estimated to number eight hundred thousand in the United States, bore "sufficiently high risk for CAD [coronary artery disease] so that the net cost of lovastatin therapy can be favorably compared with other widely used medical interventions." The metric of cost per life-year saved had become a way of sizing the market in which the use of Mevacor could be promoted as cost-effective.[106] As cost-effectiveness studies became useful in the marketing of statins, many authors who had originally written critical studies set up a small industry of cost-effectiveness research firms that fielded contract work for the pharmaceutical industry.[107] Soon after the 4S trial, the *New York Times* reported a Merck-based study claiming that Zocor could significantly lower hospital bills when used in secondary prevention. Similar data was gathered during Pravachol's subsequent CARE trial, and more detailed cost-effectiveness outcomes were incorporated into the primary-prevention trials of Pravachol and Mevacor, arguing that the drugs were an "economically attractive" remedy averaging $30,000 per life-year saved.[108]

These cost-effectiveness data collected by industry-funded trials were contested by other analysts, who argued that the widespread use of statins was not cost-effective.[109] But even as this critical pharmaco-economic literature persisted, its voice was attenuated by the multiplicity of possible metrics and methodologies of costing life available by the end of the century.[110] In addition to the intricate definitions of the DALY and the QUALY, a range of other pharmaco-economic indicators came to include the WTP (willingness to pay), the WTGT (willingness to give up leisure time), the MAR (maximum acceptable risk), and others.[111] Collectively known as "contingent evaluation approaches," these metrics focused on individual preference, rather than systematic public health prioritization, as the base unit of proper economic evaluation, and they helped to generate a more industry-friendly perspective on cost-effectiveness. This literature was further supported by surveys of American cardiologists, which suggested a tendency to recommend the use of lipid-lowering therapy even when it was estimated to cost well over $100,000 per life-year saved.[112] Although cost could suggest a bottom threshold for cholesterol treatment, that threshold would prove be easily contested and impossible to enforce.[113]

In addition to utilitarian critiques of pharmaco-preventive practice, a set of diffusely deontological critiques have also continued to oppose the expanded prescription of statins and loosely work to maintain a lower limit on treatable cholesterol levels. Based on a priori moral principles rather than any standardized calculus of benefit, this family of arguments start with the premise that routinely medicating a population that is not egregiously sick represents a fundamental moral breach that devalues human life and dignity. For the purposes of this discussion, such arguments against the wide-scale medication of a population can be divided into arguments of "medicalization" on the one hand and what Mickey Smith has dubbed "pharmacologic Calvinism" on the other.[114]

The medicalization critique is typically a top-down approach, which accuses a powerful and interested organization—most frequently the medical profession, the state, or the pharmaceutical industry—of manufacturing a disease and producing populations of patients to consolidate control over power and resources.[115] Less frequently, critics of medicalization instead lament more generally the loss of wellness and the broader social costs (beyond the risk of adverse events and the dollars spent) incurred by defining the human body as inherently diseased instead of inherently healthy.[116] Critics of medicalization have often been able to deftly untangle and expose the links between interest and the definition of disease; however, such analyses rarely trump clinical trial data when decisions are being made at the level of drug regulation, formulary acceptance, or clinical prescription. In the case of the statins, most clinicians are more concerned with the underutilization of the drugs than with any notion that widespread statin prescription might be culturally or psychologically harmful.

Critiques of medicalization are rooted in political economy, but critiques of "pharmacologic Calvinism" focus instead on the morality of the individual consuming the drug.[117] Such arguments reflect suspicions that the decision to seek a pharmaceutical solution as a replacement for some other, more individually responsible solution (e.g., diet, exercise, existential reckoning) reflects a corrosive moral laxity, a short-circuiting between effort and result. In some popular and medical literatures, statins are not seen as agents curing a disease state but rather as technologies of enhancement, physiological crutches that are used to support an immoral lifestyle.[118] The image of the overfed, underexercised American consumer who takes a statin with his cheeseburger has swiftly become something of a cultural cliché.[119] But in a culture of irony, the moral valence of such clichés is easily twisted. Thus a food critic could, by 2000, offer her highest praise to a restaurant's pâté de fois gras, Chateaubriand, or crème brûlée by advising her readers to be sure to take their Lipitor before the meal.[120] In this twist of morality, the cure for the latter-day ailments of excess consumption lies, cleverly, not in limiting consumption but in consuming additional products. Were it approved for the general marketplace, Mevacor OTC would represent the ideal extension of a culture of consumption, which finds its ultimate solution to the morbid consequences of overconsumption in the production of another consumable.

Conclusion

In examining the relationship between cholesterol, Mevacor, and the other statins, we see the negotiation of disease at its most abstractly market-oriented: bounded at the bottom by the cost of the product and bounded at the top by a series of large, expensive, and persuasive clinical trials that have influenced guidelines and practice. Ultimately, what shields the remainder of the adult population from being eligible for statin consumption is a very loosely tied network of economic and moral arguments that might yet be altered by another large study or by a shift in regulatory and consumer practices. Although the possibility of adding Mevacor to the drinking water still seems quite remote, the possibility of over-the-counter statins has recently been rekindled. By early 2004 newspapers announced that statins would soon be made available over the counter in Great Britain, and the American pharmaceutical trade press noted that Merck and Bristol-Myers Squibb were likely to petition the FDA once more to seek OTC status for Mevacor and Pravachol.[121] After all, it is not impossible in a product's life cycle for a resurrection to occur. Perhaps, years after its death, Mevacor will rise again.

We have seen that the various intersecting institutions that support and in some ways constitute the "life" of a pharmaceutical include, among others, the approval and indication-granting practices of the Food and Drug Administration, the diagnostic threshold and therapeutic recommendations of the National Cholesterol Education Program, the conduct of clinical trials, and the marketing and clinical research practices of the American pharmaceutical industry. The observation that the goals of the public health advocates of the NCEP easily merged with the goals of the pharmaceutical industry in a shared expansive tendency toward pharmaceutical prevention and a shared marketing of risk to the general population does not suggest that any scandalous influence was exerted on the part of either. Instead, the sequence of events chronicled here illustrates precisely how much the pharmaceutical industry has managed to accomplish through the construction of a means of product promotion that seems transparently *licit* at the highest administrative levels.

The recent history of elevated-cholesterol-as-disease uniquely demonstrates the fluid contemporary boundaries between physiology and pathology. Elevated cholesterol is a disorder of pure number, in which the diagnostic process is now as much a negotiation between the pharmaceutical industry and guideline-setting committees as it is a negotiation between doctor and patient. Once the bodily perceived symptom is no longer necessary for the delineation of disease—once the number and the guideline themselves effectively become symptoms—the arbitration of normality floats free from the individual body into the broader logics of bureaucratic systems and the marketplace.

The Therapeutic Transition

Iconoclastic defenders of nontreatment must also be expected to defend
their position, foreign as it is to the spirit of American medicine.
—Committee on the Care of the Diabetic, 1980

In the spring of 2003, the medical public was advised to purchase several
copies in advance of a particular issue of the *British Medical Journal* that was
predicted to become a collector's item. In surprisingly exhortative terms for the
British publication, the editor proclaimed that the lead article might represent
"the most important piece of medical news for the past 50 years."[1] This article,
by two British epidemiologists, pronounced the deliverance of humanity from
the nemesis of heart disease through the Polypill, a salvo of preventive medications compressed into a single tablet that all adults over a particular age
would be encouraged to take, daily, with no need for screening or doctor's visits. Ignoring the undertones of Orwell and Huxley that its name suggested, the
authors hailed the Polypill as the final solution to the epidemic of cardiovascular disease that had ravaged the industrialized world for the better part of a
century.

The Polypill was not a joke.[2] In the eyes of its creators—who had, indeed,
already applied for a patent on the formula and for a trademark on the name
Polypill—it was a logical and evidence-based strategy to eliminate cardiovascular disease.[3] Using a meta-analysis of more than 750 published clinical trials

that summed the experience of over four hundred thousand research subjects, the authors had set out to determine and model the combination and dosage of drugs that would yield "a single daily pill to achieve a large effect in preventing cardiovascular disease with minimal adverse effects."[4] The resulting tablet combined a statin, three blood-pressure-lowering agents (a thiazide diuretic, a beta-blocker, and an angiotensin-converting enzyme inhibitor), folic acid, and aspirin. The mixture of drugs applied to the general, healthy population, was projected to reduce coronary events by 88 percent and reduce stroke by 80 percent. One out of every three people over the age of fifty-five taking this pill would benefit, the authors claimed, gaining an average of eleven more years of life free from heart disease or stroke.[5]

For an entirely theoretical intervention, the Polypill attracted an impressive degree of international attention and controversy: its results were translated into several languages, posted on nearly three thousand Web sites worldwide, and drew more than one hundred letters to the editor from across the globe.[6] Although the majority of respondents were critical of one or more aspects of the paper's claims, citing methodological, epistemological, rhetorical, logistical, and moral lapses on the part of the authors, most respondents thought that in one form or another, the Polypill represented an innovation of the most frustratingly obvious kind: "There are many remarkable things about these papers," the *British Medical Journal*'s editor summarized, "and one is that you could almost have thought of them yourself." Indeed, as several respondents pointed out, since the average citizen over fifty-five in most developed nations was already effectively committed to multiple forms of pharmaco-prevention, putting them all together in one pill hardly seemed like a drastic intervention.[7] Other respondents wondered in their letters whether the Polypill should be limited to the prevention of heart disease when so many other conditions were now preventable. As almost all middle-aged men were at risk for benign prostatic disease and almost all women at risk for osteoporosis, why not make a male Polypill that contained prostate-protecting Cardura (doxazosin) and Proscar (finasteride) and a female Polypill that added bone-protecting Fosamax (alendronate) and calcium in addition to the other six ingredients?[8] And although many commentators suggested that the calculations of efficacy were predicated on impossible levels of compliance, a CNN poll conducted shortly after the announcement indicated that 95 percent of viewers over fifty-five would take the Polypill, if it became available on the market immediately.[9]

The Polypill represents the ultimate extension of the school of preventive pharmacotherapy traced through the history of Diuril, Orinase, and Mevacor. And yet, by favoring the mass treatment of an entire population over screening for specific physiological markers of risk, the Polypill also suggests a possible end to the risk factor as a category of diagnosis and treatment.[10] The blurring of individual risk factors is not seen only in the Polypill: by the first years of the twenty-first century, cardiovascular researchers began to suggest that millions of people diagnosed with hypertension, diabetes, or lipid abnormalities were still dying of heart attacks precisely *because* their risk factors were being treated in isolation. Instead, some critics claimed, the "number one predictor of heart disease" was a combination of risk factors that they termed Syndrome X.[11] Alternately known as the metabolic syndrome, Syndrome X combined the contemporary prediabetes of insulin resistance, hypertension or borderline prehypertension, a variety of lipid abnormalities, and obesity. In addition to those already diagnosed with hypertension, diabetes, or hypercholesterolemia, this amalgamated disorder was projected to include 33 to 44 million more Americans, or an additional 25–30 percent of the U.S. population.[12] Although exercise and diet were listed as the preferred modes of therapy—indeed, the Stanford research group responsible for publicizing the disorder had already registered their own Syndrome X Diet to combat it—experts admitted that drugs like antihypertensives, oral hypoglycemics, and statins would often be necessary.[13]

Taken together, Syndrome X and the Polypill may mark the closing of an era of diagnosis and treatment of discrete risk factors: the end of the historical moment this book has chronicled. The risk factor began to take shape in the mid-twentieth century when, as a result of epidemiological, technological, and pharmaceutical developments, chronic disease categories became preventable through the delineation and targeting of discrete symptomless precursor states like hypertension, hypercholesterolemia, and asymptomatic diabetes. As these categories, so carefully teased apart by health care professionals and pharmaceutical marketers at midcentury, are now reamalgamated into a generalized, more holistic species of risk, perhaps we will witness the death of the individual risk factor as a meaningful category.[14]

It is probably more reasonable, however, to interpret utopian visions of Polypills as a demonstration of how thoroughly the risk-factor concept now guides our basic assumptions of health and well-being. The chemotherapeutics of risk reduction seem to be everywhere expanding: in the management of osteoporosis and breast and prostate health, we see the advocacy of specific

CARDIOVASCULAR RISK SCREENING, INTERVENTION, & REPORTING TOOL   Patient Name:_____

(Return completed form to: DMH-Cambridge/Somerville, 2400 Mass Ave., Cambridge, MA 02140)   Date of Birth:_____

Risk Factor	Diagnostic/Intervention Criteria			Intervention	Results
HTN[1]	>135/85 on three different days			☐ Diet & exercise counseling	BP = / Date:
	follow JNC V1 guidelines			☐ Drug therapy	
LIPIDS[2]	Risk Category	LDL Goal (mg/dl)	LDL Level at Which to Initiate Therapeutic Lifestyle Changes (mg/dl)	LDL Level at Which to Consider Drug Therapy (mg/dl)	☐ Diet & exercise counseling	Chol_____
	CHD or CHD risk equivalents (10-year >20%)	<100	≥100	≥130 (100–129: drug optional)	☐ Nutrition referral	HDL_____ LDL _____
	2+ Risk factors (10-year risk ≤20%)	≤130	≥130	10-year risk 10–20%: ≥130 — 10-year risk <10%: ≥160	☐ Drug therapy	TG _____ Date:
	0-1 Risk factor	≤160	≥160	≥190 (160–189 LDL-lowering drug optional)		
DIABETES MELLITUS[3]	Fasting glucose >126 mg/dl on two or more tests on different days, or Random glucose ≥200 mg/dl reconfirmed with a fasting glucose >126			☐ Screening completed ___/___/___	Glucose____ Date:
SMOKING[4]	ASK, ADVISE, ASSESS, ASSIST, & ARRANGE...			☐ Assessment ☐ Advise to quit ☐ Help with quit plan and provide practical counseling	Y N Smoker? Advised? Notified MH Prov. of plan?
OBESITY				☐ Diet and exercise counseling	Weight: Height:

(See other side for care coordination information and for references.)

_____

Medical Provider signature & date

Fig. C.1. Numerical diagnosis in cardiovascular risk screening and intervention. Charts like this one are commonly used in early-twenty-first-century medical practice as screening and management tools. Note the role of numerical diagnosis and pharmaceutical intervention in most categories. HTN = hypertension. *Source:* Windsor Street Clinic, Cambridge Hospital, 2003. Courtesy of Cambridge Health Alliance.

pharmaceutical agents to be consumed in perpetuity by the asymptomatic. The pharmacotherapy of risk is now expanding into other Framingham risk factors earlier found resistant to drug therapy, including obesity (the target of Meridia [orlistat] and a topic of speculation for several pharmaceutical company pipelines) and cigarette smoking (now prescription-treatable with Zyban tablets [bupropion]). The continuing role of the cardiovascular risk factor in medical practice can be summed up in the numerical diagnosis charts (see fig. C.1) commonly found in clinics across the country by the early twenty-first century. Whatever the Polypill may signify—if it does eventually emerge as a consumable product—the legacy of Diuril, Orinase, and Mevacor is abundantly evident in the landscape of contemporary health care.

What does it mean, now, that the line between normal and pathological has become a numerical abstraction? What does it matter that pharmaceuticals have become central to public health and prevention efforts? This book has focused on a specific set of cardiovascular risk factors and preventive therapeutics in late-twentieth-century America to map out the central relationships between drugs and diseases in contemporary understandings of health. Stretching from the postwar drug boom to current concerns over increasing pharmaceutical expenditures, the narratives of Diuril, Orinase, and Mevacor overlap to provide a unique perspective on the growth of asymptomatic disease categories and the role of the pharmaceutical in their emergence.[15] This final chapter offers three sets of thematic conclusions regarding the relations of drug and disease, the debated boundaries of the normal and the pathological, and the role of pharmaceuticals in the economy of medical knowledge. Although I have taken pains to make the book more descriptive than prescriptive, my final paragraphs include reflections on some of the more unsettling problems that contemporary relationships of drug and disease bring to the fore.

Drugs and the Definition of Disease

An immediately evident conclusion to be drawn from this book is that pharmaceuticals have become central agents in the definition of disease categories. Neither drugs nor drug marketers can single-handedly define disease, though: the process involves patients, physicians, families, consumer groups, insurance companies, diagnostic technologies, expert committees, regulatory bodies, and the material basis of pathology itself, and all of these are engaged with one another in a constantly shifting system of meaning and bodily consequence.[16] However, as the cases presented in this book powerfully demonstrate, pharmaceuticals have become increasingly important to how we live our lives and how we understand both chronic disease and healthy living. In addition to changing our conceptions of disease, the widespread practice of risk reduction through long-term pharmaceutical consumption has reshaped the experience of patienthood, the ethical priorities of medical practice, and the political economy of health and medicine. Briefly, drugs define diseases in seven ways: as agents of therapeutic safety and efficacy, as bearers of therapeutic pragmatics and convenience, as technologies of enhancement, as research tools, as functional diagnostic tests, as sites of regulation and political activism, and as marketing vehicles.

Perhaps the most readily visible mode in which pharmaceuticals come to define diseases is by safely and effectively treating them, thus transforming previously deadly scourges into readily manageable conditions. This received narrative is tightly bound up with the stereotypical heroes of twentieth-century medicine: antibiotics and vaccines. Here disease eradication, though only truly achieved in the case of smallpox via selective vaccination, is the idealized relationship between drug and disease. More commonly, drugs have tamed or defanged a disease: for example, antibiotics have removed tuberculosis and pneumonia, formerly the leading causes of death in the United States, from the roster of leading causes of national mortality and made them into more curable and preventable conditions.[17] In the case of chronic conditions for which cure is not a possibility, safe and effective drugs offer a means of management, seen in the influence of antiretroviral therapies on HIV/AIDS, for example.

In all of these relationships, the drug serves as a technology of control, reshaping the formerly unruly contours of disease into forms more acceptable to human life and livelihood. Certainly this claim can be made for the impact of Diuril on hypertension, Orinase on diabetes, and Mevacor on elevated cholesterol. Each of these drugs helped to make their associated conditions more manageable, and in the process the drugs themselves became defined in terms of the related diseases: Diuril was no longer just a diuretic but an antihypertensive, Orinase an antidiabetic, Mevacor and the other statins cholesterol-reducing agents. As these terms become fused in the regulatory practice of the therapeutic indication, drug and disease become formally (and legally) understood in terms of each other.

And yet these agents enabled their associated conditions as much as they eroded them. As we have seen, asymptomatic hypertension, diabetes, and hypercholesterolemia became broadly mobilized public health concerns precisely at the moment when acceptable drugs emerged to treat them. Moreover, the three drugs studied in this book possessed a quality beyond safety and efficacy that proved crucial for such a transformation: palatability. Orinase was not necessarily safer or more effective than insulin, but it was a far more attractive option than the daily injection, and it made the treatment of asymptomatic diabetes a much more feasible option to a population that consequently allowed itself to become known as diabetic patients. Similarly, although cholestyramine had been deemed both medically safe and effective in lowering serum cholesterol, the inconvenience of swallowing several grams of a fishy gravel every day made it extraordinarily difficult to recruit patients and collect long-term data regarding its efficacy in the primary prevention of heart disease. In contrast, the odorless and easy-to-swallow tablet of Mevacor became a means of recruiting thousands of clinical subjects to demonstrate broader and still broader claims of efficacy and safety. In the history of these three drugs, the relationship of palatability, efficacy, and safety is shown to be far more interactive than is conventionally understood.

Palatability has its flip side, however, and negative connotations of moral laxity were often conflated with conceptions of therapeutic ease to complicate the cultural reception of these drugs. Both Orinase and Mevacor shortcircuited the link between effort, responsibility, and reward in the arena of health. Just as diabetologists worried about the untoward consequences of obese diabetics taking a pill instead of following a rigorous diabetic diet, later cardiologists and general commentators lamented a nation of consumers that might dose themselves with Mevacor so that they could continue to enjoy cheeseburgers. Such complaints are less apparent in connection with hypertension: for some reason it seems less morally questionable to manage one's blood pressure with a pill, just as it raises fewer eyebrows if a person with dyspepsia takes Prilosec so he can eat spicy food and drink coffee, or an allergic individual takes Claritin so she can own a pet, or an asthmatic uses Advair so he can engage in competitive sports. As technologies of enhancement, drugs make us uneasy, but it is nonetheless extraordinarily difficult to define the exact distinction between enhancement and treatment, or between ethical and unethical modes of enhancement.[18] Whether the use of pharmaceuticals to enable lifestyle choices is to be praised or criticized, it appears, is entirely dependent on moral tensions still lurking deep within our general understandings of disease.

Pharmaceuticals have many more functions in relation to disease besides efficacy, safety, palatability, and convenience, however. The role of the pharmaceutical in propping up the "diseaseness" of a putative pathological category can also involve the production of a model or mechanism of action that makes the disease more plausible or offers a nucleus for laboratory investigation in relation to a disease. In the case of Diuril, the suggestion of a unitary mechanism behind the drug's action in hypertension—though never found—helped to produce an optimistic sense that hypertension itself might indeed have a single molecular mechanism that could be elucidated. We see a similar process at work in the role of Mevacor in verifying Brown and Goldstein's Nobel Prize–winning research on the mechanistic importance of the LDL-receptor in atherosclerosis, work that supported a mechanistic understanding of disease that was later crucial to the promotion of Mevacor's efficacy and to the NCEP's promotion of high cholesterol itself as a condition worthy of broad detection and treatment.

Pharmaceutical agents also support disease categories by acting as diagnostic devices; clinicians use drugs this way when they treat patients empirically in a form of ongoing diagnosis. If your diffuse symptoms of unwellness are alleviated by an antidepressant, then most clinicians would say, retroactively, that you were depressed; if your stomach pain recedes after treatment with Prevacid, then there is a good chance that you had a gastritis or an esophageal reflux disease and not a severe ulcer. The distinction often writes itself back into the medical understanding of the disease itself, as, for example, the separation between Orinase-responsive and Orinase-unresponsive diabetes helped to contribute to the delineation of insulin-dependent and non–insulin dependent variants of diabetes mellitus. The role of the pharmaceutical as a diagnostic agent became even more overt in the later career of Orinase, when Upjohn unsuccessfully sought to promote a new formulation of tolbutamide, Orinase Diagnostic, as the gold-standard test in the diagnosis of diabetes. That Orinase Diagnostic failed in this bid by no means suggests that it represented a transgression of the place of a pharmaceutical agent in medical diagnostics.

However, Orinase's broader transgressions during the UGDP crisis—in which the agent crossed from an agent of putative risk reduction to an agent of putative harm—highlight how drugs influence the definition of disease by becoming key sites for political activism and regulation. The politics of drugs and disease is always historically contingent. Whereas the political usage of Diuril as a chemical Sputnik in the cold war helped the pharmaceutical industry successfully defend its pricing during the Kefauver investigations in the late 1950s and early 1960s, the crisis that erupted around Orinase in the 1970s brought to light a set of formerly invisible struggles over the definition of disease, tensions between the federal government and practicing physicians, between consumer-rights advocates and the pharmaceutical industry, and between older paternalistic approaches to medical information and the new preference for a more open market of medical knowledge. As we see in the public hearings on Mevacor OTC and the morning-after pill, in the history of pharmaceutical-centered AIDS activism, and in countless other examples, pharmaceuticals have become key sites for political activism around disease categories and public health.

One further relationship between drug and disease now permeates and makes possible all the other relations: namely, the role of the pharmaceutical as product and the disease as its consequent market. One way for a pharmaceutical product to increase its market share is to beat out other competitors in the field in terms of safety, efficacy, or convenience. Another is to pursue new indications in other established disease categories. A third strategy to grow a drug's market—the mechanism most noticeable in the stories of Diuril, Orinase, and Mevacor—is for the drug to alter and expand the definition of the disease category itself. As this narrative has illustrated, it is difficult to redefine a disease category when the disease is limited by symptoms. But in the treatment of risk, where there are no symptoms except for the numbers themselves, the prospects for therapeutic expansion appear limitless.

Numerical Diagnosis and the Contested Boundary of Normal

As much as market expansion may be a principal goal for pharmaceutical marketers, many aspects of a drug's career will always lie outside of their immediate control. Even when a therapeutic indication is successfully expanded, this process does not follow a single course but instead tends to proceed in a manner highly contingent on the material profile of both the disease and the drug in question. Furthermore, these material profiles are never fully knowable in advance; several physical properties of drug and disease are revealed only in the interaction of the two. The numerical diagnoses discussed in this book hinge upon debated boundaries between normal and abnormal that are pushed outward by pharmaceutical companies and public health advocates alike in a manner that defines an increasing number of health states as sufficiently abnormal to warrant treatment. These three case studies demonstrate different ways in which this boundary gets relocated: shifts in a single set of numbers (hypertension), fragmentation of multiple risk groups (cholesterol), and attempts to translate borderline test results into incipient forms of disease (diabetes).

Hypertension offers the most purely linear model of the expansion of a pathological category. The consolidation of consensus around the National High Blood Pressure Education Program in the 1970s focused a cacophony of divergent thresholds into a single pair of numbers, agreed upon by committee, which divided pathological from normal blood pressures and which have since slid smoothly down the column of mercury to gradually enfold larger and larger populations within their boundaries. Hypertension's retreating threshold and the dramatic population growth of the group defined as hypertensive is checked, ultimately, by the acutely dangerous condition of pathologically low blood pressure, or hypotension. Although the threshold for high blood pressure may continue to move lower until there is very little normal left between the pathologically high and the pathologically low, it is clear that the downward slide of this threshold cannot continue forever and that the dosing of blood-pressure-lowering medications can never be entirely casual.

The expansive definition of abnormally high cholesterol has followed a far less linear course. Instead of focusing the attention of diagnosis on a single mobile threshold, the pathological career of hypercholesterolemia has involved a multiplication of thresholds and flowcharts that have come to include more and more Americans in the ranks of the treatable. Whereas the primary thresholds of high, borderline, and normal cholesterol have remained constant since the 1985 NIH Consensus Conference, several subthresholds, including LDL cholesterol, HDL cholesterol, coronary heart disease and its equivalents, and more recently C-reactive protein levels, have proved to be mobile and have successively widened the total population deemed eligible for preventive statin therapy. This nonlinear proliferation of biological targets is a conceptually postmodern counterpoint to hypertension's linear modernism. Unlike the case of hypertension, these targets do not appear to be bounded on the lower extremity by any pathological condition of low cholesterol or low LDL, and hence the gradual expansion of pathological categories may meet no necessary physiological boundary.[19] This one-tailed nature of cholesterol as a pathological variable has allowed cardiologists to make cavalier comments about adding statins to the drinking water and has prompted drug makers to present a series of arguments for making cholesterol-lowering medications available to the entire U.S. population on an over-the-counter basis.

In contrast to the twentieth-century syndromes of hypertension and hypercholesterolemia, the expanding definition of an ancient disease such as diabetes offers particular insight into what happens when the site of diagnosis shifts from perceived symptom to pathognomonic sign to numerical marker. For diabetes to become an asymptomatic disease, it needed to expand not down in line (as hypertension did), nor outward into a proliferation of targets (as cholesterol did), but rather laterally, across different metrics, from symptoms (polydipsia/polyphagia/polyuria), to the pathognomonic sign of the clinical laboratory (sugar in the urine with high blood sugar), to the numerical thresholds of the finger-stick glucose test and the glucose tolerance test. As it traced its own unique trajectory of expansion, the career of asymptomatic diabetes required the buffer space of the borderline and the predisease to mediate between thresholds of disease and health and to stabilize both categories. As we saw, the sliding definitions of prediabetes, protodiabetes, chemical diabetes, and other prediseases were progressively folded into diabetes to make room for new buffer states that further expanded the population of hidden diabetics and the Orinase market.

If the threshold is a thin dividing line, a precise curtain marking the break between normal and abnormal, the category of the borderline as commonly used in medicine refers not to the line itself but to a space or territory made up of individuals or populations who sit, problematically, within a reasonable margin of error from the line. As a space around a threshold, the borderline soon develops its own boundaries, with new thresholds on either side. The space of the borderline simultaneously acts as a buffer zone between the territories of pathology and normality, while at the same time offering itself as a colonizable land, a fertile area for future disease expansion. The experimental treatment of a borderline state makes the treatment of nonborderline states seem less experimental and thus less controversial as a form of therapeutics. Additionally, the presence of the border tempers the potential for conflict between individually minded practicing physicians and the standardizing intent of guideline-producing bodies by offering a sanctioned arena in which idiosyncratic patterns of practice are still permitted. Although failure to treat a clearly pathological individual is grounds for malpractice litigation, failure to treat a borderline case is no failure at all; in these grey realms, some individualization of medical practice remains possible. Thus the borderline functions as a pressure valve for those who might dissent from therapeutic guidelines by maintaining a sanctioned space for therapeutic libertarianism.

Finally, although the two terms often seem to denote the same category, it is important to delineate some subtle but increasingly significant distinctions between the description of the borderline state and the predisease state; the latter became increasingly important in discussions of asymptomatic disease over the course of the period studied. The importance of the difference is perhaps most evident in the recent interest in the description, detection, and treatment of the prehypertensive—formerly "borderline hypertensive"—individual.[20] Whereas the borderline hypertensive existed on the fringes of a diagnosis, with uncertain prognostic significance, the prehypertensive now exists in direct temporal relation to the hypertensive, placing the linear gradient of the sphygmomanometric number in apposition to the linear gradient of time-course and pathogenesis. The prehypertensive state may not be defined as pathological, but it implies a clear warning of future pathology in a way that the borderline does not. Indeed, the current public health mobilization of prehypertension detection efforts places health professionals in the unusual situation of prioritizing an asymptomatic condition that is itself only a predictor of another asymptomatic condition.[21] As a pathological category, prehypertension exists two degrees of separation away from any symptoms or other "hard" disease outcomes. As these categories take on more concreteness in clinical practice, the fundamental concept of disease becomes as abstract and spaceless as the marketplace itself.

Pharmaceuticals and the Economy of Medical Knowledge

As Pfizer president John McKeen noted in a 1959 talk to fellow pharmaceutical executives, prescription drugs participate in an economy of medical knowledge that is every bit as important as their participation in the global economy of currencies and commodities: "The world is more than a market for pharmaceutical commodities; it is a marketplace of knowledge . . . With this industry's world growth, it has fashioned a pipeline through which medical knowledge flows from the farthest corners of the globe."[22] A half century later, McKeen's remarks appear self-evidently true. By the end of the twentieth century, the American pharmaceutical industry had become an institution with fully global circulation, and its research, development, and marketing funds were crucial to the production and circulation of medical knowledge at all levels: from industry labs to academic medical centers; from bench research to clinical trials; from the education of medical students, residents, and practicing physicians to the forging of national treatment guidelines and health policy. The influence of the pharmaceutical industry now permeates the global economy of medical knowledge, a currency exchange that has no functional firewalls to insulate its transactions from the more earthly economy of branded products and sales revenues.

Perhaps the greatest risk in the practice of evidence-based medicine lies in treating the available pool of medical knowledge as a balanced reservoir of facts that emerge from dedicated scientists and circulate solely through the convective currents of peer-reviewed journals and enlightened discussion. Medical knowledge neither arises spontaneously nor diffuses passively. Rather, in the fluid dynamics of medical knowledge there are deep, undisturbed trenches and there are continually pumping vents of activity, such as pharmaceutical corporations, which act as engines for the development and promotion of forms of knowledge they find useful. The pharmaceutical industry is not the only driver of medical knowledge: in similar fashion the insurance, device, and biotechnology industries, the various medical specialties, patient-activists, foundations, and the politics of federal funding all help to shape what sort of knowledge is produced and publicized. For reasons we have explored over the course of this book, however, pharmaceuticals have become uniquely important to this process. As individual pharmaceutical agents encourage action in the spheres of clinical research, clinical practice, and medical marketing, they bring the economies of medical knowledge and the economies of hard currency into close apposition.

The principal means by which they bring these worlds together is the clinical trial. Historians have described the long process by which the randomized clinical trial achieved preeminence as a prime vehicle for therapeutic rationalism.[23] Large trials structured to provide definitive answers are now conducted almost continuously and swiftly linked to guideline-producing bodies and publicity mechanisms intended to transform clinical practice to better reflect available evidence. Clinical trials constitute a key historical punctuation of the contemporary period: every month another new era of therapy is ushered in based on new results accumulated around a pharmaceutical agent, facts that alter the ethics of living with or without medication. In the second half of the twentieth century, it became apparent to most pharmaceutical companies that this mechanism for promoting a more rational therapeutics could be skillfully adapted to promote larger and larger markets for their products. Now that the public reporting of clinical trials is as likely to be found in the business sections of newspapers as in the science or health sections, it becomes hard for physicians or patients to know with what balance of skepticism and sincerity they should consume such knowledge.

The fact that the average clinical trial today is part of a larger industrial marketing engine does not mean, however, that such research can be dismissed as merely marketing. The group of postmarketing statin trials conducted in the 1990s—funded by pharmaceutical companies in the hopes of expanding the market of treatable cholesterol levels and extending the market share of their designated products—were at the same time ideal models of large-scale, long-term, double-blind, randomized, placebo-control prevention trials. Even though most of these $50 million trials were feasible only because they suited the marketing goals of the products involved, there are no grounds to accuse CARE, 4S, WOSCOPS, or AFCAPS/TexCAPS of being "bad science." For the most part, it is of vital interest to the pharmaceutical industry for its research to be regarded as "good science," not only to obtain FDA approval but also so that the industry can more generally promote itself as the most legitimate and scientifically based means of therapeutics. The occasional case of fraudulent research notwithstanding (e.g., MER/29), the industry is cautious not to let its zeal for expansive markets overtake its own legitimacy in the eyes of the medical profession and the general public. Proof lies in the recent PROVE-IT trials, in which a large, expensive, and well-publicized clinical trial funded by Bristol-Myers Squibb to promote the additional use of its cholesterol-lowering Pravachol instead demonstrated the superiority of the competing Lipitor—much to the detriment of Pravachol's market share.[24] Bounded by the dual bottom lines of scientific rigor and fiscal responsibility, the process by which pharmaceuticals influence the landscape of available medical knowledge is always a careful tightrope between marketing ambitions and the realm of the clinically provable.

Although, as Diuril demonstrates, this relationship between marketing and clinical research already existed at midcentury, there were significant changes in the economy of pharmaceutical knowledge between 1950 and 2000. One unmistakable shift regards the definition of knowledge consumers. The physician was addressed as the primary consumer of medical knowledge in the 1950s (both in journal literature and in pharmaceutical promotion), but that role, along with other vestiges of medical paternalism, was successfully challenged in the 1960s and 1970s and a much broader, more egalitarian circulation of medical information was common by the end of the century. The past fifty years have witnessed the resulting increase in health reporting in newspapers and magazines, the rise of popular medical information technologies, the transformation of pharmaceutical public relations into direct-to-consumer advertising, and the proliferation of many other participants in health care debates. These new participants—including patient advocates, formulary committees, managed care organizations, lawyers, ethicists, lawmakers, and other "strangers at the bedside"—now constitute a multitude of consumers in the medical information market.

Generally speaking, the intellectual economy of medicine and health has shifted from a Keynesian system to a free-market approach, and this broad shift has clearly multiplied the venues available for the promotion of pharmaceuticals. While the structure of drug and disease promotion described in the launch of Diuril still stands as a scaffolding for pharmaceutical marketing, it is now supplemented by a dramatically expanded corps of sales representatives, new marketing technologies such as drug Web sites and handheld PDA (personal digital assistant) devices with access to real-time market research data, extensive direct-to-consumer advertising budgets, celebrity endorsements, funding of public sporting events, the establishment and promotion of disease-identity groups, and the rising trend of "ghostwritten" scholarly articles subsequently attributed to prominent academic physicians.

It is tempting to muse that most medical knowledge is itself prescribed; that is to say, pharmaceutical marketing forces now visibly drive our lay and professional consumption of medical knowledge. The present saturation of pharmaceutical promotion has recently begun to provoke a backlash against the industry. Promotional mechanisms that went unquestioned in the 1950s are now the subject of increased public scrutiny; the omnipresence of consumer-oriented pharmaceutical advertising has helped to fuel a new market for pharmaceutical muckraking literature. An increasingly vocal segment of the medical profession, as well as a significant group of lay activists, are now criticizing the Food and Drug Administration as an entity too cowed by its financial dependence on the companies it is supposed to be reviewing to properly regulate their promotional activities.

Much of the recent attention to pharmaceutical promotion has been prompted by scandals or obvious ethical breaches on the part of a pharmaceutical company, of which there have been many.[25] I have deliberately avoided tales of the scandalous in this book, because of the tendency of scandals to suggest that undue influence is a result of extraordinary practice, bad science, or a breach of normal operating ethics. As the preceding chapters have demonstrated, the greater part of the influence that pharmaceutical firms wield over medical knowledge and clinical practice operates in the everyday occurrence of rigorously conducted science and rational therapeutics. This influence is both more insidious and more morally ambiguous. The fact that an industry with a product to sell should seek all means available to sell it, within the boundaries set for it by the constraints of scientific practice, regulatory practice, and clinical practice, should not be surprising. Perhaps it should be surprising, however, that the industry finds such willing and able promotional allies in the body of cardiologists, epidemiologists, and public and private institutions, whose relative duties to the public are more clearly defined.

Beyond the Therapeutic Transition

As a mainstream movement within American medicine, the pharmaceutical approach to primary prevention has shown tangible benefits. After decades of widespread use of antihypertensives, antidiabetics, and cholesterol-lowering agents, deaths of Americans from cardiovascular disease and stroke are less frequent than they were at midcentury. But another result of the project of prevention is that the American population now consumes an enormous amount of pharmaceuticals and bears direct and indirect costs that have not necessarily been balanced in the interests of the general population. The long-term consumption of preventive pharmaceuticals can carry direct costs for the bodies and pocketbooks of those who consume them, as well as indirect costs in the broader de-prioritization of other forms of prevention and other public health goals. In addition, recent studies have begun to document the ecological costs of pharmaceutical consumption, as high levels of pharmaceuticals excreted by American populations have medicated local riverine and wetland ecosystems with a notably toxic effect.[26] Three ruptures have emerged in the otherwise smooth synthesis of pharmaceutical prevention, and I would classify these major problem areas as economic cost, substituted risk, and distraction.

First, increased pharmaceutical consumption and pharmaceutical expenditures have come at a time of increasing health care costs and swelling ranks of the U.S. population for whom health insurance is unaffordable and health care inaccessible. In the year 2000, more than $40 billion was spent on brand-name cholesterol-reducing agents, oral antidiabetic agents, and antihypertensive agents; by 2003 the global cholesterol drug market alone was $26 billion, and one out of every two dollars spent on cholesterol-reducing drugs was spent by the U.S. population.[27] For the patients who can afford to pay for their medications and suffer no adverse effects from them, the pharmacopoeia of prevention is likely a win-win situation whether they ultimately benefit from their probabilistic risk reduction or not. For many others who face serious economic decisions surrounding their ability to pay for their medications, or who must bodily countenance adverse effects of such medications in order to achieve their numerical targets of physiological normality, or both, the calculus becomes far more complex. In the absence of a guiding symptom, it is often difficult for the physician or the patient to know when the benefits are worth the costs.

The second problem is substituted risk. On the level of the individual patient, the application of population-based data regarding the appropriateness of risk reduction can be difficult; risk reduction is not necessarily risk-free. Even if clinical trials are internally rigorous, the exclusion requirements of efficacy trials often make it hard to generalize from a trial population to the case of an individual patient. Moreover, the results of even the most successful risk-reducing trials still indicate that the majority of patients treated will not receive a direct improvement in morbidity or mortality from taking their course of treatment. The tendency to collapse the complexities of population-level study results into a "soundbite" fact (e.g., "statins prevent heart disease in people with high cholesterol") that is digestible and seemingly applicable to the individual patient who "has high cholesterol" often distorts the relative benefits and risks of any clinical decision. Should one continue to take cholesterol-lowering drugs in the face of mild liver damage? Does the benefit of maintaining one's blood pressure at approved levels justify experiencing side effects such as impotence? Do the long-term gains of prevention balance out the immediate costs? These questions cannot be answered by clinical guidelines. They require a localized understanding of the ethics of clinical decision-making, a complex humanistic practice that the integrated promotion of pharmaceutical products and numerical treatment thresholds can, unfortunately, obscure.[28]

Third, the synergistic "win-win" collusion of public health and private wealth runs the risk of distracting global public health efforts away from those diseases that do not represent attractive markets. The narrative in this book deals with a narrow period in American history in which national public health campaigns have focused increasingly on prevention via the detection and treatment of asymptomatic diseases. In other parts of the world, frankly symptomatic diseases dominate public health concerns, diseases that present themselves forcefully on populations without need for any promotional efforts or consumer education. Symptomatic cases of malaria are estimated at 300–500 million per year, causing well over 1 million deaths annually.[29] Roughly 90 percent of the cases of malaria occur in sub-Saharan Africa, where they are joined by roughly half a million symptomatic cases of African trypanosomiasis, or "sleeping sickness," each year.[30] Symptomatic trypanosomiasis is almost always fatal without proper treatment, but it is estimated that only 10 percent of those afflicted receive medication. These indisputably diseased populations, though numbering millions, do not translate into profitable markets for a drug. They are, in the fiscal perspective of the global pharmaceutical industry, simply not visible. Although several antimalarial and anti-trypanosomiasis compounds emerged from the American pharmaceutical industry in the 1940s and 1950s, the half century that produced our pharmacopoeia of risk reduction was a largely stagnant period for developing agents to fight these global epidemics. Only very recently, after years of concerted pressure and financial challenges from various independent public health institutions and foundations, has the pharmaceutical industry slowly become a partner in developing drugs for these conditions.[31]

The responsibility for this mismatch of supply and demand—quaintly known as "market failure"—does not rest solely on the shoulders of the pharmaceutical industry, but it greatly undermines the industry's claims that free-market approaches are the best way to meet public health goals.[32] The examples of malaria and trypanosomiasis make evident, in a way that hypertension, diabetes, and elevated cholesterol may obscure, that there is and should be a divergence between the interests of the pharmaceutical industry and the interests of public health. We should be rightly wary of arguments that simply present the priorities of one as a proxy for the other.

Numerical diagnoses offer the promise of controlling our own future health, but their careless application runs the risk of dehumanizing the medical experience and fueling public skepticism about the growing influence of market forces on medical practice. Over the course of this book I have taken pains to avoid prescriptive polemic in the narration of a controversial set of events—the interaction between the marketing of pharmaceuticals and the changing definitions of chronic disease. As the cases in this book suggest, it is simply misguided to ask whether market forces should be allowed to exist in medicine. For they have always been there in some form, certainly since the often nostalgically misremembered 1950s, even in the production and circulation of "cheap" drugs like thiazides. Instead, the lessons from this book suggest a more significant question: Which forms of engagement between marketing, public health, and medical practice are productive, and which are deceptive? Just because diseases now represent markets does not mean that we need to embrace a laissez-faire approach to the economy of medical knowledge. Left solely to the whim of the invisible hand, the discourse of health will be dominated by high-margin and high-profit issues at the expense of other vital issues. If we want to maintain public health as a public good, it will be important for the public to invest in it and to prioritize those efforts that address the needs of the populace.

The pharmaceutical industry did not simply buy the influence it now wields over the production and circulation of medical knowledge: it has developed its relevance in the production and circulation of medical knowledge because it was allowed, even encouraged, to do so. The industry has subtly and very effectively filled a void in the production and circulation of medical knowledge, and it now occupies a relative advantage in this asymmetric economy. But neither cynicism nor nostalgia can effectively blame the pharmaceutical industry for all the present ills of the health care system. If we are to restore trust in the balanced pool of medical knowledge, the intersecting efforts of many institutions—from professional associations to advocacy groups to medical journals to federal agencies—will be required. To confront the economic problems posed by the spread of preventive pharmaceuticals, we will also need to develop metrics to think beyond statistical significance to clinical significance and social significance, so that merely demonstrating a divergence between two mortality curves does not mandate a mass prescription. We will also need to confront difficult questions in defining the boundary between essential medicine and enhancement and consider how much of our resources we should be spending trying to live a little bit longer, as opposed to engaging in other, perhaps more noble, pursuits.

At the dawn of the twenty-first century, disease is still more than a numerical threshold. A vast amount of human suffering is encountered each year in struggles with unremitting pain, depression, stigma, loss of functionality, and death, providing ample testimony that we are far from any real eradication of the symptom. For the multitudes who bear no symptoms, however, the popularization and widespread pharmaceutical treatment of asymptomatic diseases has irreversibly altered the nature of health and disease in contemporary American society in ways no one could have predicted at midcentury. This new philosophy of health and medicine has generated new experiences of patienthood and has reoriented both the practice of medicine and the ethical priorities of the doctor-patient relationship. The pharmaceutical-centered program of risk reduction has ushered in a new economy of health values, a new approximation of public health and consumerism, and an expanding set of surveillance structures by which not only patients but also clinicians, policymakers, and even pharmaceutical executives find themselves constrained in their abilities to make decisions about the proper means of promoting good health and quality of life.

Preventive pharmaceuticals and risk factors are instruments of therapeutic rationalism. Of many possible therapeutic rationales, however, they tend to support a particular kind: a structure of risk-based knowledge that must be followed by therapy. Reinforced by marketing and public relations, this rationale inserts itself seamlessly into the thoughts and practices of physicians, policymakers, and patients as a commonsense partnership between clinical science and the marketplace. These drugs and the numerical conditions they treat thus exemplify in microcosm the links more broadly forged between bureaucracy, technology, and capital in contemporary society, and they show no sign of diminishing.

Notes

PREFACE

Epigraph: This quotation has been attributed to Osler by many sources, though it does not seem to come from any of his published works. Curiously, the same quotation has also been attributed to Mark Twain and Oliver Wendell Holmes and has been labeled a Chinese proverb, all without adequate documentation. See, e.g., William J. Hall, "The Doctors of Time," *Annals of Internal Medicine* 132 (2000): 18–24.

INTRODUCTION: The Pharmacopoeia of Risk Reduction

Epigraph: E. E. Evans-Pritchard, *Witchcraft, Oracles, and Magic among the Azande* (Oxford: Clarendon Press, 1937), 161. Margaret Lock pointed out this quotation.

1. Charles M. Mottley, "Operations Research Looks at Long-Range Planning," *Proceedings of the American Drug Manufacturer's Association* 512 (1958): 161–62.

2. Although an early description of the Framingham Study came out in 1951, the study's findings were first published in 1957; see Thomas R. Dawber, Felix Moore, and George V. Mann, "Coronary Heart Disease in the Framingham Study," *American Journal of Public Health* 47 (1957): 3–24; see also Thomas R. Dawber, Gilcin F. Meadors, and Felix Moore, "Epidemiologic Approaches to Heart Disease: The Framingham Study," *American Journal of Public Health* 41 (1951): 279–86. For a concise retrospective on the goals and proceedings of the Framingham Study, see Thomas R. Dawber, *The Fram-*

ingham Study: The Epidemiology of Atherosclerotic Disease (Cambridge: Harvard University Press, 1980); Merwyn Susser, "Epidemiology in the United States after World War II: The Evolution of Technique," *Epidemiologic Reviews* 7 (1985): 147–77; Daniel Levy and Susan Brink, *A Change of Heart: How the People of Framingham, Massachusetts, Helped Unravel the Mysteries of Cardiovascular Disease* (New York: Knopf, 2005).

3. George V. Mann, "The Epidemiology of Coronary Heart Disease," *American Journal of Medicine* 23 (1957): 478.

4. William B. Kannel, Thomas R. Dawber, Abraham Kagan, Nicholas Revotskie, and Joseph Stokes III, "Factors of Risk in the Development of Coronary Heart Disease: Six Year Follow-Up Experience—The Framingham Study," *Annals of Internal Medicine* 55, no. 1 (1961): 33–50.

5. Jeremiah Stamler, "Epidemiological Analysis of Hypertension and Hypertensive Disease in the Labor Force of a Chicago Utility Company," *Hypertension* 7 (1958): 23–52, quotation on 48–49.

6. The categories of cholesterol- and triglyceride-reducing agents, calcium-channel blockers, ACE-inhibitors (the latter two are the most highly marketed antihypertensive medications), and oral antidiabetic agents accounted for a cumulative $38.9 billion in global sales in the year 2000. These four categories of drugs represented more than 12% of total pharmaceutical sales. Data from IMS Health, *2001 World Review,* online at www.ims-global.com/insight/news_story/0103/news_story_010314.htm.

7. Susser, "Epidemiology after World War II"; Allan Brandt, "The Cigarette, Risk, and American Culture," *Daedalus* 119, no. 4 (1990): 155–76; Robert Aronowitz, *Making Sense of Illness: Science, Society, and Disease* (Cambridge: Cambridge University Press, 1998).

8. *Epidemiologic transition* is a familiar term used by historians, sociologists, epidemiologists, and demographers to describe the overarching shift from acute, infectious causes of death to more chronic, noninfectious causes. Although Omran did not formalize the theory of epidemiologic transition until the mid-1970s, the American epidemiological data relating the decline of infectious disease mortality and the concomitant rise in mortality from heart disease and cancer was evident as early as the 1920s and had been robustly articulated by the 1940s; see J. E. Gordon, "The Twentieth Century—Yesterday, Today, and Tomorrow"; public address, c. 1920, reprinted in *The History of American Epidemiology,* ed. C.-E. A. Winslow (St Louis: Mosby, 1957); Charles-Edward Amory Winslow, *The Conquest of Infectious Disease: A Chapter in the History of Ideas* (Princeton, NJ: Princeton University Press, 1943); Abdul R. Omran, "The Epidemiologic Transition: A Theory of the Epidemiology of Population Change," *Milbank Quarterly* 49, no. 4 (1971): 509–38.

9. On the agency of nonhuman actors, see Michel Callon, "Towards a Sociology of Translation: Domestication of the Scallops and the Fishermen of St. Brieuc Bay," in *The Science Studies Reader,* ed. Mario Biagioli (New York: Routledge, 1999); on pharmaceuticals as nonhuman actors, see Jordan Goodman and Vivien Walsh. *The Story of Taxol: Nature and Politics in the Pursuit of an Anti-Cancer Drug* (Cambridge: Cambridge University Press, 2001). The historical "sampling device" is derived from Charles Rosenberg's usage of successive epidemics over the course of the nineteenth century in *The Cholera Years* (Chicago: University of Chicago Press, 1962) to trace subtle shifts in American cultural history.

10. A critical literature on "medicalization" was energized in the 1970s by Eliot

Friedsen's sociology of the medical profession, the strident antimedical polemics of Ivan Illich, and the translated work of Michel Foucault, who described the classification of pathology as the subtlest police function by which the modern state inscribes order onto the bodies of its subjects. See Friedsen, *Profession of Medicine* (New York: Dodd, Mead, 1970); Illich, *Medical Nemesis: The Expropriation of Health* (London: Calder and Boyars, 1975); Foucault, *Discipline and Punish: The Birth of the Prison,* trans. Alan Sheridan (New York: Vintage, 1979). The subject of medicalization persists as a subfield of medical sociology. See, e.g., Peter Conrad and Joseph W. Schneider, *Deviance and Medicalization: From Badness to Sickness* (Philadelphia: Temple University Press, 1981); for the example of hypertension, see Ichiro Kawachi and Peter Conrad, "Medicalization and the Pharmaceutical Treatment of Blood Pressure," in *Contested Ground: Public Purpose and Private Interest in the Regulation of Prescription Drugs,* ed. Peter Davis (New York: Oxford University Press, 1996). Although other historians of medicine have attempted to salvage the term *medicalization* by expanding the field of historical actors— see, e.g., Maren Klawiter, "Risk, Prevention, and the Breast Cancer Continuum: The NCI, the FDA, Health Activism, and the Pharmaceutical Industry," *History of Technology* 18, no. 4 (2002): 309–53—I have intentionally avoided use of the term to avoid the simple power dynamics it suggests and to present an analysis of disease-naming as a more overdetermined and interactive phenomenon.

11. Max Weber, *The Protestant Ethic and the Spirit of Capitalism,* trans. Talcott Parsons (1903; New York: Scribner, 1958), 181; for discussion of the "iron cage" in relation to history, science, and technology, see Terry Maley, "Max Weber and the Iron Cage of Technology," *Bulletin of Science and Technology* 24, no. 1 (2004): 69–86; John Patrick Diggens, *Max Weber: Politics and the Spirit of Tragedy* (New York: Basic Books, 1996). For a more general argument on modernity and risk reduction, see Ulrich J. Beck, *Risk Society: Towards a New Modernity,* trans. Mark Ritter (London: Sage, 1992).

12. Not all risk factors have become treatable conditions or proper diagnoses in the same sense as hypertension, early diabetes, and high cholesterol; many others merely remain risks. For recent scholarship on the history of the risk factor, see Aronowitz, *Making Sense of Illness;* and William Rothstein, *Public Health and the Risk Factor: A History of an Uneven Medical Revolution* (Rochester, NY: University of Rochester Press, 2003).

13. H. Franz Messrli, "This Day 50 Years Ago," *New England Journal of Medicine* 332 (1995): 1038–39; H. G. Bruenn, "Clinical Notes on the Illness and Death of President Franklin D. Roosevelt," *Annals of Internal Medicine* 72 (1970): 579–91.

14. Although the insertion of numbers into medical diagnosis is often credited to the mid-nineteenth-century Parisian "numerical school" of Pierre Louis, the origins of the practice can be traced back at least to medieval uroscopy. See J. Rosser Matthews, *Quantification and the Quest for Medical Certainty* (Princeton, NJ: Princeton University Press, 1995); Stanley J. Reiser, *Medicine and the Reign of Technology* (Cambridge: Cambridge University Press, 1978).

15. Richard C. Cabot: "The Historical Development and Relative Value of Laboratory and Clinical Methods of Diagnosis," *Boston Medical and Surgical Journal* 157 (August 1907): 150–53, quotation on 151. On the historiography of technology as it affects the doctor-patient relationship, see Reiser, *Medicine and the Reign of Technology;* Hughes Evans, "Losing Touch: The Controversy over the Introduction of Blood Pressure Instruments into Medicine," *Technology and Culture* 34 (1993): 784–808; Keith

Wailoo, *Drawing Blood: Technology and Disease Identity in Twentieth-Century America* (Baltimore: Johns Hopkins University Press, 1997).

16. Beginning in the 1950s, a set of studies documented the effect of laboratory and radiological diagnostics in precipitating symptoms in previously asymptomatic patients; see, e.g., E. O. Wheeler and C. R. Williamson, "Heart Scare, Heart Surveys, and Iatrogenic Heart Disease," *Journal of the American Medical Association* 167 (1958): 1096–1102. More recent studies include Arthur Barsky, *Worried Sick: Our Troubled Quest for Wellness* (Boston: Little, Brown, 1988); Dorothy Nelkin and Laurence Tancredi, *Dangerous Diagnostics: The Social Power of Biological Information* (New York: Basic Books, 1989); Robert T. Croyle, ed., *Psychosocial Effects of Screening for Disease Prevention and Detection* (New York: Oxford University Press, 1995).

17. For recent sociological work on the role of industry funding in publication bias, see Sheldon Krimsky, *Science in the Private Interest* (New York: Rowan and Littlefield, 2003), 141–76.

18. Barbara Alving, "Cholesterol Guidelines: The Strength of the Science Base and the Integrity of the Development Process, press release, September 24, 2004, National Heart, Lung, and Blood Institute, Bethesda, MD.

19. The thesis that the medical profession was able to isolate itself from the influence of the marketplace for most of the twentieth century is most forcefully stated by Paul Starr in *The Social History of American Medicine* (New York: Basic Books, 1982). The model of medicine in relation to the marketplace that I present challenges Starr's guiding narrative of insulation; nonetheless, as a rough heuristic Starr's thesis correctly points out a that, by 1982, the relative autonomy of the medical profession had already declined markedly from its midcentury peak.

20. Susser "Epidemiology after World War II."

21. George C. Shattuck, *Principles of Medical Treatment*, 6th ed. (Cambridge: Harvard University Press, 1926), 3–4.

22. Ibid., 7. On the long history of digitalis as a drug that persisted through several different therapeutic identities, see J. K. Aronson, *An Account of the Foxglove and Its Medical Uses, 1775–1985* (New York: Oxford Medical Publications, 1985).

23. On the role of rheumatic heart disease in the early public health activity of American cardiologists, see W. Bruce Fye, *American Cardiology: History of a Specialty and Its College* (Baltimore: Johns Hopkins University Press, 1996). On pellagra and silicosis, respectively, see Joseph Goldberger, G. A. Wheeler, Wilford I. King, William S. Bean Jr., R. E. Dyer, J. D. Reichard, P. M. Stewart, M. C. Edmunds, R. E. Tarbett, Dorothy Wiehl, and Jennie C. Goddard, *A Study of Endemic Pellagra in Some Cotton-Mill Villages of South Carolina*, Hygienic Laboratory Bulletin no. 153 (Washington, DC: Government Printing Office, 1929); David Rosner and Gerald Markowitz, *Deadly Dust* (Princeton, NJ: Princeton University Press, 1991).

24. In the mid-1920s, the New Jersey Department of Institutions and Agencies began its first statewide survey of the chronically ill with an eye toward policy recommendations. New York and Massachusetts followed suit; and nationwide surveys began in the early 1930s. Mary C. Jarret, *Chronic Illness in New York City* (New York: Columbia University Press, 1933); George H. Bigelow and Herbert L. Lombard, *Cancer and Other Chronic Diseases in Massachusetts* (New York: Houghton-Mifflin, 1933); Commission on Chronic Illness, *Chronic Illness in the United States*, vol. 1, *Prevention of Chronic Illness* (Cambridge, MA: Commonwealth Fund, 1957); National Health Survey,

1935–36, *The Magnitude of the Chronic Disease Problem in the United States,* Preliminary Reports, Sickness and Medical Care Series, Bulletin no. 6 (Washington, DC: Public Health Service, 1938). It is worth noting that the idea of the asymptomatic precursor state is not unique to twentieth-century preventive medicine. It has ties to the nineteenth-century concepts of constitutions or proclivities that predisposed certain sorts of people to develop chronic illness. Nonetheless, the ontological status of the risk factor in late-twentieth-century America represents a distinct rupture with previous, more bodily rooted notions of predilection.

25. Although the asymptomatic diagnosis of hypertension became routine practice for the insurance physical in the first half of the twentieth century, the diagnosis remained an actuarial category rather than a clinical entity, with risk discussed in financial terms (i.e., the solvency of the insurance company) rather than in terms of individual health and prevention; see Audrey Davis, "Life Insurance and the Physical Examination," *Bulletin of the History of Medicine* 55 (1981): 392–406.

26. Asymptomatic diagnosis through periodic physical examination was suggested as a public health measure in early-twentieth-century medical literature, but the practice did not spread widely until it was explicitly tied in to insurance reimbursement practices. See George Rosen, *Preventive Medicine in the United States* (New York: Science History Publications, 1975), 59–60; for early publications calling for the identification of asymptomatic disease states, see Alexander M. Campbell, "The Necessity for a Periodical Examination of the Apparently Healthy," *Detroit Medical Journal* 4 (1904): 193–95; Haven Emerson, "The Protection of Health by Periodic Medical Examinations," *Journal of the Michigan State Medical Society* 21 (1922): 158–71.

27. Senate Subcommittee of the Committee on Labor and Public Welfare, *Hearings of the Subcommittee of the Committee on Labor and Public Welfare,* 20th Cong., 2nd sess., May 5–6, 1948, 29–32.

28. The periodical literature of the 1950s and 1960s was saturated with the subject of heart disease and attempts to control it: e.g., "Last Ten Years: Giant Steps against Heart Disease," *Today's Health,* July 1959, 30; "Can Heart Attacks Be Predicted?" *Today's Health,* December 1960, 36; J. Stuart, "Any Man Can Have a Heart Attack," *Today's Health,* November 1961, 12–13; L. Kavaler, "Will There Soon Be a Drug That Might Ultimately Prolong Your Husband's Life?" *Good Housekeeping,* October 1969, 112; "Drug to Prevent Heart Attacks?" *U.S. News & World Report,* July 8, 1968, 13.

29. Clarence G. Lasby, *Eisenhower's Heart Attack: How Ike Beat Heart Disease and Held on to the Presidency* (Lawrence: University Press of Kansas, 1997).

30. David Seegel, "Proceedings of the Conference on Preventive Aspects of Chronic Disease," March 12–14, 1951, as cited in *Chronic Illness in the United States,* by the Commission on Chronic Illness, vol. 1.

31. For an example of the "discovery" genre of narrative focused on a particular agent, see Albert N. Brest, "Milestones in Clinical Pharmacology: Antihypertensive Drug Therapy: A 30-Year Retrospective," *Clinical Therapeutics* 14, no. 1 (1992): 78; Daniel P. Steinberg and Antonio M. Gotto, "Preventing Coronary Artery Disease by Lowering Cholesterol Levels: Fifty Years from Bench to Bedside," *Journal of the American Medical Association* 282 (1999): 2043–50. Relevant corporate histories include Tom Mahoney, *The Merchants of Life: An Account of the American Pharmaceutical Industry* (New York: Harper, 1959); Leonard Engel, *Medicine Makers of Kalamazoo* (New York: McGraw-Hill, 1961); Robert D. B. Carlisle, *A Century of Caring: The Upjohn Story* (Elmsford, NY: Ben-

jamin, 1987); Louis Galambos and Jeffrey Sturchio, *Values and Visions: A Merck Century* (Rahway, NJ: Merck, 1991).

32. E.g., Samuel Hopkins Adams, "The Great American Fraud," *Collier's Weekly*, February 17, 1906, 22; Morton Mintz, *The Therapeutic Nightmare: A Report on the Roles of the United States Food and Drug Administration, the American Medical Association, Pharmaceutical Manufacturers, and Others in Connection with the Irrational and Massive Use of Prescription Drugs That May Be Worthless, Injurious, or Even Lethal* (Boston: Houghton Mifflin, 1965). Recently a wave of well-researched critiques of the pharmaceutical industry have emerged from clinician-authors associated with Harvard Medical School, most notably Marcia Angell, *The Truth about the Drug Companies: How They Deceive Us and What to Do about It* (New York: Random House, 2004); Jerome Avorn, *Powerful Medicines: The Benefits, Risks, and Costs of Prescription Drugs* (New York: Knopf, 2004); and John Abramson, *Overdo$ed America: The Broken Promise of American Medicine* (New York: HarperCollins, 2004). Of the three, Angell's critique is the most strident; Avorn takes care to describe the relationship of dependency between medical practice and the pharmaceutical industry, while Abramson provides a narrative of disenchantment that laments a loss of trust in the doctor-patient relationship and structures of medical knowledge. See also Ray Moynihan and Alan Cassels, *Selling Sickness: How the World's Biggest Pharmaceutical Companies Are Turning Us All into Patients* (New York: Nation Books, 2005).

33. E.g., Thomas Szasz, *Pharmacracy: Medicine and Politics in America* (London: Praeger, 2001); Linda Marsa, *Prescription for Profits: How the Pharmaceutical Industry Bankrolled the Unholy Marriage between Science and Business* (New York: Scribner's, 1999); and Tonda R. Bian, *The Drug Lords: America's Pharmaceutical Cartel* (Kalamazoo, MI: No Barriers, 1997).

34. Perhaps the most successful overview of the social roles of pharmaceuticals remains Harry F. Dowling, *Medicines for Man: The Development, Regulation, and Use of Prescription Drugs* (New York: Knopf, 1970); on the regulatory and investigative networks around pharmaceuticals, see Harry M. Marks, "Cortisone, 1949: A Year in the Political Life of a Drug," *Bulletin of the History of Medicine* 66 (1992): 419–39; Marks, *The Progress of Experiment: Science and Therapeutic Reform in the United States, 1900–1990* (Cambridge: Cambridge University Press, 1997); John P. Swann, *Academic Scientists and the Pharmaceutical Industry: Cooperative Research in Twentieth-Century America* (Baltimore: Johns Hopkins University Press, 1988); Nicholas Rasmussen, "The Moral Economy of the Drug Company–Medical Scientist Collaboration in Interwar America," *Social Studies of Science* 34, no. 2 (2004): 161–86; on the social lives of pharmaceuticals in popular culture, see Elizabeth Siegel Watkins, *On the Pill: A Social History of Oral Contraceptives, 1950–1970* (Baltimore: Johns Hopkins University Press, 1998); Susan L. Speaker. "From Happiness Pills to National Nightmare: Changing Cultural Assessment of Minor Tranquilizers in America, 1955–1980," *Journal of the History of Medicine* 52 (1997): 338–77; Andrea Tone, *Devices and Desires: A History of Contraceptives in America* (New York: Hill and Wang, 2001); Jonathan M. Metzl, *Prozac on the Couch: Prescribing Gender in the Era of Wonder Drugs* (Durham, NC: Duke University Press, 2003).

35. Although a substantial literature in recent American history has focused on corporations, advertising, marketing, and the consumer, few accounts have connected this scholarship to the history of medicine. An exception is Nancy Tomes, "Merchants of

Health: Medicine and Consumer Culture in the United States: 1900–1940," *Journal of American History* (2001): 519–45; key texts from the broader field include Lizbeth Cohen, *A Consumer's Republic: The Politics of Mass Consumption in Postwar America* (New York: Knopf, 2003); Alfred Chandler, *The Visible Hand* (Cambridge, MA: Belknap Press, 1977); Olivier Zunz, *Making America Corporate: 1870–1920* (Chicago: University of Chicago Press, 1990); Susan Strasser, *Satisfaction Guaranteed: The Making of the American Mass Market* (New York: Pantheon, 1989); and Roland Marchand, *Advertising the American Dream: Making Way for Modernity, 1920–1940* (Berkeley: University of California Press, 1985).

36. Early texts in pharmaceutical anthropology include Sjaak van der Geest, Susan Reynolds-Whyte, Anita Hardon, "The Anthropology of Pharmaceuticals: A Biographical Approach," *Annual Review of Anthropology* 25 (1996): 153–78; Adriana Petryna, Andrew Lakoff, and Arthur Kleinman, eds., *Global Pharmaceuticals: Ethics, Markets, Practices* (Durham, NC: Duke University Press, 2006); Joseph Dumit, *Drugs for Life: Managing Illness through Facts and Pharmaceuticals* (forthcoming, 2006); Jennifer Fishman, "Manufacturing Desire: The Commodification of Female Sexual Dysfunction," *Social Studies of Science* 34 (2004): 187–218; Nathan Greenslit, "Pharmaceutical Branding: Identity, Individuality, and Illness," *Molecular Interventions* 2 (2002): 342–45. For relevant work in economic anthropology, see Arjun Appadurai and Igor Kopytoff, *The Social Life of Things: Commodities in Cultural Perspective* (New York: Cambridge University Press, 1986); Marcel Mauss, *The Gift: The Form and Reason for Exchange in Archaic Societies* (1925; New York: Norton, 1990).

37. E.g., Peter Kramer, "The Message in the Capsule," in *Listening to Prozac*, by Kramer (New York: Penguin, 1997); David Healy, "The Luke Effect," in *The Antidepressant Era*, by Healy (Cambridge: Harvard University Press, 1995). A more interactive approach to the relations between somatic and mental illness categories and pharmaceutical treatments is offered by Joseph Dumit, "Drugs for Life," *Molecular Interventions* 2, no. 3 (2002): 124–27.

38. Existing historical works examining the role of pharmaceuticals and the negotiation of somatic disease include Wailoo, *Drawing Blood;* Chris Feudtner, *Bittersweet: Diabetes, Insulin, and the Transformation of Illness* (Chapel Hill: University of North Carolina Press, 2003); Robert Kaiser, "The Introduction of the Thiazides: A Case-Study in 20th Century Therapeutics," in *The Inside Story of Medicines,* ed. Greg Higby and Elaine Stroud (Madison, WI: American Institute of the History of Pharmacy, 1997), 121–37; Rein Vos, *Drugs Looking for Diseases: Innovative Drug Research and the Development of the Beta Blockers and the Calcium Antagonists* (Dordrecht: Kluwer Academic Publishers, 1991).

39. Theophrastus Von Hohenheim (Paracelsus), "On the Miners' Sickness and Other Miners' Diseases," in *Paracelsus: Four Treatises,* trans. George Rosen (Baltimore: Johns Hopkins University Press, 1941), book 3, tractate 4, chap. 4, 117–18. Although the *Antidotarium Nicolai* was more widely used, a more visually striking *antidotarium* is Armengaud's *Tabula Antidotarii,* which presented relations of drugs and associated conditions in tabular form. See Michael McVaugh, "The *Tabula Antidotarii* of Armengaud Blaise and Its Hebrew Translation," *Transactions of the American Philosophical Society* 90, no. 6 (2000): 1–10; also see Tony Hunt, *Popular Medicine in Thirteenth-Century England: Introduction and Texts,* (London: D. S. Brewer, 1990), 13–16.

40. In framing questions in terms of normal and pathological, my own work has been greatly influenced by Georges Canguilhem, *The Normal and the Pathological*, trans. Carolyn R. Fawcett (1943; New York: Zone Books, 1991).

41. Feudtner, *Bittersweet*.

CHAPTER 1: Releasing the Flood Waters

Epigraph: Diuril News Report 2, April 25, 1958, folder 7, Diuril Papers, Merck Archives, Whitehouse Station, NJ.

1. On antibiotics and psychotropic development, see Harry F. Dowling, *Fighting Infection: Conquests of the 20th Century* (Cambridge: Harvard University Press, 1977); Gladys Hobby, *Penicillin: Meeting the Challenge* (New Haven, CT: Yale University Press, 1985); Judith Swazey, *Chlorpromazine in Psychiatry: A Study of Therapeutic Innovation* (Cambridge: MIT Press, 1974); Mickey Smith, *Small Comfort: A History of the Minor Tranquilizers* (New York: Praeger Scientific, 1985); Susan L. Speaker, "From Happiness Pills to National Nightmare: Changing Cultural Assessment of Minor Tranquilizers in America, 1955–1980," *Journal of the History of Medicine* 52 (1997): 338–77.

2. On the postwar surge in the productivity and profitability of the pharmaceutical industry, see Harry F. Dowling, *Medicines for Man: The Development, Regulation, and Use of Prescription Drugs* (New York: Knopf, 1970); Milton Silverman and Philip R. Lee, *Pills, Profits, and Politics* (Los Angeles: University of California Press, 1974).

3. W. D. McAdams, "Three Major Marketing Problems on the Desks of Pharmaceutical Management Today," *Proceedings of the American Pharmaceutical Manufacturer's Association Midyear Meeting,* December 17, 1947, 272–80. The first published volume of the *Physicians' Desk Reference (PDR)* in 1947 was a slim thing of some three hundred pages. By 1960 the *PDR* had swelled to nearly one thousand pages, and it was dominated by several classes of pharmacological agents unheard of before the war. *Physicians' Desk Reference to Pharmaceutical Specialties and Biologicals* (Oradell, NJ: Medical Economics, 1946); cf. *Physicians' Desk Reference to Pharmaceutical Specialties and Biologicals,* 14th ed. (Oradell, NJ: Medical Economics, 1960).

4. *Facts about Pharmacy and Pharmaceuticals* (New York: Health News Institute, 1958), 37. Sales figures represent wholesale sale values; retail figures can be expected to have been significantly larger.

5. These therapeutic agents are described in more detail in chapter 2; also see Milton J. Chatton, Sheldon Margen, and Henry Brainerd, *Handbook of Medical Treatment,* 5th ed. (Los Altos, CA: Lange Medical Publications, 1956), 157–60; also see J. Earle Estes, "Hypertension in 1958: A Tale of Pills, Philosophy, and Perplexity," *Medical Clinics of North America* 42 (1958): 899–915; *Physicians Desk Reference to Pharmaceutical Specialties and Biologicals,* 12th ed.(Oradell, NJ: Medical Economics, 1958).

6. Irvine H. Page, *Hypertension Research: A Memoir* (New York: Pergamon, 1988), 129–40; "Treating Hypertension: The Four Periods of Antihypertensive Treatment," in *A Century of Arterial Hypertension, 1896–1996,* ed. Nicolas Postel-Vinay (New York: Wiley, 1996), 123–26.

7. Estes, "Hypertension in 1958," 906.

8. Richard G. Kedersha, ed., *Pharmaceutical Marketing Orientation Seminar* (New Brunswick, NJ: Rutgers, the State University College of Pharmacy, 1959). In the decade of the 1950s alone, the total number of prescriptions filled in American pharmacies rose

from less than 350 million to 550 million, while the number of new prescriptions per doctor rose from 1,400 in 1950 to nearly 2,000 by 1958; see "Prescription Trends," *New Material Medica,* April 1963, 31.

9. As Hugh Hussey, president of the AMA, testified before the Kefauver Committee in 1961: "By 1950 it was apparent that the acceptance program, which fostered the consideration of multiple brands of older drugs, left the council little time for the consideration and early publication of information on new drugs which was desired of physicians." Although Hussey would officially deny such claims under interrogation, it appears that financial reasons were also a significant driver of the demise of the AMA Seal of Acceptance program. Quotation from the statement of Hugh H. Hussey, Senate Subcommittee on Antitrust and Monopoly of the Committee of the Judiciary, *Hearings before the Subcommittee on Antitrust and Monopoly of the Committee of the Judiciary, Pursuant to S. 1552,* 87th Cong., 1st sess., 1961, 104; also see Harry F. Dowling, "The American Medical Association's Policy on Drugs in Recent Decades," in *Safeguarding the Public: Historical Aspects of Medicinal Drug Control,* ed. John B. Blake (Baltimore: Johns Hopkins Press, 1968).

10. Richard J. Hull, director of marketing research at Smith, Kline & French, noted in 1959, "We are operating in an increasingly fragmented market . . . we offer more and more specific medications and each product competes, or will compete, with a limited number of other products in its own area." Hull, "Marketing Concepts," in *Workings and Philosophies of the Pharmaceutical Industry,* ed. Karl Reiser, (New York: National Pharmaceutical Council, 1959), 53–63, quotation on 63.

11. Pharmaceutical branding and marketing had necessarily existed in some form since the inception of the American pharmaceutical industry in the mid-nineteenth century. *Marketing* as an explicit subject in the business literature, however, did not emerge in the American business landscape until the early twentieth century. Marketing was well established as a discipline in most business schools by the 1920s, and many pharmaceutical firms had explicitly developed marketing divisions by the mid-1930s. But the 1950s saw the advent of the "marketing concept" in the pharmaceutical industry, that is, the integration of marketing at earlier stages of product development and at higher levels of long-term corporate strategy. See Hull, "Marketing Concepts"; also see Robert Bartels, "Influences on the Development of Marketing Thought, 1900–1923," *Journal of Marketing* 16, no. 1 (1951): 1–17. For general works on the history of marketing in America, see Alfred Chandler, *The Visible Hand* (Cambridge, MA: Belknap Press, 1977); Susan Strasser, *Satisfaction Guaranteed: The Making of the American Mass Market* (New York: Pantheon, 1989).

12. By 1957 Merck could boast six Nobel Prize winners among its network of in-house and externally supported researchers; see Max Tishler, "The Search for Better Drugs: Diuril Symposium," November 11, 1957, folder 2, Diuril Papers, R3-2.74, Public Affairs Subject Files, Merck Archives (hereafter cited as Diuril Papers). For a brief history of the research tradition at Merck, see Louis Galambos and Jeffrey Sturchio, *Values and Visions: A Merck Century* (Rahway, NJ: Merck, 1991), 27–28.

13. In the 1930s, 75% of all prescriptions were compounded by pharmacists; by 1950 that figure had dropped to 25%, by 1960 only one in twenty-five prescriptions required compounding skill, and by 1970 only one out of every hundred prescriptions were compounded. Gregory J. Higby, "Evolution of Pharmacy," in *Remington: The Science and Practice of Pharmacy,* 20th ed., ed. Alfonso R. Gennaro (Philadelphia: Lippincott Wil-

liams and Wilkins, 2000), 14. Also see Raymond Gosselin, "Massachusetts Prescription Survey" (master's thesis, Massachusetts College of Pharmacy, 1950).

14. By 1959 adrenocortical steroids were indicated as specific therapy for nearly eighty specific diseases and saw much broader empirical use in rheumatologic, allergic, metabolic, dermatologic, dental, gastrointestinal, hematopoietic, pulmonary, renal, ocular, neurological, neoplastic, and infectious conditions. See Harry M. Marks, "Cortisone, 1949: A Year in the Political Life of a Drug," *Bulletin of the History of Medicine* 66 (1992): 419–39; Nicolas Rasmussen, "Steroids in Arms: Science, Government, Industry, and the Hormones of the Adrenal Cortex in the United States, 1930–1950," *Medical History* 46 (2002): 299–324.

15. In 1949 Merck held a monopoly over the steroids market; by 1958 competitors had pushed Merck to a mere 18% market share. "Steroids: Confidential Summary," November 1959, 42–44, Merck Archives.

16. In the 1951 *Physicians' Desk Reference* (Oradell, NJ: Medical Economics, 1950), Sharp & Dohme's forty-five listed products included various formulations of amphetamines, analgesics, vitamins, sulfas, antispasmodics, sedatives, and a few biologicals and vaccines. Of these, only one—hexylresorcinol—was a compound unique to Sharp & Dohme, and all except sulfamerizine were known by brand, not chemical, names. In contrast, the 1951 *PDR* lists a total of 227 chemical agents produced by Merck, of which only *Cortone* and a handful of others were listed by brand name.

17. As the *New York Times* reported, the merger "required a new word in the financial vocabulary," properly reflecting these broad changes in the pharmaceutical industry; see Robert E. Bedingfield, "Proposed Sharp & Dohme, Merck Merger Regarded as Deal of New, 'Encircling' Type," *New York Times,* April 14, 1953, 37, 41. As Merck Sharp & Dohme's president, John T. Connor, later recalled, "In 1953—just three years after we achieved the first mass production of cortisone—our profits were sharply reduced . . . It became quite obvious to us that we, as a company, had to compete directly with the leading pharmaceutical companies, such as Lilly, Parke-Davis, and Upjohn. We had to bring new products from our research directly to the attention of the 200,000 physicians in the country . . . Sharp & Dohme had the pharmaceutical marketing staff and know-how that Merck did not possess, as well as an equally fine reputation with the profession. Today we are beginning to capitalize on this change of the direction of our business. DIURIL is an outstanding example." John T. Connor, "Merck's Changing Times," *Merck Review* 19 (1959): 20–21. Ironically, the research leading to Diuril's development originated in Sharp & Dohme's laboratories—not Merck's.

18. Although Karl Beyer subsequently claimed in his memoirs that he had been guided all along by a preformed concept of a saluretic agent that would work in hypertension, there is no other evidence in Diuril's preclinical or early clinical trials suggesting that Beyer or the company encouraged the usage of Diuril in hypertension at all before the independent reports of external researchers such as Robert Wilkins and Edward Freis.

19. Paraminohippuric acid (PAH), the first compound tested, required a high dose of 200 g/day to maintain penicillin levels; a subsequent compound, carinamide, required a lower dose of 20 g/day but was still deemed inefficient for practical purposes. Probenecid, the final compound, effectively maintained penicillin levels at a dosage of only 2 g/day. See Karl Beyer, "New Concept of Competitive Inhibition of Renal Tubular Excretion of Penicillin," *Science,* January 24, 1947.

20. As a subsequent internal document noted, by the time probenecid had been demonstrated to be safe and efficacious as a potential consumer product, "penicillin was so cheap and plentiful that retention was no longer a problem." Probenecid was never launched as a penicillin adjunct. Rather, it languished in the Sharp & Dohme Research Laboratories until 1952, when it found use as an antigout agent and was launched with the product name Benemid. See "'Diuril': Confidential Product Summary," November 1959, 3, 10, Merck Archives.

21. Morris Fishbein, "The Story of a New Drug," *Postgraduate Medicine*, April 1960; also see W. B. Schwartz, "The Effect of Sulfanilamide on Salt and Water Excretion in Congestive Heart Failure," *New England Journal of Medicine* 240 (1949): 173–79.

22. Milton Moskowitz, "Diuril Creates a New Market," *Drug and Cosmetic Industry* 87 (1960): 841–42.

23. E.g., Carroll Handley and John Moyer, *The Pharmacology and Clinical Use of Diuretics* (Springfield, IL: Charles C. Thomas, 1959).

24. The diuretic project, or Project no. 1208 of the Sharp & Dohme Research Laboratories, was authorized on May 23, 1949, and involved the screening of a number of sulfa compounds to develop bases for synthetic development. The first promising compound, Dirnate, was found to increase secretion of sodium and chloride during clinical trials in 1950 but suffered from poor oral absorption and low potency. Daranide, developed in March 1955, showed promise of higher potency with heterocyclic sulfonamides, and in May of 1955 chlorothiazide was synthesized and immediately named Diuril. Preclinical evaluations were conducted in dogs in late 1955 and 1956 and showed tolerance at high dosages. "'Diuril': Confidential Product Summary," 14–17, 18–19.

25. John H. Moyer, "Historical Aspects of the Development of Chlorothiazide," in *A Decade of Diuril*, ed. Charles Lyght (Rahway, NJ: Merck Sharp & Dohme, 1968), 8–9.

26. "Hypertension: Dual Role Found for Chemical in High Blood Pressure," *Boston Globe*, September 15, 1957; "New Drug Tested for Hypertension," *New York Times*, September 15, 1957; "Hypertension, New Hope," *Newsweek*, November 25, 1957, 82.

27. R. G. Denkewalter to Marketing Area, "Resume of Meeting on Diuril," memorandum, May 14, 1957, folder 3, Diuril Papers.

28. George Schott to Beyer, Edmonston, Gibson, Heath, Horan, Jennings, Klodt, Krieger, Kuryloski, Novello, Sprague, Thomas, "Press Information on 'Diuril,'" memorandum, October 23, 1957, folder 3, Diuril Papers (emphasis added).

29. Leon Gortler and Jeffrey L. Sturchio, "Karl Henry Beyer, Jr.: An Interview for the Merck Archives," 1988, 33–34, oral history transcript, Merck Archives. The initial papers referred to are Edward D. Freis, Annmarie Wanko, and I. M. Wilson, "Potentiating Effect of Chlorothiazide (Diuril) in Combination with Other Antihypertensive Agents: Preliminary Report," *Medical Annals of the District of Columbia* 26 (1957): 468–516; W. Hollander and R. W. Wilkins, "Chlorothiazide: A New Type of Drug for Treatment of Arterial Hypertension," *Boston Medical Quarterly* 8 (1957): 69–75; R. W. Wilkins, "New Drugs for Hypertension with Special Reference to Chlorothiazide," *New England Journal of Medicine* 257 (1957): 1026–30; E. D. Freis, A. Wanko, I. M. Wilson, and A. E. Parrish, "Treatment of Essential Hypertension with Cholorothiazide (DIURIL)," *Journal of the American Medical Association* 166 (1958): 137–40.

30. *Statistics from the U.S. National Health Survey: Heart Conditions and High Blood Pressure Reported in Interviews, July 1957-June 1958*, series B, no. 13, February 1960

(Washington, DC: U.S. Department of Health, Education, and Welfare, Public Health Service, 1960), 3.

31. Jeffrey L. Sturchio and Louis Galambos, "Gordon R. Klodt: An Interview for the Merck Archives," 1992, oral history transcript, Merck Archives.

32. George Schott, memorandum, n.d., attached to "Press Information on 'Diuril,'" folder 3, Diuril Papers (emphasis added).

33. Before the thiazides were developed, diuretic therapy was used primarily in the treatment of congestive heart failure, premenstrual edema, edema of pregnancy, nephritic syndrome, hepatic disease, and iatrogenic steroid edema. Handley and Moyer, *Pharmacology and Clinical Use of Diuretics,* 150–62.

34. Instead, 1940s texts on drug marketing listed the functions of research and the functions of marketing as distinct entities within the pharmaceutical firm. See, e.g., Louis Bader and Sidney Picker, *Marketing Drugs and Cosmetics* (New York: Van Nostrand, 1947); also Paul Olsen, *Marketing Drug Products* (New Brunswick, NJ: Rutgers University Press, 1948).

35. Paul de Haen, *Development Schedule of New Drug Products* (New York: Romaine Pearson, 1949), 16.

36. This executive was careful to preserve some independence for basic science research, though: "*Total* acceptance of the Marketing Concept, or of marketing as the most dominant function in the pharmaceutical industry could have many implications, and I am not sure that they are all desirable ones. I would be most concerned with the possibility of restricting or circumscribing basic research in our laboratories." Hull, "Marketing Concepts," 1–2.

37. John Huck, director of marketing at Merck Sharp & Dohme soon after the merger, recalls, "It was very difficult to get the research people to listen to us. When Max Tishler [head of Merck Sharp & Dohme Research Laboratories] was here, they didn't really feel that the sales people had much to offer . . . Well, management prevailed, and they said, 'All right, he can come and sit in the meetings so long as he doesn't say anything.' [laughter] So we had a little bit of that when I came here: 'That's all right, you can sit in, but don't say anything.'" Jeffrey L. Sturchio and Louis Galambos, "John Lloyd Huck: An Interview for the Merck Archives," 1990, 23, oral history transcript, Merck Archives.

38. Economic Research Area, "Pre-Market Recognition and Usage of Chlorothiazide or 'Diuril,'" January 15, 1958, folder 16, Diuril Papers.

39. Abstract and applied study of the "influential" had become prominent in sociological literature and marketing research journals by the 1950s; see, e.g., Elihu Katz and George Menzel, "Social Relations and Innovation in the Medical Profession: The Epidemiology of a New Drug," *Public Opinion Quarterly* (Winter 1955): 337–72; James Coleman, Herbert Menzel, and Elihu Katz, "Social Processes in Physicians' Adoption of a New Drug," *Journal of Chronic Disease* 9, no. 1 (1959): 337–72. Merck marketers measured requests for Diuril in the months preceding release to compare the marketing influence of their researchers in various geographic regions. Economic Research Area, "Pre-Market Recognition."

40. F. K. Heath, "Final 1957 Report on Diuril Promotional Program," folder 3, Diuril Papers. Merck budgeted twenty-five thousand dollars for the NYAS meeting; although the NYAS represented itself as a neutral space for the advancement of scientific

progress, the institution and its journal, the *Annals of the New York Academy of Sciences,* were frequent sites for the promotion of new drugs in the 1950s.

41. F. K. Heath, *Diuril Meetings,* n.d., folder 3, Diuril Papers.

42. Denkewalter to Marketing Area, "Resume of Meeting on Diuril."

43. Medical Publications Department, Merck Sharp & Dohme, *Chlorothiazide and Hydrochlorothiazide: Annotated Bibliography* (Rahway, NJ: Merck Sharp & Dohme Research Laboratories, 1960).

44. Freis, Wanko, and Wilson, "Potentiating Effect of Chlorothiazide"; Hollander and Wilkins, "Chlorothiazide"; the quotation is from "'Diuril': Confidential Product Summary," 22.

45. C. Cray Williams, *The Pharmaceutical Manufacturers Association: The First 30 Years* (Washington, DC: Pharmaceutical Manufacturers Association, 1989), 6.

46. Ian Maclean Smith, "Recent Contributions to Antibacterial Therapy," *Archives of Internal Medicine* 106, no. 3 (1959): 459–60. Annotated reproduction, MS C 372, "Private Regulatory Agencies" folder, box 11, Harry Fillmore Dowling Papers, National Library of Medicine, Bethesda, MD (hereafter cited as Dowling Papers).

47. See, e.g., Eli Lilly's *Physician's Bulletin,* Parke Davis & Company's *Therapeutic Notes,* Lederle's *Bulletin,* Ciba's *Symposium,* Winthrop's *Clinical Excerpts,* McNeil's *McNeil-O-Gram,* Hoffman-LaRoche's *Roche Review,* Upjohn's *Scope,* and Squibb's *Memoranda.*

48. Kenneth Avis, "House Organs as Important Sources of Information," *American Journal of Pharmacy* 117, no. 11 (1945): 400–408.

49. John H. Moyer, "Diuretics," *Merck Sharp & Dohme Seminar-Report* (Spring 1958): 2–9; Joseph C. Edwards, "Management of Patients with Hypertension," *Merck Sharp & Dohme Seminar-Report* (Summer 1958): 2–8; Harriet P. Dustan, "Chlorothiazide in the Treatment of Hypertension," *Merck Sharp & Dohme Seminar-Report* (Winter 1958): 2–9.

50. Bader and Picker, *Marketing Drugs and Cosmetics,* 194, estimated the cost of maintaining a good house organ at forty thousand dollars in 1947 dollars.

51. Economic Research Area, "Pre-Market Recognition."

52. As late as 1947, market research was still a novel concept to many drug firms. Bader and Picker noted in the first (1947) edition of *Marketing Drugs and Cosmetics,* 295, "In the drug field most of the principal producers have made extensive use of research in the physical sciences and more particularly in the field of the relations of their products to the needs of consumers, but of marketing research there has not been much."

53. The methodology section of the Economic Research Area report, "Pre-Market Recognition and Usage of Chlorothiazide or 'Diuril,'" was explicit: "The sample was drawn in a manner that allows it to be duplicated and repeated. It is intended to make subsequent studies with parallel samples."

54. De Haen, *Development Schedule of New Drug Products,* 28.

55. George Morris Piersol, medical director of the National Disease and Therapeutic Index, to unnamed recipient, n.d., box 11, Dowling Papers. Also see *Lea Associates, Inc.,* undated pamphlet, box 11, Dowling Papers. Within a few years, Lea Associates would be producing regular reports as well as drug- and disease-specific analyses of its data for thirty-seven major pharmaceutical firms, including Merck Sharp & Dohme.

56. Economic Research Area, "Pre-Market Recognition." Although references to other prelaunch market research studies on Diuril exist, "Pre-Market Recognition and Usage of Chlorothiazide or 'Diuril'" is the only market-research document currently available in the Merck Archives. It is worth noting that 25% of physicians who claimed to recognize the drug at the time of the study already had had experience using it as an experimental agent before its launch. It was also estimated that thirty-four hundred physicians had already written prescriptions for the drug (which had obviously gone unfilled) months before it was available in pharmacies.

57. "Press Releases," January 30 1958, folders 9 and 10, Diuril Papers.

58. "There are now quite a number of highly respected science writers who can be trusted with significant news stories on a confidential basis. By briefing responsible members of the press in advance of the release date, the quality of the stories that appear later can be greatly improved." *A Primer of Public Relations for the Pharmaceutical Industry* (Washington, DC: Pharmaceutical Manufacturers of America, 1953), 25.

59. "Press Information on 'Diuril,'" 2, 3.

60. G. R. Klodt to J. J. Horan, E. L. Kuryloski, and J. A. Wells, n.d., folder 13, Diuril Papers.

61. Paul de Kruif, "New Hope for Overloaded Hearts," *Reader's Digest,* April 1959, 44–46.

62. Ben Pearse, "The Pill with the Built-In Surprise," *Saturday Evening Post,* October 4, 1958, 26, quotation on 57.

63. Daniel Starch and staff to John T. Connors, "Consumer Readership: The Pill with the Built-In Surprise," November 26, 1958, folder 13, Diuril Papers.

64. *Reader's Digest* to John T. Connors, "1,500,000 Patients a Month Now Use Merck-Developed Diuril," 1959, folder 13, Diuril Papers.

65. "Press Information on 'Diuril,'" 3, Diuril Papers. The delicacy of managing lay publicity around a new drug and complying with the industry-wide restriction of marketing to physicians was discussed by William D. Jenkins of American Cyanamid in a pharmaceutical marketing seminar published in 1959: "Lay publicity has seemed—I want to emphasize 'seemed'—inconsistent with ethical advertising and promotion. This line of thought would hold that if you are going to advertise and promote only to the medical profession, the less you tell the general public, the better . . . [Nonetheless] the doctor's attitude toward publicity has changed considerably . . . the doctor no longer regards publicity as the work of the devil." In Kedersha, *Pharmaceutical Marketing Orientation Seminar,* 146.

66. *Primer of Public Relations,* 26.

67. *Diuril News Report* 11 (June 27, 1958): 1, in folder 7, Diuril Papers. The article referred to was "A Triple-Threat Drug That Fights High Blood Pressure," *Coronet,* July 1958, 45–48.

68. The pharmacist was also an important target for pharmaceutical marketing, but with the passage of the Durham-Humphrey Act in 1951 requiring prescriptions for the sale of ethical drugs, and the shift in pharmaceutical production from standardized chemicals toward "specialty" products, pharmaceutical marketing came to focus more centrally on the physician.

69. Jonathan Liebenau, *Medical Science and Medical Industry* (Baltimore: Johns Hopkins University Press, 1987), 4.

70. Bader and Picker, *Marketing Drugs and Cosmetics*, 234.

71. On the history of the detail man, see Arthur F. Peterson, "The Professional Care Pharmacist," *Journal of the American Pharmaceutical Association* 12, no. 4 (1951): 212–13, 251; Jeremy A. Greene, "Attention to Details: Etiquette and the Pharmaceutical Salesman in Postwar America," *Social Studies of Science* 34, no. 2 (2004): 271–92.

72. Hull, "Marketing Concepts," 58–59. Hull continues: "As more and more physicians become specialists it is increasingly important that we pinpoint our promotion through special mailing lists, by advertising in specialty journals, and by having our professional service representatives use discrimination in the selection of products to be detailed."

73. Economic Research Area, "Pre-Market Recognition"; Medical Publications Department, *Information for the Physician on Diuril (Chlorothiazide), a New Diuretic and Antihypertensive Agent* (Philadelphia: Merck Sharp & Dohme, 1958).

74. Economic Research Area, "Pre-Market Recognition"; also see Mark Kenyon Dresden Jr., "Are Physicians Receptive to Pharmaceutical Promotion by Direct Mail?" pamphlet, c. 1958, box 11, Dowling Papers.

75. Theodore Caplow, "Market Attitudes: A Research Report from the Medical Field," *Harvard Business Review*, November 1952; Robert Ferber and Hugh G. Wales, "The Effectiveness of Pharmaceutical Marketing: A Case Study," *Journal of Marketing* 22, no. 4 (1956): 398–407; Robert Ferber and Hugh G. Wales, *The Effectiveness of Pharmaceutical Promotion* (Urbana: University of Illinois Press, 1958); Coleman, Menzel, and Katz, "Physicians' Adoption of a New Drug."

76. Walter O. Wegner, "Trends in Pharmaceutical Advertising," *Journal of Marketing*, January 1960, 65–67.

77. Moskowitz, "Diuril Creates a New Market."

78. "Telling the DIURIL Story," *Merck Review* 19 (1958): 7, Merck Archives.

79. "A Story with Heart," *MSD Sales Dispatch*, April 1960, 10, Merck Archives.

80. Marketers themselves were very aware of the value of the gift in building relationships; see, e.g., "Does It Pay to Present Gifts through Detail Men?" *Medical Marketing*, October 1941, 4–7, 12. Some midcentury marketers had extensive social science training, and it is tempting speculate whether a few were familiar with the anthropological literature on the social value of the gift, especially the popular work of Marcel Mauss, *The Gift: The Form and Reason for Exchange in Archaic Societies*, trans. W. D. Hall (1925; New York: Norton, 1990).

81. Moskowitz, "Diuril Creates a New Market."

82. Ibid., quotation in Moskowitz, "The Ethical Drug Hit Parade," *Drug and Cosmetic Industry* 89, no. 5 (1961): 575.

83. See Carsten Timmermann, "A Matter of Degree: The Normalisation of Hypertension, circa 1940–2000," in *The Normal and the Abnormal*, ed. Waltraud Ernst (London: Routledge, forthcoming). Also see "The Question of Norms: The Fallacy of the Dividing Line between the Normal and the Pathological," in *A Century of Arterial Hypertension, 1896–1996*, ed. Nicolas Postel-Vinay (New York: Wiley, 1996), 133–42.

84. For an overview of pharmaceutical response to the 1960s revival of "medical muckraking," see Mickey C. Smith, "The Medical Muckrakers," *Pharmaceutical Marketing and Media* 1 (1966): 19.

85. Barbara Mintzes and Catherine Hodgkin, "The Consumer Movement: From Single-Issue Campaigns to Long-Term Reform," in *Contested Ground: Public Purpose and Private Interest in the Regulation of Prescription Drugs*, ed. Peter Davis (New York: Oxford University Press, 1996), 76–91.

86. Senate Subcommittee on Antitrust and Monopoly of the Committee of the Judiciary, *Hearings before the Subcommittee on Antitrust and Monopoly of the Committee of the Judiciary, Pursuant to S. 1552*, 87th Cong., 1st sess., 1961. The history of the Kefauver hearings has been documented by participants such as Estes Kefauver, *In a Few Hands: Monopoly Power in America* (New York: Pantheon, 1965); Richard Harris, *The Real Voice* (New York: Macmillan, 1964); and Dowling. *Medicines for Man*. Recent accounts of the hearings include Arthur Daemmrich's *Pharmacopolitics: Drug Regulation in the United States and Germany* (Chapel Hill: University of North Carolina Press, 2004).

87. Samuel Hopkins Adams, "The Great American Fraud," *Collier's Weekly*, February 17, 1906, 22.

88. J. T. Connor, *An Early Skirmish* (Rahway, NJ: Merck, 1958), collected publications of Merck Sharp & Dohme, QV 772 M5565 U67, National Library of Medicine, Bethesda, MD, quotation on 12–13. By 1958 27% of Merck's revenues came from its overseas division, Merck Sharp & Dohme International.

89. John T. Connor, "Drug Profits," *Drug and Cosmetic Industry* 86, no. 4 (1960): 563.

90. "DIURIL is an outstanding example, for it is a product of our research, our manufacturing know-how, and our ability to bring it to market through our own sales staff throughout the major sections of the Free World." "Merck's Changing Times," 20–21.

91. "US, Soviet 'Longevity Race' Proposed by Merck President to Gauge Progress in Health," *Oil, Paint, and Drug Reporter* 173 (1958): 5. A reader wrote back in a letter to Connor: "You are to be commended for your suggestion that we engage Russia in a medical race and I sincerely trust that such a wish will come to pass in years to come." Anton Hogstad to John T. Connor, June 19, 1958, folder 13, Diuril Papers.

92. "PMA Leader Replies to Critics: More Drugs for the Average Man," *Oil, Paint, and Drug Reporter* 177 (1960): 4.

93. Williams, *Pharmaceutical Manufacturer's Association*, 26.

94. The Kefauver bill had been severely attacked by opponents and was essentially dead in the water as a price-control measure until the thalidomide scandal broke out and Congress revived the bill as an FDA-strengthening measure. See Harris, *Real Voice*.

95. P. Roy Vagelos, "Are Prescription Drug Prices High?" *Science* 252 (1991): 1080–84.

96. E.g., Charles D. May, "Selling Drugs by 'Educating' Physicians." *Journal of Medical Education* 36 (1961): 1.

97. A scandal around another dual figure had been instrumental in generating the momentum behind Kefauver's initial investigations. Henry Welch was the director of the antibiotics division of the FDA until Francis Engelstad of the General Accounting Office exposed documents demonstrating that Welch had received compensation from two journals he edited on the side, *Antibiotics and Chemotherapy* and *Antibiotic Medicine and Clinical Therapy;* this compensation was explicitly tied to the pharmaceutical advertising of the journals, at a 7.5% commission. Welch's kickbacks totaled nearly three hundred thousand dollars over a six-year period, and later evidence surfaced demon-

strating that Pfizer-authored promotional materials appeared verbatim within Welch's official speeches. "FDA Fires Dr. Welch after Probe, Says Ads Were Basis for Compensation," *Advertising Age,* May 23, 1959, 2; John Lear, "Public Health at 7½%," *Saturday Review,* June 4, 1959; "Pfizer Ad Copy Planted in Welch Talk, Probe Told," *Advertising Age,* June 6, 1959; "Welch Letters Show How He Pushed Sales of Reprints to Advertisers," *Advertising Age,* June 20, 1959.

98. Lawrence Galton, "The Amazing Drug That Helps High Blood Pressure," *Pageant,* April 1958, 96–99, quotation on 99; American Heart Association, *Hypertension,* vol. 6, *Drug Action, Epidemiology, and Hemodynamics: Proceedings of the Council for High Blood Pressure Research* (November 1958).

99. M. Moser, M. D. Blaufox, E. Freis, R. W. Gifford, W. Kirkenhall, H. Langford, A. Shapiro, and S. Sheps, "Who Really Determines Your Patients' Prescriptions?" *Journal of the American Medical Association* 265 (1991): 336–43.

100. Prominent hypertension researchers Edward D. Freis and Marvin Moser mounted a vigorous campaign to reverse the perceived taint of the thiazide diuretics, with little success. Edward D. Freis and Vasilios Papademetriou, "Thiazides Do Not Increase Cardiovascular Risk," *Drug Therapy,* February 1986, 41–49; Edward D. Freis, "The Cardiovascular Risks of Thiazide Diuretics," *Clinical Pharmacology and Therapeutics* 39, no. 3 (1986): 239–44; Edward D. Freis, "The Cardiotoxicity of Thiazide Diuretics: Review of the Evidence," *Journal of Hypertension* 8, supp. 2 (1990): 23–31; Edward D. Freis, "The Efficacy and Safety of Diuretics in Treating Hypertension," *Annals of Internal Medicine* 122, no. 3 (1995): 223–26; Marvin Moser, "Why Are Physicians Not Prescribing Diuretics More Frequently in the Management of Hypertension," *Journal of the American Medical Association* 279 (1998): 1813–16.

101. Lawrence J. Appel, "The Verdict From ALLHAT: Thiazide Diuretics Are the Preferred Initial Treatment For Hypertension," *Journal of the American Medical Association* 288 (2002): 3039–42.

CHAPTER 2: Shrinking the Symptom, Growing the Disease

Epigraph: Edward D. Freis, "The Changing Outlook in Essential Hypertension," *Hypertensive Cardiovascular Disease* 1, no. 1 (1967): 285.

1. The metallic ink was explicitly linked to Diuril's tenth (or "aluminum") anniversary.

2. Charles Lyght, ed., *Decade with Diuril* (West Point, PA: MSD, 1968), iv–v. Also see "Creativity Using 'A Decade with Diuril,'" *MSD Frontline,* May–June 1968, 6, Merck Archives, Whitehouse Station, NJ.

3. Ray W. Gifford, "Chlorothiazide in Hypertension," in Lyght, *A Decade with Diuril,* 42.

4. Correspondence between Max Tishler and the Lasker Foundation, 1967, folder 12, Diuril Papers, Merck Archives (hereafter cited as Diuril Papers).

5. Lasker Prize Citation, Edward Freis, 1971, in a collection of correspondence, manuscripts, publications, and artifacts maintained by Susan Freis, Bluemont, VA (hereafter cited as Freis Papers).

6. Carsten Timmermann, "A Matter of Degree: The Normalisation of Hypertension, circa 1940–2000," in *The Normal and the Abnormal,* ed. Waltraud Ernst (London:

Routledge, in press); also see "The Question of Norms: The Fallacy of the Dividing Line between the Normal and the Pathological," in *A Century of Arterial Hypertension, 1896–1996*, ed. Nicolas Postel-Vinay (New York: Wiley, 1996), 133–42.

7. William Goldring and Herbert Chassis, "Antihypertensive Drug Therapy: An Appraisal," in *Controversy in Internal Medicine*, ed. Franz J. Inglefinger and Arnold S. Relman (Philadelphia: Saunders, 1966), 83.

8. See Ian Hacking, "The Looping Effects of Human Kinds," in *Causal Cognition: A Multidisciplinary Debate*, ed. D. Sperber, D. Premack, and A. J. Premack (New York: Oxford University Press, 1995).

9. Paul Dudley White, *Heart Disease* (New York: MacMillan, 1931), 400.

10. John Hay, "The Significance of a Raised Blood Pressure," *British Medical Journal* 2 (1931): 43–47, quotation on 44.

11. For a collection of primary sources tracing the history of high blood pressure (or "hard pulses") as a category from early Chinese and Hippocratic texts on to the twentieth century, see A. Ruskin's *Classics in Arterial Hypertension* (Springfield, IL: Charles C. Thomas, 1956).

12. Harvey Cushing, the iconoclastic neurosurgeon of the newly formed Brigham and Women's Hospital, is credited with introducing the Korotkoff technique into American hospital practice in 1906. By World War I, although blood pressure measurement had become routine in hospital practice, physicians observing the sphygmomanometer tended to be more concerned with readings of low blood pressures than any fear that their patients might be chronic hypertensives. See Hughes Evans, "Losing Touch: The Controversy over the Introduction of Blood Pressure Instruments into Medicine," *Technology and Culture* 34 (1993): 784–808.

13. Fisher's hypertension projects began in 1907, when Northwestern Mutual began using sphygmomanometric readings in all of its initial insurance physicals. A 1922 publication, based on fifteen years of experience with the Northwestern Mutual applicant pool, concluded that across populations, high blood pressure exhibited a dose-response curve with early cardiovascular mortality but that an "apparently healthy person may have high arterial tension extending over a considerable period of time without a discoverable impairment to account for same." John Walton Fisher, *The Diagnostic Value of the Use of the Sphygmomanometer in Examinations for Life Insurance* (Milwaukee: Northwestern Mutual Life Insurance Co., 1922), 7.

14. Robert M. Daley, Harry E. Ungerleider, and Richard S. Gurner, "Prognosis and Insurability of Hypertension with Particular Reference to the Electrocardiogram," *Transactions of the Association of Life Insurance Medical Directors of America* 28, no. 18 (1942): 18–48, quotation on 19. Irvine Page, an early advocate and promoter of blood pressure management, complained in 1951 that "in recent years, army, navy, and public health examinations have discovered vast numbers of men with hypertension. Insurance and employment examinations have uncovered thousands of other cases . . . the finding of hypertension or prehypertension often makes a man unemployable and nearly always uninsurable . . . And yet hypertension as such is not even officially recognized as a disease by the Boards of Health and insurance companies." Page, *Hypertension: A Manual for Patients with High Blood Pressure* (Springfield, IL: Charles C. Thomas, 1951), 85–87.

15. Allen G. Rice, *The Value of Blood Pressure in the Diagnosis and Prognosis of Dis-*

ease (Providence, RI: Snow and Farnham, 1916), 36–37; as cited in Evans, "Losing Touch," 797.

16. G. W. Norris, H. C. Bazett, and T. M. McMillan, *Blood-Pressure: Its Clinical Applications* (Philadelphia: Lea and Febiger, 1926).

17. For secondary sources on antipyretic faddism, see Charles C. Mann, *The Aspirin Wars: Money, Medicine, and 100 Years of Rampant Competition* (New York: Knopf, 1991); Anne A. J. Audermann, "Physicians, Fads, and Pharmaceuticals: A History of Aspirin," *McGill Journal of Medicine* 2 (1996): 115–20.

18. Theodore Janeway, "A Clinical Study of Hypertensive Cardiovascular Disease," *Annals of Internal Medicine* 12 (1913): 755–98; D. Ayman and J. Pratt, "Nature of the Symptoms Associated with Essential Hypertension," *Archives of Internal Medicine* 47 (1931): 675–87. Ayman and Pratt conducted a descriptive search of symptoms among one hundred hypertensive patients (higher than 160 mm Hg systolic); they described early symptoms including headache, nervousness, fatigability, irritability, and dizziness.

19. N. M. Keith, H. P. P. Wagener, and N. W. Barker, "Some Different Types of Essential Hypertension: Their Course and Diagnosis," *American Journal of the Medical Sciences* 196 (1939): 332–43.

20. See, e.g., A. M. Butler, "Chronic Pyelonephritis and Arterial Hypertension," *Journal of Clinical Investigation* 16 (1937): 889–97.

21. Following M. C. Pincoffs's initial demonstration of pheochromocytoma in 1929, it became the first distinct form of "surgically curable hypertension" and acquired a specific etiological mechanism, which remained elusive for essential hypertension. See M. C. Pincoffs, "A Case of Paroxysmal Hypertension Associated with a Suprarenal Tumor," *Transactions of the Association of American Physicians* 44 (1929): 295–99, cited in "Pheochromocytoma and the Concept of Surgically Curable Hypertension," and "Conn and Aldosteronism," in Postel-Vinay, *Century of Arterial Hypertension,* 79–85; also see W. Conn, "Primary Aldosteronism, a New Clinical Syndrome," *Journal of Laboratory and Clinical Medicine* 4 (1955): 661–64.

22. An analogous problem is described in the field of psychiatry by Gerald Grob in *From Asylum to Community: Mental Health Policy in Modern America* (Princeton, NJ: Princeton University Press, 1991), 125–26: "When a clear causal relationship between pathology and behavior was demonstrated, jurisdiction for the illness was invariably transferred from psychiatry to another medical specialty . . . By surrendering authority over diseases of known somatic origins marked by disturbed behavior, [psychiatrists] were in the somewhat odd position of retaining jurisdiction over all the mental diseases of unknown etiology."

23. R. Platt, "The Nature of Essential Hypertension," *Lancet* 2 (1959): 1092.

24. For analysis of the Platt-Pickering controversy as a case study in gradualist versus Manichean approaches to defining normality and pathology, see Timmermann, "Matter of Degree"; J. D. Swales, ed., *Platt versus Pickering: An Episode in Recent Medical History* (Cambridge, U.K.: Keynes Press, 1985).

25. Irvine H. Page, "Pathogenesis of Arterial Hypertension," *Journal of the American Medical Association* 140 (1949): 451.

26. Society of Actuaries Committee on Mortality, *Build and Blood Pressure Study, 1959* (Chicago: Society of Actuaries, 1960).

27. G. W. Pickering, "The Genetic Factor in Essential Hypertension," *Annals of Internal Medicine* 43 (1955): 457–64, quotation on 463.

28. S. Robinson, "Range of Normal Blood Pressure: A Statistical and Clinical Study of 11,388 Persons," *Archives of Internal Medicine* 64 (1939): 409–44.

29. See "Question of Norms."

30. Excision of sympathetic ganglia was first performed in Germany, by Bruening in 1923; the practice of sympathectomy, as popularized by Smithwick and Peet in the United States, did not rise into prominence until the mid-1940s. The adverse effects typically resulting from surgical sympathectomy included postural hypotension, syncope, impotence, and the inability to sweat. R. H. Smithwick, "Splanchniectomy for Essential Hypertension: Results in 1,266 Cases," *Journal of the American Medical Association* 152 (1953): 1501–4. also see Marvin Moser, *The Treatment of Hypertension: A Story of Myths, Misconceptions, Controversies, and Heroics* (Darien, CT: Le Jacq Communications, 2002), 8; "An Assorted Catalog of Treatments: Radiotherapy, Electrotherapy, Pyrotherapy, Surgery, and Pharmacology," in Postel-Vinay, *Century of Arterial Hypertension.*

31. J. Earle Estes, "Hypertension in 1958: A Tale of Pills, Philosophy, and Perplexity," *Medical Clinics of North America* 42(1958): 899–915. In lower doses, hydralazine later became a very popular antihypertensive. It is still widely used today in hospital practice and in some cases of intractable hypertension in the outpatient context, particularly when angiotensin-converting enzyme inhibitors are contraindicated.

32. Briefly, even diuretics had been considered as therapeutics for high blood pressure; R. S. Megibow and co-workers demonstrated in 1948 that frequent high-dose administration of intravenous mercurials could reduce blood pressure, but this finding was not followed over the ensuing decade because it involved a very time- and material-intensive approach with a highly toxic formulation of mercury. R. S Megibow, H. Pollack, G. H. Stollerman, E. H. Roston, and J. J. Bookman, "The Treatment of Hypertension by Accelerated Sodium Depletion," *Journal of Mount Sinai Hospital* 15 (1948): 233–39; historical perspective was provided a decade later by Gifford, "Chlorothiazide in Hypertension."

33. Edward D. Freis, interview by author, Bethesda, MD, August 2003; Marvin Moser, telephone interview by author, June 2003.

34. Although mention of the low-salt diet as a therapeutic modality for hypertension can be found in the earlier work of L. Ambard and E. Beaujard, "Causes de l'hypertension arteriole," *Archives of General Medicine* 1 (1904): 520–23, its broad implementation in the United States owes much to the writings and public demonstrations of W. Kempner, e.g., Kempner, "Treatment of Kidney Disease and Hypertensive Vascular Disease with Rice Diet," *North Carolina Medical Journal* 5 (1944): 125, 273.

35. Charles E. Lyght, ed., *The Merck Manual of Diagnosis and Therapy,* 9th ed. (Rahway, NJ: Merck, 1956), 234.

36. As discussed by Marvin Moser in *Treatment of Hypertension,* 8.

37. Harriet P. Dustan, "The Usefulness of Drugs in the Treatment of Hypertension," *Hypertension* 6 (1958): 107–9, quotation on 109.

38. Milton J. Chatton, Sheldon Margen, and Henry Brainerd, *Handbook of Medical Treatment,* 5th ed. (Los Altos, CA: Lange Medical Publications, 1956), 156.

39. H. P. Dustan, Roland E. Schneckloth, A. C. Corcoran, and Irvine H. Page, "The Effectiveness of Long-Term Treatment of Malignant Hypertension," *Circulation* 18

(1958): 644–51; H. Mitchell Perry Jr. and Henry A. Schroeder, "The Effect of Treatment on Mortality Rates in Severe Hypertension: A Comparison of Medical and Surgical Regimens," *Archives of Internal Medicine* 102 (1958): 418–25; P. A. Restall and F. H. Smirk, "The Treatment of High Blood Pressure with Hexamethonium Iodide," *New Zealand Medical Journal* 49 (1950): 206.

40. Lyght, *Merck Manual*, 9th ed., 232.

41. Arthur M. Master, Charles I. Garfield, and Max B. Walter, *Normal Blood Pressure and Hypertension: New Definitions* (Philadelphia: Lea and Febiger, 1952), 124.

42. William A. Brams, *Your Blood Pressure and How to Live with It* (Philadelphia: Lippincott, 1956), 100.

43. Merck Sharp & Dohme, Medical Publications Department, *Information for the Physician on Diuril (Chlorothiazide), a New Diuretic and Antihypertensive Agent* (Philadelphia: Merck Sharp & Dohme, 1958), 19; the quote continues, "The mode of action by which DIURIL exerts an antihypertensive effect appears to differ from that of other antihypertensive agents and has not been clearly elucidated."

44. The only adverse effect of Diuril immediately perceived to be relevant was a reversible risk of potassium depletion (hypokalemia) that could easily be detected and corrected by supplement; see Leon Gortler and Jeffery L. Sturchio, "Karl Henry Beyer Jr.: An Interview for the Merck Archives," 1988. Also see Jeffrey L. Sturchio and Louis Galambos, "John Lloyd Huck: An Interview for the Merck Archives," 1990, 36; both are oral history transcripts of interviews conducted in Penllyn, PA, stored in the Merck Archives. For a full side-effect profile, see Merck, *Information for the Physician (Chlorothiazide)*, 18–19.

45. E. D. Freis, "Should We Treat Hypertension Early?" *GP* 14 (1956): 72–73, quotation on 73.

46. E. D. Freis, "Rationale and Methods for the Treatment of Early Essential Hypertension" *Journal of the National Medical Association* 50, no. 6 (1958): 405–12, quotation on 405.

47. F. K. Heath, "Diuril Meetings," memorandum, 1958, folder 3, Diuril Papers. The parenthetical "rarely adequate" refers to the importance of Diuril in combination therapy relative to monotherapy.

48. John H. Moyer, "Summary: Today's Recommendations for Drug Therapy of Hypertension," in *Hypertension: The First Hahnemann Symposium on Hypertensive Disease*, ed. John H. Moyer (Philadelphia: Saunders, 1959), 735–74, Schroeder quoted on 737. As Corcoran elaborated, "If hypertension is a reversible process and if, once reversed, the blood pressure does tend to self-sustain at normal levels, I think that there is every indication for whatever is required to restore blood pressure to or toward normal, even if there is some small risk involved" (736).

49. Ibid., 735–36.

50. Even before the thalidomide disaster, there were highly publicized cases of large-scale pharmaceutical iatrogenesis, such as the Cutter polio vaccine incident and the sulfanilamide tragedy, both well detailed in Phillip J. Hilts, *Protecting America's Health: The FDA, Business, and One Hundred Years of Regulation* (New York: Knopf, 2003).

51. Moyer, "Summary," 737–38.

52. Ibid., 755–60.

53. Estes, "Hypertension in 1958," 915.

54. Perry and Schroeder, "Effect of Treatment," 418.

55. A. W. D. Leishman, "Hypertension—Treated and Untreated, a Study of 400 Cases," *British Medical Journal* 15 (1959): 1361–68, quotation on 1367.

56. Institute for Motivational Research, *Research Study on Pharmaceutical Advertising* (Croton-on-Hudson, NY: Pharmaceutical Advertising Club, 1955).

57. Milton Moskowitz, "Diuril Creates a New Market," *Drug and Cosmetic Industry* 87 (1960): 841. In comparison, five years earlier Merck had spent only $70,000 on yearly advertising in the *Journal of the American Medical Association.* The next-largest advertiser in *JAMA* in 1960 was Schering, with a budget of $316,000. Senate Subcommittee on Antitrust and Monopoly of the Committee of the Judiciary, *Hearings before the Subcommittee on Antitrust and Monopoly of the Committee of the Judiciary, Pursuant to S. 1552,* 87th Cong., 1st sess., 1961, 132.

58. The patent on hydrochlorothiazide was ultimately shared between Ciba and MSD. As Karl Beyer recalls, "An avalanche of publications attesting to the efficacy and safety of chlorothiazide evoked a chemical effort in the worldwide pharmaceutical industry that could not be kept up with. Every company seemed to want its very own patentable thiazide." Karl Beyer, "Chlorothiazide: How the Thiazides Evolved as Antihypertensive Therapy," *Hypertension* 22 (1993): 390.

59. Hodge and Smirk in 1961 published a series of hypertensive deaths from 1959 to 1964 that demonstrated that coronary heart disease accounted for half of all hypertensive deaths, while stroke accounted for one-quarter. J. V. Hodge, E. G. McQueen, and H. Smirk, "Results of Hypotensive Therapy in Arterial Hypertension," *British Medical Journal* 1 (1961): 1–7; also see Freis, "Changing Outlook in Essential Hypertension," 280–85.

60. This shift has been discussed by Merwyn Susser, "Epidemiology in the United States after World War II: The Evolution of Technique," *Epidemiologic Reviews* 7 (1985): 147–77; and by Robert Aronowitz, *Making Sense of Illness: Science, Society, and Disease* (Cambridge: Cambridge University Press, 1998), 111–44.

61. Freis, "Changing Outlook in Essential Hypertension," 285.

62. Shortly after the early publications of the Framingham Study, Framingham researchers had been quick to suggest the possible application of their findings to the treatment of hypertension. See Abraham Kagan, Tavia Gordon, William B. Kannel, and Thomas R. Dawber, "Blood Pressure and Its Relation to Coronary Heart Disease in the Framingham Study," *Hypertension* 7 (1958): 53–81.

63. Jeremiah Stamler, "Epidemiological Analysis of Hypertension and Hypertensive Disease in the Labor Force of a Chicago Utility Company," *Hypertension* 7 (1958): 23–52, quotation on 48–49.

64. W. Goldring and H. Chasis, "Antihypertensive Drug Therapy: An Appaisal," *Archives of Internal Medicine* 11 (1965): 523–25, quotation on 523. Goldring and Chasis's textbook, *Hypertension and Hypertensive Disease* (New York: Commonwealth Fund, 1944), had become an important reference in the field.

65. Goldring and Chassis, "Antihypertensive Drug Therapy," in *Controversy in Internal Medicine,* 83.

66. Goldring and Chassis, Antihypertensive Drug Therapy," *Archives of Internal Medicine,* 525.

67. Goldring and Chassis, Antihypertensive Drug Therapy," in *Controversy in Internal Medicine,* 83, 85.

68. In his retrospective account of the VA study, Freis cites Goldring and Chassis's repeated critiques of antihypertensive therapy as a prime motivating factor behind the conceptualization of the VA study as a randomized controlled trial. Edward D. Freis, "Reminiscences of the Veterans Administration Trial of the Treatment of Hypertension," *Hypertension* 16, no. 4 (1990): 472–75. A similar account appears in Irvine H. Page, *Hypertension Research: A Memoir* (New York: Pergamon, 1988).

69. See F. W. Wolff and R. D. Lindeman, "Effects of Treatment in Hypertension: Results of a Controlled Study." *Journal of Chronic Disease* 19 (1966): 227–40.

70. Edward D. Freis, interview by author, June 2003, Bethesda, MD.

71. Veterans Administration Cooperative Study Group on Antihypertensive Agents (VACSGAA), "A Double Blind Control Study of Antihypertensive Agents: I. Comparative Effectiveness of Reserpine and Hydralazine, and Three Ganglionic Blocking Agents, Chlorisondamine, Mecamylamine, and Pentolinium Tartrate," *Archives of Internal Medicine* 106 (1960): 81–96; VACSGAA, "A Double Blind Control Study of Antihypertensive Agents: II. Further Report on the Comparative Effectiveness of Reserpine, Reserpine and Hydralazine, and Three Ganglionic Blocking Agents, Chlorisondamine, Mecamylamine, and Pentolinium Tartrate," *Archives of Internal Medicine* 110 (1962): 222–29.

72. VACSGAA, "Effects of Treatment on Morbidity in Hypertension: Results in Patients with Diastolic Blood Pressures Averaging 115 through 129 mm Hg," *Journal of the American Medical Association* 202, no. 11 (1967): 116–22; VACSGAA, "Effects of Treatment on Morbidity in Hypertension: II. Results in Patients with Diastolic Blood Pressure Averaging 90 through 114 mm Hg," *Journal of the American Medical Association* 213 (1970): 1143–52.

73. Excerpted from the "Citation of the Clinical Research Award of the Albert and Mary Lasker Foundation," as recorded by Congressman Olin E. Teague in the *Congressional Record* (November 17, 1971): E 12340.

74. Jeffrey L. Sturchio and Louis Galambos, "Eugene L. Kuryloski: An Interview for the Merck Archives," oral history transcript, 44, Merck Archives.

75. George DiDomizio, Merck Sharp & Dohme Advertising Department, to Edward D. Freis, May 1, 1973, bound correspondence files, Freis Papers.

76. "What Effect Does the Treatment of Essential Hypertension Have on Morbidity?" proof copy of advertisement, 1972; "When Is Blood Pressure High Enough for Drug Treatment?" proof copy of advertisement, 1973, accompanied by a memo from John P. Burns, Merck Sharp & Dohme, to Freis, October 23, 1973. All in the Freis Papers.

77. Jane E. Brody, "Drive on High Blood Pressure Urged," *New York Times*, November 22, 1971.

78. Brian Blades to Edward Freis, November 10, 1971, Freis Papers.

79. Mary Lasker to Edward D. Freis, December 2, 1971; further memos from the Lasker Foundation to Freis included press clippings and scripts of TV news programs that had reported on the award ceremony, including *NBC News*, the *New York Times*, *Daily News*, *Women's Wear Daily*, and *Medical World News*. See Maier to Freis, November 17, 1971; Maier to Lasker, Fordyce, and Freis, August 15, 1972; Ruth Maier to Lasker, Fordyce, and Freis, September 20, 1972, all in bound correspondence files, 1971–73, Freis Papers.

80. Evidence of the Lasker Foundation's intent to utilize the occasion of Freis's award to lobby for the creation of the National High Blood Pressure Education Pro-

gram is found in correspondence between Lasker's press secretary Ruth Maier and Campbell Moses, the president of the American Heart Association. Meier to Freis, November 17, 1971, Freis Papers.

81. Lasker's activities were a crucial factor in the expansion of the NIH budget from $2.4 million in 1945 to $5.5 billion by 1986. See, e.g., Sana Siwolup, "The Fairy Godmother of Medical Research: Mary Lasker Has Wrested Money Out of Congress for Nearly Half a Century and She's Still Going Strong," *Business Week,* June 5, 1986, 67.

82. As NHLBI director Robert I. Levy recalled, in early 1972 "former Secretary of Health, Education, and Welfare, Elliot Richardson, on the very strong advice of Mrs. Mary Lasker, Mr. Gorman and others, called for the initiation of the National High Blood Pressure Education Program." Robert Levy to Lyman J. Olson, 1978, NHBPEP Files folder, from the collections of Edward J. Roccella, director, National High Blood Pressure Education Program, NHLBI, NIH, Building 32, Bethesda, MD (hereafter cited as Roccella Papers). See also Claude Lenfant and Edward J. Roccella, "Beginnings," in *National High Blood Pressure Education Program: 20 Years of Achievement* (Bethesda, MD: National High Blood Pressure Education Program, 1992), 3–6.

83. Ruth Maier to Mrs. Albert D. Lasker, August 15, 1972, Freis Papers.

84. Transcript, *Today Show,* August 11, 1972, 7:00 AM, WNBC and the NBC Television Network, New York, NY, Freis Papers.

85. Mike Gorman, "A Combined Federal, State, and Citizens Campaign against Hypertension," unpublished manuscript, in Mike Gorman Papers, MS C 462, box 14, folder 32, National Library of Medicine, Bethesda, MD (hereafter cited as Gorman Papers).

86. Citizens was described in 1973 by Elliot Richardson as "a group of distinguished individuals who will marshal private and voluntary resources for professional and public education on the importance of early detection and treatment of high blood pressure. These undertakings will complement, supplement, and expand existing and planned governmental and voluntary efforts in these areas." Unpublished manuscript, addressed to the National Conference on High Blood Pressure Education, January 15, 1973, Washington Hilton Hotel, Washington, DC, Roccella Papers. Also see Stuart Auerbach, "Hypertension Fight Launched by Richardson," *Washington Post,* January 16, 1972, A3.

87. As chief of the NHLBI's health education branch Graham Ward noted in 1981, "The NHBPEP, although coordinated by the NHLBI, is not a single agency activity. Rather, it is an ever growing coalition of some 15 Federal agencies; 150 national programs, voluntary and trade organizations; nearly all state and many local public health departments; and over 2,000 organized community programs. Insurance companies, pharmaceutical firms, and many other private industries play active and increasing roles in this program as do many organized labor associations." Unpublished (draft) manuscript entitled "Statement by Graham Ward before the Committee on Labor and Human Resources, United States Senate, 16 July 1981," NHBPEP Files folder, Roccella Papers.

88. Elliot explicitly referred to the NHBPEP as a model of New Federalism in several interviews and speeches, e.g., Auerbach, "Hypertension Fight Launched by Richardson," A3. The initial members of the Coordinating Committee of the NHBPEP were representatives of the NHLBI, the American Heart Association, the National Kidney Foundation, the American College of Cardiology, and Citizens for the Treatment of

High Blood Pressure, Inc. See unpublished manuscript, "Background of the National High Blood Pressure Coordinating Committee, 1977, folder 5, box 20, Gorman Papers.

89. Cooper to Richardson, March 1, 1973, unfiled documents, Roccella Papers.

90. *The Hypertension Handbook* (Westpoint, PA: Merck Sharp & Dohme, 1974). A parenthetical admonition on the cover stated, "This does not constitute an endorsement by the program of any of the drug products mentioned in this book."

91. Promotional efforts of the pharmaceutical industry on behalf of the NHBPEP were cataloged in a memorandum written by the assistant secretary for health of the Department of Health, Education, and Welfare, Charles C. Edwards; one or more of these advertisements appeared on the pages of the *Atlantic Monthly, Harper's*, the *Intellectual Digest, Newsweek, Psychology Today*, the *Saturday Evening Post*, the *Saturday Review, Time, U.S. News and World Report, Hospital Medicine, Postgraduate Medicine*, the *American Journal of Cardiology*, the *Archives of Internal Medicine*, the *American Journal of Medicine, Circulation, Lancet*, and the *American Heart Journal;* see Edwards to Theodore Cooper, "National High Blood Pressure Education Program-INFORMATION MEMORANDUM," March 12, 1974, unfiled document, Roccella Papers. In addition the NHBPEP maintained its own press secretary, who encouraged the publication of articles on hypertension in newspapers and periodicals such as *McCall's, Harper's Bazaar, Vogue, Women's Day*, and others; see the "Articles in Publications" subsection of a memorandum, "Quarterly Report," NHBPEP Coordinator Graham Ward to the NHBPEP Coordinating Committee, June 30, 1977, box 5, folder 20, Gorman Papers.

92. Mike Gorman to Mrs. Albert D. Lasker, November 1, 1977, box 5, folder 17, Gorman Papers. The role of the guideline in late-twentieth-century practice has been the subject of an extensive clinical literature and recently has attracted attention as a subject of medical sociology; see, e.g., the recent work of Stefan Timmermans and Marc Berg, *The Gold Standard: The Challenge of Evidence-Based Medicine and Standardization in Health Care* (Philadelphia: Temple University Press, 2003).

93. "Fact Sheet #1—New Frontiers in Lowering Cholesterol," Mike Gorman to Antonio Gotto, box 2, folder 3, Gorman Papers; also see Roger A. Rosenblatt, D. Cherkin, R. Schnelweiss, and G. Hart, "The Content of Ambulatory Care in the United States," *New England Journal of Medicine* 309 (1983): 892–97.

94. On the development of the JNC, see Marvin Moser, "The First Decade: 1972–1982," in *National High Blood Pressure Education Program*, 45–52. The expansive changes in the chronology of the JNC recommendations are also noted by William Rothstein, *Public Health and the Risk Factor: A History of an Uneven Medical Revolution* (Rochester, NY: University of Rochester Press, 2003), 270. See also Aram V. Chobanian, George L. Bakris, Henry R. Black, William C. Cushman, Lee A. Green, Joseph L. Izzo Jr., Daniel W. Jones, Barry J. Materson, Suzanne Oparil, Jackson T. Wright Jr., and Edward J. Roccella, "The Seventh Report of the Joint National Committee on Prevention, Detection, Evaluation, and Treatment of High Blood Pressure," *Journal of the American Medical Association* 289 (2003): 2560–71.

95. "The 1978 Albert Lasker Public Service Award, to the Honorable Elliott L Richardson," press release from the Albert and Mary Lasker Foundation via Ruth Maier, November 21, 1978, unfiled document, Roccella Papers.

96. The generalizability of the VA study—which included only middle-aged men—to other populations was called into question by critics, and subsequent studies varied

in their ability to demonstrate the applicability of the JNC guidelines beyond this core population. Moreover, because the VA conclusions were not statistically significant for "mild" hypertension, a continuing debate over the treatability of mild hypertensives persisted well into the 1980s even after a general consensus had accumulated around the treatment of asymptomatic "moderate" hypertensives. Nonetheless, the joint actions of the JNC, the NHBPEP, and the expanded armamentarium of antihypertensive drugs functioned to make hypertension the leading reason for office visits to physicians by 1980. See Ichiro Kawachi and Peter Conrad, "Medicalization and the Pharmacological Treatment of Blood Pressure," in *Contested Ground: Public Purpose and Private Interest in the Regulation of Prescription Drugs,* ed. Peter Davis (New York: Oxford University Press, 1996), 26–41.

CHAPTER 3: Finding the Hidden Diabetic

Epigraph: Elliott P. Joslin, *Diabetes, Its Control by the Individual and the State* (Cambridge: Harvard University Press, 1931), 47.

1. Milton Moskowitz, "Diuril Creates a New Market," *Drug and Cosmetic Industry* 87 (1960): 841.

2. Claude Bernard, *Introduction à la médicine expérimentale* (Paris: J. B. Baillere, 1865), 236.

3. Georges Canguilhem, *The Normal and the Pathological,* trans. Carolyn R. Fawcett (1943; New York: Zone Books, 1991).

4. Histories of diabetes include N. S. Papaspyros, *The History of Diabetes Mellitus* (London: Robert Stockwell, 1952); Michael Bliss, *The Discovery of Insulin* (Chicago: University of Chicago Press, 1982); Dietrich von Englehardt, *Diabetes: Its Medical and Cultural History* (Berlin: Springer Verlag, 1989); James Wright Presley, "A History of Diabetes Mellitus in the United States, 1880–1990" (PhD diss., University of Texas at Austin, 1991); Chris Feudtner, *Bittersweet: Diabetes, Insulin, and the Transformation of Illness* (Chapel Hill: University of North Carolina Press, 2003).

5. Claude Bernard, *Leçons sur le diabète et la glycogenèse animale* (1877), as translated in Canguilhem, *Normal and the Pathological,* 70.

6. In other fields (e.g., the diagnosis of anemia in hematology) diseases defined on deviant laboratory values appear to have taken root earlier; see Keith Wailoo, *Drawing Blood: Technology and Disease Identity in Twentieth-Century America* (Baltimore: Johns Hopkins University Press, 1997).

7. Chris Feudtner, "The Want of Control: Ideas, Innovations, and Ideals in the Modern Management of Diabetes Mellitus," *Bulletin of the History of Medicine* 69 (1995): 66–90; D. A. Pyke, "The History of Diabetes," in *The International Textbook of Diabetes Mellitus,* ed. R. A. DeFronzo, 2nd ed. (New York: Wiley, 1997).

8. On the discovery of insulin, see Michael Bliss, *The Discovery of Insulin* (Chicago: University of Chicago Press, 1982).

9. Feudtner, *Bittersweet,* 21.

10. Joslin, *Diabetes,* 14.

11. Paul Kimmelstein and Clifford Wilson, "Intercapillary Lesions in the Glomeruli of the Kidney," *American Journal of Pathology* 12 (1936): 83–98.

12. Rachmiel Levine, "Editorial Statement," *Metabolism* 5 (1956): 723–26.

13. The self-maintenance practices involved in being a diabetic patient after the ad-

vent of insulin are well described in Feudtner, *Bittersweet*, 89–120. See also the numerous contemporary guidebooks written for newly diagnosed diabetics, e.g., R. D. Lawrence, *The Diabetic Life: Its Control by Diet and Insulin* (Philadelphia: P. Blakiston's Son, 1928); Anthony M. Sindoni Jr., *The Diabetic's Handbook: How to Work with Your Doctor* (New York, Ronald Press, 1948); T. S. Danowski, *Diabetes as a Way of Life* (New York: Coward-McCann, 1957).

14. On the divergent histories of the British and American associations, see Leo P. Krall, "The History of Diabetes Lay Associations," *Patient Education and Counseling* 26 (1995): 285–91.

15. Feudtner, "Want of Control."

16. In the early twentieth century, the diagnosis of diabetes with blood sugar was made only as an adjunct to clinical suspicion; see G. C. Shattuck, *Principles of Medical Treatment*, 6th ed. (Cambridge: Harvard University Press, 1926), 157: "The patient's story, if the onset has been acute, is almost pathognomonic; otherwise, we must depend upon the examination of the urine. Blood-sugar estimations may help us in determining the degree of diabetes, but are of more aid in the treatment."

17. Patricia A. Rosales, "The History of the Hypodermic Syringe: 1850s–1920s" (PhD diss., Harvard University, 1997).

18. Pyke, "History of Diabetes," 7.

19. Martin Goldner, "Oral Hypoglycemic Agents, Past and Present," *Archives of Internal Medicine* 102 (1058): 830–41. These vegetable substances included glucokinin (a yeast extract), phaseolin (from bean pods), myrtillin (from blueberries), and hypoglycin (from chestnuts).

20. See Walter Sneader, *Drug Prototypes and Their Exploitation* (New York: Wiley, 1996), 736. Guanidine derivatives had the unfortunate distinction of emerging as therapeutic agents immediately following the widespread success of insulin; the product was marketed by Schering AG as an oral hypoglycemic under the trade name Synthalin in 1926. Synthalin was followed by the supposedly less toxic Synthalin B, though both were found to have unfortunately high rates of liver toxicity and were withdrawn from production in the early 1940s.

21. These early agents were produced by Rhone Poulenc under licensing agreement with Schering AG. One of these compounds was a homologue of Schering's Globucid and was known in Germany as VK 57. See Hans Kleinsorge, "Carbutamide—The First Oral Antidiabetic," *Endocrinology and Diabetes* 106 (1998) 149–51.

22. Auguste Loubatieres, "The Hypoglycemic Sulfonamides: History and Development of the Problem from 1942 to 1955," *Annals of the New York Academy of Sciences* 71, no. 11 (1957): 4–11.

23. Sneader, *Drug Prototypes and Their Exploitation*, 616; Kleinsorge, "Carbutamide," 151; George E. Farrar Jr., Charles R. Shuman, and Arthur Krosnick, "Milestones in Clinical Pharmacology: Oral Hypoglycemic Agents: Sulfonylurea," *Clinical Therapeutics* 13, no. 2 (1991): 319–23.

24. See "Agreement between the Upjohn Co. and Farbwerke Hoechst of August 30, 1949, and Subsequent Correspondence and Memos," and "Agreement between the Upjohn Co. and Farbwerke Hoechst of August 6, 1956, and Subsequent Correspondence," reproduced in the Senate Subcommittee on Antitrust and Monopoly of the Committee on the Judiciary, *Hearings before the Subcommittee on Antitrust and Monopoly of the Committee on the Judiciary, Part 20: Administered Prices in the Drug Industry (Oral*

Antidiabetic Drugs), 1960, 11253–82. Also see "Orinase: Its Development within the Up-john Company," *Overflow*, June 1957, 245–49, Upjohn Collections, Kalamazoo Public Library, Kalamazoo, MI (hereafter cited as Upjohn Collections). For a brief history of the pharmaceutical component of I. G. Farben, see Charles C. Mann, *The Aspirin Wars: Money, Medicine, and 100 Years of Rampant Competition* (New York: Knopf, 1991).

25. "Orinase: Its Development." For a listing of sites of clinical research, see "Symposium on Clinical and Experimental Effects of Sulfonylureas in Diabetes Mellitus," *Metabolism* 5 (1956): 721–972.

26. Characterizations of oral hypoglycemics as "oral insulins" were common in early news reports surrounding the development of the sulfonylureas; see, e.g., A. L. Blakeslee, "Insulin Substitute?" *Today's Health*, May 1956, 17.

27. "A.M.A. Studies Use of Diabetic Drugs," *New York Times*, April 15, 1956, 42.

28. "Symposium on Sulfonylureas in Diabetes," 847–63; "Diabetes Identified as Two Distinct Types," *Science News Letter* 72 (November 9, 1957): 296.

29. The Brook Lodge, once the ancestral home of the Upjohn family, had by the early 1950s been converted into a conference center that was fully controlled by the Upjohn Company, yet not connected to it by name. It was a venue for research conferences, press releases, corporate picnics, and sales training seminars. See "Brook Lodge in Transition," *Overflow*, February 1957, 91–94; "Brook Lodge: Meeting of the Minds," *Upjohn News*, July–August 1957, 1–5, both in Upjohn Collections. The circulation of *Metabolism* in 1956 was fifteen thousand; by December of 1956 Upjohn had made plans to ship another thirty-five thousand copies of the issue to its salesmen for distribution to physicians. This distribution more than tripled the journal's circulation. "Symposium on Sulfonylureas in Diabetes"; see also "Orinase: A Progress Report," *Upjohn Overflow*, June 1956; "Orinase Symposium," *Upjohn Overflow*, September 1956, both in the Upjohn Collections.

30. "Symposium on Sulfonylureas in Diabetes," 860 (emphasis added).

31. Ibid., 725.

32. Ibid., 910, 952.

33. R. Camerini-Davalos, A. Marble, and H. Root, "Clinical Experience with Orinase: A Preliminary Report," *Metabolism* 5 (1956): 904–10. Also see Rafael Camerini-Davalos, Howard F. Root, and Alexander Marble, "Clinical Experience with Carbutamide (BZ-55): A Progress Report," *Diabetes* 6, no. 1 (1957): 74–77; William B. Hadley, Avedis Khachadurian, and Alexander Marble, "Studies with Chlorpropamide in Diabetic Patients," *Annals of the New York Academy of Sciences* 74, no. 3 (1959): 621–24.

34. "Pills for Diabetics," *Time*, October 29, 1956.

35. The average size for an NDA in 1956 was a single volume of less than a thousand pages; "Good News for Diabetics: ORINASE* the Long-Sought Oral Antidiabetic," *Upjohn News*, January 1957, 160–65, Upjohn Collections.

36. "Orinase: Its Development," 245–49.

37. "Clinic Test Given to Diabetes Drug," *New York Times*, February 16, 1957, 14.

38. "Orinase Symposium," *Upjohn Overflow*, January 1957, Upjohn Collections.

39. "Clinic Test Given to Diabetes Drug," 14. Also see Alexander Marble and Rafael Camerini-Davalos, "Clinical Experience with Sulfonylurea Compounds in Diabetes," *Annals of the New York Academy of Sciences* 71, no. 1 (1957): 239–48; Henry Dolger, "Ex-

perience with the Tolbutamide Treatment of Five Hundred Cases of Diabetes on an Ambulatory Basis," *Annals of the New York Academy of Sciences* 71, no. 1 (1957): 275–79.

40. Donald S. Gilmore, "How Are We Doing?" *Upjohn Overflow*, March 1957, 144–55, quotation on 148, Upjohn Collections.

41. The first mentions of Orinase in Upjohn's in-house publications were dampening efforts during Orinase's clinical evaluation, which reported findings largely in the negative: e.g., "Probable release date is anybody's guess. In spite of the recent rash of publicity on product, target date is still next year." *Upjohn Overflow*, June 1956, Upjohn Collections.

42. "Orinase Conference," *Upjohn Overflow*, April 1957, Upjohn Collections.

43. Levine, "Editorial Statement, 725, 726.

44. Levine continued, "Whatever clinical advantages these drugs may afford they do not constitute a more effective treatment than that already available and cannot be considered a substitute for insulin. For these reasons and, when it is a matter of convenience that is at stake, it would appear logical that elements of doubt should be removed from long term sulfonylurea therapy, before trials on the entire diabetic population are embarked upon." Ibid., 726.

45. Pete van Haften, "Sales Education: Background: Diabetes," *Upjohn Overflow*, October 1956, 393; van Haften, "Science Information: Part II: Diabetes Background," *Upjohn Overflow*, November 1956, 439–71; van Haften, "Science Information: Part III: Diabetes Background," *Upjohn Overflow*, December 1956, 490–521; van Haften, "Science Information: Part IV: Diabetes Background," *Upjohn Overflow*, January 1957, 45–47, all in the Upjohn Collections.

46. "Hope for Diabetics," *Time*, February 27, 1956, 98.

47. "Good News for Diabetics," 161.

48. "Coming—Diabetic Pills," *Time*, April 23, 1956, 88–89; "Topics of the Times," *New York Times*, November 19, 1957, 32; "Pills for Diabetics," *Time*, February 10, 1958, 76; "Pills for Diabetes," *Time*, April 20, 1959, 72.

49. "Caution," *Newsweek*, June 10, 1957, 66.

50. "Oral Hypoglycemic Agents," *Diabetes* 7, no. 1 (1958): 53–58, quotations on 54, 57.

51. R. M. Royle, "The Orinase Profile," *Upjohn Overflow*, January 1959, 4–6.

52. Ibid., 6.

53. Dolger continues: "There is no doubt that for handicapped diabetic patients, those visually impaired, those whose problems of intelligence preclude taking insulin properly, in the socio-economic situation that exists in a fairly significant part of our population—if such diabetics can be taken off insulin and put on *Orinase* safely a great boon would be achieved." "Oral Hypoglycemic Agents," 58.

54. Upjohn responded that the simplicity of administration of Pfizer's drug was compromised by increased rates of side effects; see "Pfizer Uses Big Ad to Introduce New Antidiabetic," *Advertising Age* 29 (1958): 3.

55. Ibid.

56. Lois J. Rolland, "Upjohn Goes Public," *Financial World*, February 4, 1959, 22.

57. Ibid.

58. Royle, "Orinase Profile," 4.

59. Initial screening efforts seem to have been largely limited to children; see Joslin, *Diabetes*, 33.

60. "Diabetes Drive," *Newsweek,* April 3, 1950, 46.

61. Russell M. Wilder, "Presidential Address," ADA Meetings, June 1947, as cited in American Diabetes Association, *The Journey and the Dream: A History of the American Diabetes Association* (American Diabetes Association, 1990), 39–40.

62. "The Hunt for Hidden Diabetics," *Saturday Evening Post,* December 6, 1958, 20.

63. "Diabetes Drive," 46; "Diabetes Detection," *Today's Health,* November 1950, 22.

64. Hugh L. C. Wilkerson and Leo P. Krall, "Diabetes in a New England Town," *American Journal of Public Health* 135, no. 4 (1947): 209–16.

65. "Hunt for Hidden Diabetics," 79.

66. Ibid., 75.

67. Paul de Kruif, "A Million Hidden Diabetics," *Today's Health,* November 1958, 22.

68. The text of the advertisement read: "No more needles for many U. S. sufferers? Yes, tests among 250,000 people have already proved that swallowing a pill—Orinase—can eliminate insulin shots for certain types of diabetics. In the April Readers' Digest, Paul de Kruif reports on why this means a happier, healthier life for many diabetics." "A New Day for Diabetics," advertisement, *New York Times* March 28, 1958, 14.

69. H. E. LeBrecht, "Products: Diabetes Mellitus—1960," *Upjohn Overflow,* October 1960, 334–37, quotation on 337, Upjohn Collections.

70. L. C. Hoff, "The Diabetic Deserves Early Diagnosis," *Upjohn Overflow,* October 1961, 382–85, Upjohn Collections. Note that by 1961 the estimated number of hidden diabetics had increased from 1 million to 1.5 million.

71. "A Diabetes Detection Film to Help Find the Million and a Half Unknown Cases," *Upjohn Overflow,* October 1961, 382–83, Upjohn Collections.

72. American Diabetes Association, *Journey and the Dream,* 113. Also see *Upjohn Overflow,* October 1960, Upjohn Collections.

73. Clinistix advertisement, *Today's Health,* May 1963.

74. Dextrostix advertisement, *Diabetes* 31 (1964): 5. Unlike Clinistix, Dextrostix was marketed only to physicians. The practice of home blood sugar assessments was not endorsed by the ADA until 1981. ADA, *Journey and the Dream,* 106.

75. "Detecting Diabetes Early," *Time,* June 26, 1964, 54.

76. Orinase promotional pamphlet, c38(a)I Upjohn, Kremers Files, American Institute of the History of Pharmacy, Madison, WI (hereafter cited as the Kremers files).

77. Upjohn Company, *Diabetes: Research, Detection, Therapy,* collected volumes located in the National Library of Medicine, Bethesda, MD.

78. Upjohn Company, *Finding the Hidden Diabetic,* (Kalamazoo, MI: Upjohn Co., 1965), filmstrip, copy in the National Library of Medicine, Bethesda, MD.

79. As Alexander Marble explained in the Upjohn film, "In screening procedures one hopes to identify those individuals in which diabetes is probable, so as to follow up with more exact testing procedures." Ibid.

80. ADA, *Journey and the Dream,* 102–3.

81. "New Look at Diabetes," *Time,* June 25, 1965, 79.

82. Statement of Gerald Kent, in Upjohn, *Finding the Hidden Diabetic.*

83. Before 1940 diabetics do not appear to have been eligible for life insurance; by the 1950s group life insurance was increasingly available to diabetics at dramatically higher premiums. See R. D. Montgomery, "Insurance and Diabetes," *ADA Forecast,* 1949; "The Truth about Life Insurance for Diabetics," *ADA Forecast,* 1957; Groff Conklin, "Facts about Health Insurance for Diabetics," *ADA Forecast,* 1957.

84. ADA, *Journey and the Dream*, 44–47. Also see Jon J. Pouts, "Industry and Diabetes," *ADA Forecast*, 1949; ADA Committee on Employment, "Analysis of a Survey concerning Employment of Diabetics in Some Major Industries," *Diabetes* 6 (1957): 6.

85. E. Tolstoi, "Treatment of Diabetes with the 'Free Diet' during the Last Ten Years," in *Progress in Clinical Endocrinology*, ed. Samuel Soskin (New York: Grune and Stratton, 1950), 292–302; Edward Tolstoi, *The Practical Management of Diabetes* (Springfield, IL: Charles C. Thomas, 1953).

86. L. H. Newburgh and J. W. Conn, "New Interpretation of Hyperglycemia in Obese Middle-Aged Patients," *Journal of the American Medical Association* 112 (1939): 7–11, quotation on 11.

87. Quotations from Mosenthal, Boyd, Jackson, and Allen, as cited in ADA, *Journey and the Dream*, 16.

88. LeBrecht, "Products," 336.

89. Ibid.

90. Upjohn, *Finding the Hidden Diabetic*.

91. Ibid.

92. Joslin's "susceptible" populations in 1931 were "according to the statistics . . . middle-aged women, fat women, fat Jewish women, and particularly those in whose families diabetes has existed previously, because diabetes is to some degree hereditary." Joslin, *Diabetes*, 47.

93. Charles Best, "The Prevention of Diabetes," keynote address, First Annual Meeting of the American Diabetes Association, Cleveland, OH, June 1, 1941, as cited in ADA, *Journey and the Dream*, 11–12.

94. Newburgh and Conn, "New Interpretation of Hyperglycemia," 11.

95. E. Perry McCullagh, William N. Fawell, and Fenton J. Lane, "Significance of Hyperglycemia without Glycosuria: A Ten to Twenty-Eight Year Study," *Journal of the American Medical Association* 156 (1954): 925–29; "The Natural History and Identification of Diabetes," *Diabetes* 1, no. 5 (1955): 381–88, quotation on 381.

96. E. Allen, "The Glycosurias of Pregnancy," *American Journal of Obstetrics and Gynecology* 38 (1939): 98; H. C. Miller, "The Effect of the Prediabetic State on the Survival of the Fetus and the Birth Weight of the Newborn Infant," *New England Journal of Medicine* 233 (1945): 376; J. M. Moss and H. B. Mulholland, "Diabetes and Pregnancy with Special Reference to the Prediabetic State," *Annals of Internal Medicine* 34 (1951): 678; W. P. U. Jackson, "Studies in Prediabetes," *British Medical Journal* 2 (1952): 690; E. R. Carrington, H. S. Reardon, and C. R. Schuman, "Recognition and Management of Problems Associated with Prediabetes during Pregnancy," *Journal of the American Medical Association* 166 (1958): 245.

97. W. P. U. Jackson, "A Concept of Diabetes," *Lancet*, September 24, 1955, 625–31, quotation on.631.

98. R. G. Sprague, H. L. Mason, and M. H. Power, *Proceedings of the American Diabetes Association* 8 (1949): 213.

99. J. W. Conn, "Banting Memorial Lecture, 18th Annual Meeting, American Diabetes Association: The Prediabetic State in Man," *Diabetes* 7 (1958): 347–57, quotation on 355.

100. John A. Osmundsen, "Early Diabetes Detected in Test," *New York Times*, April 11, 1959, 19.

101. LeBrecht, "Products," 335, 337.

102. Steven S. Fajans and Jerome W. Conn, "The Use of Tolbutamide in the Treatment of Young People with Mild Diabetes Mellitus: A Progress Report," *Diabetes* 11, supp. (1962): 123–26, quotation on 126. Fajans and Conn's enthusiasm was shared by many other diabetes researchers of the time, e.g., W. P. U. Jackson, "Present Status of Prediabetes," *Diabetes* 9, no. 5 (1960): 373–78; J. A. Tulloch and R. A. Lambert, "Latent Diabetes," *Diabetes* 10, no. 3 (1961): 207–10.

103. L. C. Hoff, "Orinase: Tolbutamide after Five Years," *Upjohn Overflow*, April 1962, 126–29, quotation on 129, Upjohn Collections.

104. Bernard A. Seeman, "What You Should Know about Diabetes," *Today's Health*, January 1960, 51, quotation on 66. By 1961 several critics had observed that the cortisone-GTT was far from an optimal test for identifying prediabetic populations: many individuals who tested positive never developed the disease (false positives), and several individuals who received negative test results later developed diabetes (false negatives). But several publications at the time suggested the possible advantages of the tolbutamide test. W. P. U. Jackson, "The Glucose-Cortisone Tolerance Test with Special Reference to the Prediction of Diabetes: Diagnosis of Prediabetes," *Diabetes* 10, no. 1 (1961): 33–40. "Tolbutamide Tolerance Test in Carbohydrate Metabolism Evaluation," *Archives of Internal Medicine* 107 (1961): 212.

105. Orinase Diagnostic was brought to market in 1961; an Upjohn in-house document described the product as "an accurate test for the diagnosis of mild diabetes. It is a sensitive test, that is, it provides a minimum of false negative answers. In addition, the *Orinase Diagnostic* test provides an excellent indication of the physiologic ability of the beta cells to elaborate insulin. This is not a test for oral *Orinase* response; only in extensive clinical trial may individual responsiveness to oral *Orinase* therapy be determined." L. C. Hoff, "Products: The Diabetic Deserves an Early Diagnosis," *Upjohn Overflow*, October 1961, 382–85, quotation on 385.

106. These conditions included (1) situations in which the patient received inadequate carbohydrate intake for the three days preceding the GTT, (2) conditions that affect glucose absorption, (3) old age and/or physical inactivity, (4) liver disease, and (5) adrenal steroid activity. In L. C. Hoff, "*Orinase Diagnostic:* The False Positive Diagnosis of Diabetes," *Upjohn Overflow*, March 1962, 82–83.

107. Hoff, "Products," 385.

108. "Oral Tolbutamide Test Aids Diagnosis," *Diabetes: Research, Detection, Therapy* 3, no. 3 (1966): 1.

109. Rafael A. Camerini-Davalos, James B. Caulfield, Searle B. Rees, Oscar Lozano-Castaneda, Santiago Naldjian, and Alexander Marble, "Preliminary Observations on Subjects with Prediabetes," *Diabetes* 12, no. 6 (1960): 508–18.

110. James M. Moss, DeWitt DeLawter, C. Robert Meloni, and Edward Gallagher, *Diabetes Mellitus: Under Good Diagnostic Protocol,* pamphlet reproduced in FDA docket 75N-0062, n.d., vol. 8, 176, Office of the Hearing Clerk, FDA, Rockville, MD.

111. By 1964 Camerini-Davalos and Alexander Marble had presented long-term evidence that asymptomatic, "chemical" diabetes was an ideal site of intervention, with capacity for reversing the course of the disease. A. Marble, J. Taton, A. Cervantes-Amezcua, and R. A. Camerini-Davalos, "Early Changes in 'Chemical' (Asymptomatic) Diabetes Including the Long-term Effect of Tolbutamide and Phenformin," Fifth International Congress of the International Diabetes Federation, Toronto, July, 1964. As a subsequent Orinase pamphlet noted, "although the effects of this therapy on the delay

of development of the complications of diabetes is still uncertain, as with other forms of antidiabetes therapy, it can reasonably be expected that improvement in the glucose tolerance or its restoration to normal may have a salutary effect on the long-term course of the disease." Upjohn Company, *Orinase (Tolbutamide): Time Confirms Decisions Founded on Reality,* promotional pamphlet, May 1965, Kalamazoo, MI, 6–11, C38(a)I Upjohn Co., Kremers files.

112. Upjohn Company, *Orinase (Tolbutamide),* 12–15; also see Upjohn Company, *Orinase (Tolbutamide): The Importance of Time in the Treatment of Diabetes,"* promotional pamphlet, March 1965, Kalamazoo, MI, 6–11, C38(a)I Upjohn Co., Kremers files.

113. Upjohn Company, *Orinase (Tolbutamide): Control of the New Diabetic,* promotional pamphlet, March 1965, Kalamazoo, MI, 6–11, C38(a)I Upjohn Co., Kremers Files.

114. Conn, "Banting Memorial Lecture," 355. W. P. U. Jackson made a similar statement in 1955: "Most people would say 'prediabetes' here; but, since some quite definite clinical features may be manifest in this state, it is evident that the disorder which will eventually lead to hyperglycemia is already making itself known. Is not this also to be called 'diabetes'? It is not pre-anything; it is already there." Jackson, "Concept of Diabetes," 625.

115. "Who Is a Diabetic? . . . Where Does He Come From? . . . Are We Meeting His Needs?" *Diabetes: Research, Detection, Therapy* 5, no. 1 (1968): 1.

116. *National Prescription Audit: Therapeutic Category Report, Ten-Year Trends* (Dedham, MA: R. A. Gosselin, 1968), unfiled item in the Kremers files.

117. "Prescriptions: Orinase/600," *Upjohn Overflow,* August 1960, 264, Upjohn Collections.

118. Although "prehypertension" has subsequently become an important category in early-twenty-first-century preventive treatment of hypertension, as is discussed in the concluding chapter.

119. See, e.g., S. M. Haffner, "Insulin Resistance, Inflammation, and the Prediabetic State," *American Journal of Cardiology* 92, no. 4A (2003): 18J–26J.

120. In a surprising reversal, the lifestyle-modification arm of the trial found a greater level of efficacy (58% reduction) than did pharmaceutical intervention; nonetheless, for the purposes of this discussion, the demonstrated benefit of Glucophage has become crucial for the continued product-life of the pharmaceutical agent and the treatment of insulin resistance. Diabetes Prevention Program Research Group, "Reduction in the Incidence of Type 2 Diabetes with Lifestyle Intervention or Metformin," *New England Journal of Medicine* 346 (2002): 393–403.

121. See, e.g., R. Giannarelli, M. Aragona, A. Coppelli, and S. Del Prato, "Reducing Insulin Resistance with Metformin: The Evidence Today," *Diabetes and Metabolism* 29, no. 4 (2003): 6S28–35; G. Slama, "The Potential of Metformin for Diabetes Prevention," *Diabetes and Metabolism* 29, no. 4 (2003): 6S104–11; E. Standl, "Metformin: Drug of Choice for the Prevention of Type 2 Diabetes and Cardiovascular Complications in High-Risk Subjects," *Diabetes and Metabolism* 29, no. 4 (2003): 6S121–22; H. Mehnert, "Metformin, the Rebirth of a Biguanide: Mechanism of Action and Place in the Prevention and Treatment of Insulin Resistance," *Experimental and Clinical Endocrinology and Diabetes* 109, no. S2 (2001): S259–64.

122. Diabetes Prevention Program Research Group, "Within-Trial Cost-Effectiveness of Lifestyle Intervention or Metformin for the Primary Prevention of Type 2 Diabetes," *Diabetes Care* 26, no. 9 (2003): 2518–23; Diabetes Prevention Program Research Group,

"Costs Associated with the Primary Prevention of Type 2 Diabetes Mellitus in the Diabetes Prevention Program," *Diabetes Care* 26, no. 1 (2003): 36–47.

CHAPTER 4: Risk and the Symptom

Epigraph: Testimony of Robert F. Bradley, transcript of proceedings, FDA Panel on Hypoglycemic Drug Labeling, August 20, 1975, 82, FDA docket 75N-0062, vol. 11, Office of the Hearing Clerk, FDA, Rockville, MD.

1. As reprinted in Harvey C. Knowles Jr., "An Historical View of the Medical-Social Aspects of the UGDP," *Transactions of the American Clinical and Climatological Association* 88 (1977): 150–56, quotation on 152.

2. "FDA Statement, Friday, May 22, 1970," *Diabetes* 19, supp. 1 (1970): 467.

3. C.P. to FDA, May 25, 1970, AF12-868, vol. 40, AF Jackets Collection, FDA, Rockville, MD. Note that all patients' names in this chapter are referred to by initials; this is a by-product of the Freedom of Information Act request process required to gain access to the FDA AF Jackets collections. Physician and manufacturer correspondence is not protected by privacy law in the same manner, so physicians' names are considered public knowledge. Because the letters submitted to the FDA hearings clerk as part of the open docket on UGDP between 1975 and 1984 were public documents, names of these letter-writers have not been blotted out in the historical record. In this chapter, I consistently use initials to identify writers of patient or consumer letters, while retaining the full names of physicians, researchers, policymakers, and lobbyists. For purposes of continuity, I have chosen to reproduce the letters faithfully, including typographical errors, without repetitive use of *sic.*

4. Morton Mintz, "Antidiabetes Pill Held Causing Early Death," *Washington Post,* May 21, 1970. Although a significant number of diabetes specialists, outside of the small circle of Upjohn-FDA-NIH, had already learned of the subject through word of mouth, the results had not been formally made available to general practitioners before Mintz's article.

5. Gina Kolata, "Controversy over Study of Diabetes Drug Continues for Nearly a Decade," *Science* 203 (1979): 990. Full publication of the study did not occur until November of 1970.

6. David L. Roberts to Charles C. Edwards, November 6, 1970, FDA AF Jackets: AF12-868, vol. 42.

7. For details of the FBI investigation into Christian Klimt and the details of Supreme Court involvement, see Kolata, "Controversy over Study of Diabetes Drugs," 986, 989.

8. As recollected by Alvan Feinstein; see James Wright Presley, "A History of Diabetes Mellitus in the United States, 1880–1990" (PhD diss., University of Texas at Austin, 1991), 447.

9. The use of ethnographic techniques to explore physician and patient perspectives in the negotiation of disease has been well established among medical anthropologists. One key crossover text in the field is Arthur Kleinman, *Patients and Healers in the Context of Culture: An Exploration of the Borderland between Anthropology, Medicine, and Psychiatry* (Berkeley: University of California Press, 1980). In spite of repeated calls for a more patient-oriented historical approach, such source material is notori-

ously difficult to find. See Roy Porter, "The Patient's View: Doing Medical History from Below," *Theory and Society* 14 (1985): 175–98; Sheila Rothman, *Living in the Shadow of Death: Tuberculosis and the Social Experience of Illness in America* (New York: Basic Books, 1994).

10. Harry M. Marks presents an account of the UGDP as a case study of epistemological conflict in *The Progress of Experiment: Science and Therapeutic Reform in the United States, 1900–1990* (Cambridge: Cambridge University Press, 1997). Accounts of the UGDP controversy also appear in Presley, "History of Diabetes Mellitus," 424–55; Curtis L. Meinert and Susan Tonascia, *Clinical Trials: Design, Conduct, and Analysis* (New York: Oxford University Press, 1986), 52–62; 197–228; and Stan N. Finkelstein, Stephen B. Schechtman, Edward J. Sondik, and Dana Gilbert, "Clinical Trials and Established Medical Practice: Two Examples," in *Biomedical Innovation,* ed. E. B. Roberts, R. I. Levy, S. N. Finkelstein, J. Moskowitz, and E. J. Sondik (Cambridge: MIT Press, 1981).

11. Elliott Joslin, *The Treatment of Diabetes Mellitus,* 10th ed. (Philadelphia: Lea and Febiger, 1959); also see Chris Feudtner, "The Want of Control: Ideas, Innovations, and Ideals in the Modern Management of Diabetes Mellitus," *Bulletin of the History of Medicine* 69 (1995): 66–90.

12. E. Tolstoi, "Treatment of Diabetes with the 'Free Diet' during the Last Ten Years," in *Progress in Clinical Endocrinology,* ed. Samuel Soskin (New York: Grune and Stratton, 1950), 292–302; Edward Tolstoi, *The Practical Management of Diabetes* (Springfield, IL: Charles C. Thomas, 1953).

13. Philip K. Bondy, "Therapeutic Considerations in Diabetes Mellitus," in *Controversy in Internal Medicine,* ed. Franz J. Inglefinger and Arnold S. Relman (Philadelphia: Saunders, 1966), 499–503.

14. Knowles, "Medical-Social Aspects of the UGDP," 150.

15. Ibid.

16. The NIAMD considered the UGDP application to be an "almost unique example of a prospective study." Special Review Committee, "General Comments on Application 06876–01," May 15, 1960, on file at Division of Research Grants, National Institute of Arthritis, Diabetes, and Digestive and Kidney Diseases. As cited in Marks, *Progress of Experiment,* 205.

17. University Group Diabetes Program, "A Study of the Effects of Hypoglycemic Agents on Vascular Complications in Patients with Adult-Onset Diabetes. I. Design, Methods, and Baseline Results," *Diabetes* 19, no. 2 (1970): 747.

18. C. R. Klimt, C. L. Meinert, M. Miller, and H. C. Knowles, "Discussion," in *Tolbutamide . . . after Ten Years,* ed. E. J. H. Butterfield and W. Van Westering (Amsterdam: Excerpta Medica Foundation, 1967), 269.

19. University Group Diabetes Program, "Effects of Hypoglycemic Agents," 750, 752.

20. M. L. Balodimos, A. Marble, J. H. Rippey, M. A. Legg, T. Kuwabara, and R. F. Gleason, "Pathological Findings after Long-Term Sulfonylurea Therapy, *Diabetes* 17, no. 8 (1968): 503–8.

21. Christopher Klimt, interview by Harry M. Marks, May 16, 1984, cited in Marks, *Progress of Experiment,* 206.

22. Theodore B. Schwartz, "The Tolbutamide Controversy: A Personal Perspective," *Annals of Internal Medicine* 75 (1971): 305–6. Also see Klimt interview, cited in Marks, *Progress of Experiment,* 206.

23. One investigator requested a reanalysis of autopsy data. See minutes of principal investigators' meeting, UGDP, June 5–6, 1969, cited in Marks, *Progress of Experiment*, 209.

24. "Report of the Committee for the Assessment of Biometric Aspects of Controlled Trials of Hypoglycemic Agents," *Journal of the American Medical Association* 321 (1975): 588.

25. Marks provides a particularly detailed account of this epistemological divide in *Progress of Experiment*, 197–228.

26. The classic sociological analysis of medical paternalism and the contractual dependency written into the early-twentieth-century doctor-patient relationship is found in Talcott Parsons, *The Social System* (Glencoe, IL: Free Press, 1951). For an analysis of the rising critique in the 1970s of the autonomous, paternalistic model of medical authority, see David Rothman, *Strangers at the Bedside: A Story of How Law and Bioethics Transformed Medical Decision-Making* (New York: Basic Books, 1991).

27. Barbara Mintzes and Catherine Hodgkin, "The Consumer Movement: From Single Issue Campaigns to Long-Term Reform," in *Contested Ground: Public Purpose and Private Interest in the Regulation of Prescription Drugs*, ed. Peter Davis (New York: Oxford University Press, 1996), 76–91.

28. Information politics was crucial to the patient-autonomy and egalitarian reforms of doctor-patient relations in the 1970s. See Jay Katz, *The Silent World of Doctor and Patient* (New York: Free Press, 1984).

29. Mintz, "Antidiabetes Pill Held Causing Early Death."

30. Harry Dowling, "The Practicing Physician and the Food and Drug Administration," in *The Impact of the Food and Drug Administration on Our Society*, ed. Henry Welch and Felix Marti-Ibanez (New York: MD Publications, 1956), 28–29.

31. Norman S. Knee to Charles E. Edwards, June 1, 1970, AF12-868, vol. 40, FDA.

32. On earlier political efforts of the organized medical profession to limit the authority of the FDA in the 1940s and 1950s, see Harry M. Marks, "Revisiting 'The Origins of Compulsory Drug Prescriptions,'" *American Journal of Public Health* 85, no. 1 (1995): 109–15.

33. Francis T. Collins to Robert H. Finch, June 4, 1970, AF12-868, vol. 40, FDA.

34. Morgan U. Stockwell to Elliot Richardson, June 9, 1970, ibid.

35. M. J. Ryan to Francis T. Collins, June 30, 1970, ibid.

36. Jesse L. Steinfeld to Morgan U. Stockwell, July 6, 1970, ibid.

37. Harold M. Schmeck, "Doubts about Oral Diabetes Drugs," *New York Times*, June 7, 1970, 157.

38. C.P. to FDA, May 25, 1970, AF12-868, vol. 40, FDA.

39. J.S.M. to FDA, May 24, 1970, ibid.

40. M.M. to Henry E. Simmons, June 1, 1970, ibid.

41. V.D. to FDA, June 6, 1970, ibid.

42. J.F. to Senator Williams, May 28, 1970, ibid.

43. Ibid.

44. R.R.J. to Senator William Proxmire, June 1, 1970, ibid.

45. Marvin Seife to J.S.M., June 5, 1970; Marvin Seife to C.C.B., June 8, 1970, both ibid.

46. The role of industry in the strategic production and maintenance of scientific

controversy is best illustrated in the tobacco industry's decades-long fight to delay consensus on the health hazards of the cigarette; that effort is amply documented in the internal documents released after the recent litigation and in Allan M. Brandt's *The Cigarette Century* (New York: Basic Books, 2006). For a more general account of the use of industrial public relations to shape public debate in the postwar era, see Elizabeth Fones-Wolf, *Selling Free Enterprise: The Business Assault on Labor and Liberalism* (Urbana: University of Illinois Press, 1994).

47. Harold M. Schmeck, "Pills for the Diabetic: Dilemma for Doctors," *New York Times,* June 21, 1970, 1.

48. Rafael Camerini-Davalos, Henry Dolger, and Arthur Krosnick, "Letters to the Editor," *New York Times,* November 13, 1970.

49. Orinase advertisement, *Journal of the American Medical Association* (1971).

50. It is ironic, given Elliot Joslin's own misgivings about oral hypoglycemic agents, that the Joslin Clinic's endorsement of physiological control of the asymptomatic diabetic had made it a strong center for promoting the broad use of oral hypoglycemic agents.

51. "In Boston: A Diabetes Tea Party Hits FDA," *Medical World News,* December 18, 1970, 13–14.

52. It should be noted that another group interested in the continued publicity of the UGDP controversy was the nascent medical news industry, with periodical titles such as *MD, Medical Tribune,* and *Medical World News,* who worked to directly translate medical controversy into sales and advertising revenues.

53. For a brief history of DESI, see the FDA History Office Web site, www.fda.gov/cder/about/history, 37. The revision of tolbutamide's label following its DESI evaluation was being finalized in late 1969 when the UGDP results were first announced to the FDA. John Jennings, Upjohn Company, to James O. Lawrence, n.d., received December 22, 1969, AF12-868, vol. 39, FDA.

54. The recall of Panalba sparked a controversy in its own right that proved to be the defining case of the DESI project. See Upjohn Company, "Drug Recall," company circular, March 19, 1970; N.H. to Richard M. Nixon, April 1, 1970; Perry C. Martineau to Nixon, March 21, 1970; Charles E. Krause to Nixon, March 24, 1970; H.E.W. to Charles E. Krause, April 29, 1970; all documents in AF12-868, vol. 40, FDA.

55. To some extent, the FDA had necessarily considered a drug's efficacy and therapeutic indication as part of its evaluation since its inception in 1938, to evaluate the drug's relative safety and the merits of its therapeutic claims. The Kefauver-Harris Act of 1962, however, set in motion a series of formal regulatory processes that increased the political importance of package inserts and the public reception of clinical trials. See Marks, *Progress of Experiment,* 71–97.

56. Ibid., 218.

57. Jennings to Lawrence.

58. Orinase package insert, 1968, as reprinted in Butterfield and Van Westering, *Tolbutamide . . . after Ten Years.*

59. See Louis Lasagna, "1938–1968: The FDA, the Drug Industry, the Medical Profession, and the Public," in *Safeguarding the Public: Historical Aspects of Medicinal Drug Control,* ed. John B. Blake (Baltimore: Johns Hopkins University Press, 1970), 171–79.

60. Harry F. Dowling, *Medicines for Man: The Development, Regulation, and Use of*

Prescription Drugs (New York: Knopf, 1970), 243–44. Dowling was also involved in the Kefauver investigation and was on the FDA advisory committee charged with the initial review of the tolbutamide findings; see FDA Bureau of Medicine, Medical Advisory Board, "Minutes: June 26–27, 1969," box 6, MS C 372, Harry F. Dowling Papers, National Library of Medicine, Bethesda, MD.

61. "FDA Statement, Friday May 22, 1970: Status of Problem of Usage of Tolbutamide: Preliminary Statements," *Diabetes* 19, no. 6 (1970): 467.

62. Harold M. Schmeck, "Diabetes Drug Use Backed by Council," *New York Times,* June 15, 1970, 42; Harold M. Schmeck, "F.D.A. Cites Doubt on Diabetes Pill: Says Two Medical Groups Share Its Misgivings about Popular Drug," *New York Times,* October 22, 1970, 1.

63. Charles C. Edwards, "Oral Hypoglycemic Agents: Report of the Food and Drug Administration, October 30, 1970," *Diabetes* 19, no. 2 (1970): viii–ix.

64. HEW news press release, November 2, 1970; "Oral Hypoglycemic Agents," *FDA Current Drug Information,* October, 1970, 1, both in AF12-868, vol. 42, FDA; AMA Council on Drugs, "Statement Regarding the University Group Diabetes Program (UGDP) Study, November 2, 1970," *Diabetes* 19, no. 2 (1970): vi–vii.

65. "In Boston," 14.

66. These clinicians' allegations that the FDA was overstepping its regulatory bounds were paired with allegations that the FDA should "think and act as physicians rather than administrators" in their regulation of other physicians. Robert F. Bradley to Henry Simmons, December 11, 1970, FDA docket 75N-0062, vol. 2.

67. Coordinating Committee of the Committee on the Care of the Diabetic (Bradley, Dolger, Forsham, Seltzer, and O'Sullivan) to Henry Simmons, March 11, 1971; also see Bradley to Simmons, July 23, 1971; both ibid.

68. Bradley to Simmons, July 23, 1971, ibid.

69. The CCD maintained that "controverting data" supported the usage of oral hypoglycemics, and it insisted that labeling must reflect a "fair balance" of scientific opinion. For a rough chronology of the court battle over labeling, see Jeremy A. Greene, "The Therapeutic Transition: Pharmaceuticals and the Marketing of Chronic Disease" (PhD diss., Harvard University, 2005), 196n82. Ironically, the two sides had nearly reached a tentative agreement when the premature publication of the Biometric Society report in early 1975 re-ignited public controversy, which led to the public set of hearings. "Notice of Public Hearing, July 7, 1975," 10, 12, FDA docket 75N-0062, vol. 1.

70. Marks, *Progress of Experiment,* 222.

71. James M. Moss, "The UGDP Scandal and Cover-up," *Journal of the American Medical Association* 232 (1975): 806–8, quotation on 806; also see John B. O'Sullivan and Ralph B. D'Agostino, "Decisive Factors in the Tolbutamide Controversy," *Journal of the American Medical Association* 232, no. 8 (1975): 825–29.

72. The Committee on the Care of the Diabetic, "Settling the UGDP Controversy?" *Journal of the American Medical Association* 232 (1975): 813–17.

73. "Notice of Public Hearing, July 7, 1975," FDA docket 75N-0062, vol. 1. This announcement opened docket 75N-0062—a public file that was not closed until 1984.

74. "Comments of the American Medical Association concerning Oral Hypoglycemic Drugs, Proposed Labeling Requirements, Published July 7, 1975," September 4, 1975, ibid., quotation on 2. The AMA further objected to extension of the study's in-

terpretation to the entire class of oral hypoglycemics, and it objected to the requirement that physicians obtain informed consent before each prescription. Parallel debates were occurring at the same time regarding package insert labeling for oral contraceptives; see Elizabeth S. Watkins, "'Doctor, are you trying to kill me?': Ambivalence about the Patient Package Insert for Estrogen," *Bulletin of the History of Medicine* 76 (2002): 84–104.

75. James E. Davis to FDA hearing clerk, August 26, 1975, FDA docket 75N-0062, vol. 1.

76. John J. McGuire to FDA hearing clerk, August 22, 1975; Nym L. Barker to FDA hearing clerk, August 22, 1975; Karl F. Moch to FDA hearing clerk, August 22, 1975, all ibid.

77. Out of the hundreds of letters the FDA received from physicians in 1975 regarding the tolbutamide controversy, only nine wrote in support of the FDA's decision. Five of the nine were identical form letters written by colleagues of one UGDP researcher at the Mayo Clinic.

78. Hans G. Engel to FDA hearing clerk, August 2, 1975, FDA docket 75N-0062, vol. 1.

79. Thomas Chalmers, "Settling the UGDP Controversy," *Journal of the American Medical Association* 213 (1975): 624.

80. David H. Walworth to FDA, August 23, 1975, FDA docket 75N-0062, vol. 1.

81. For a concise account of the expanding role of malpractice in the clinical environment of 1970s America, see Neal C. Hogan, *Unhealed Wounds: Medical Malpractice in the Twentieth Century* (New York: LFB Scholarly Publishing, 2003). The passage of the Freedom of Information Act in 1974 and the proximity of Watergate and Vietnam also clearly color the writings of both doctor and patient in the letters in the FDA files.

82. The Massachusetts Supreme Court later overturned this ruling with the proviso that "the package insert is no more and no less important than any other medical reference available to the physician." N. L. Chayet, "Power of the Package Insert," *New England Journal of Medicine* 277 (1967): 1253–54.

83. Hans G. Engel to FDA hearing clerk, August 2, 1975, FDA docket 75N-0062, vol. 1.

84. James L. Bland to HEW, September 2, 1975, ibid.

85. Clifford Emond Jr. to Division of Public Information, Office of the Commissioner, FDA, June 30, 1970, AF12-868, vol. 40, FDA.

86. "Other than the approved labeling information, information contained in the new drug application is regarded as confidential and may not be released. It is the approved labeling which sets forth the indications for use as well as appropriate cautionary and warning information made available to the prescribing physician." Carl Loustanis to Clifford Emond Jr., July 26, 1970, ibid.

87. Morse Kochtitzky to FDA, September 3, 1975, ibid.

88. L. M. Ungaro to Alexander Schmidt, September 16, 1975, ibid.

89. Testimony of James Moss, transcript of proceedings, FDA Panel, 153.

90. James L. Bland to HEW, September 2, 1975, FDA docket 75N-0062, vol. 1.

91. Testimony of T. S. Danowski, 95; testimony of Harry Keen, 139, both in transcript of proceedings, FDA Panel. Keen himself had conducted long-term studies that indicated long-term benefit of oral hypoglycemics in the prevention of diabetic complica-

tions and vascular disease, e.g. H. Keen, R. J. Jarrett, C. Chlouverakis, and D. R. Boyns, "The Effect of Treatment of Moderate Hyperglycaemia on the Incidence of Arterial Disease," *Postgraduate Medical Journal,* supp. (1968): 960–65.

92. Testimony of T.S. Danowski, transcript of proceedings, FDA Panel, 95–96.

93. Paul H. Lavietes, "Presentation at FDA Hearing on Oral Hypoglycemic Agents," August 20, 1975, FDA docket 75N-0062, vol. 11.

94. See Marks, *Progress of Experiment;* Presley, "History of Diabetes Mellitus"; Meinert and Tonascia, *Clinical Trials;* Finkelstein et al., "Clinical Trials and Established Medical Practice."

95. Testimony of Thaddeus Prout, transcript of proceedings, FDA Panel, 83–84.

96. James L. Bland to HEW, September 2, 1975, FDA docket 75N-0062, vol. 1.

97. As a practicing physician from New Jersey noted, "A doctor who was once a contributor to the UGDP study resigned from the study because one of the principals in the study was being paid off by the US Vitamin Corporation who was anxious to discredit other oral hypoglycemics and in particular the Upjohn Company. I feel that on the basis of this alone the UGDP study is not viable and should be discarded." Carl K. Friedland to FDA, October 2, 1975, ibid., vol. 8. Klimt later admitted to receiving funds from USV during the study but insisted it had not influenced his work. Ironically, the UGDP study later indicted USV's drug DBI as harmful, and the drug was subsequently pulled from the market.

98. Eugene T. Davidson to FDA, September 8, 1975, ibid.

99. B. Todd Forsyth to HEW, September 4, 1975, ibid., vol. 1.

100. J. R. Kircher to FDA hearing clerk, July 28, 1975, ibid.

101. Sudah Deuskar to FDA, July 20, 1975, ibid.

102. Ralph Nader, *Unsafe at Any Speed: The Designed-in Dangers of the American Automobile* (New York: Grossman, 1965); Mintzes and Hodgkin, "Consumer Movement."

103. Sidney M. Wolfe and Anita Johnson to Alexander M. Schmidt, June 10, 1975, 2, FDA docket 75N-0062, vol. 8.

104. Testimony of Anita Johnson, transcript of proceedings, FDA Panel, 25.

105. E.g., J.W. to FDA, August 7, 1975; G.U. to FDA, August 18, 1975; L. Latour to FDA, July 28, 1975, all in FDA docket 75N-0062, vol. 1.

106. P.G.W. to FDA, July 24, 1975, ibid.

107. A.M. to Alexander M. Schmidt, August 25, 1975, ibid.

108. Testimony of John Abeles, transcript of proceedings, FDA Panel, 121–22.

109. David H. Walworth to FDA, August 23, 1975, FDA docket 75N-0062, vol. 1.

110. T.H. to FDA, August 27, 1975, ibid.

111. A.M. to Alexander M. Schmidt, August 25, 1975, ibid.

112. H.E.J. to Edwin M. Ortiz, September 26 , 1975; B.T. to Edwin M. Ortiz, September 30, 1975; S.M.M. to Edwin M. Ortiz, November 14, 1975, all ibid., vol. 8.

113. In similar fashion, several clusters of letters with nearly identical wording were submitted by physicians, suggesting that either the letters were ghostwritten or organized letter-writing drives were behind them. The problems of pharmaceutically funded ghostwriting in prominent clinical publications has recently provoked an extended crisis of authenticity in the medical literature. See Michael Lynch, "Ghost Writing and Other Matters," 147–48; and David Healy, "Shaping the Intimate: Influences on the Everyday Experience of Nerves," 219–45, both in *Social Studies of Science* 34 (2004).

114. David Nathan, "Initial Management of Glycemia in Type 2 Diabetes Mellitus," *New England Journal of Medicine* 347 (2002): 1342–49.

115. Senate Subcommittee on Monopoly of the Select Committee on Small Business, *Competitive Problems in the Drug Industry: Oral Hypoglycemic Drugs, Continued* (testimony of Frank Davidoff), 94th Cong., 1st sess., January 31, July 9 and 10, 1975. This argument was also echoed in the presentation and testimony of Sidney M. Wolfe and Anita Johnson, transcript of proceedings, FDA Panel, 22.

116. "Thus," Davidson continues, "the physician writes a prescription for a medication that he has been led to believe is safe and effective, when it is almost certainly unsafe and is frequently ineffective." John K. Davidson, "The FDA and Hypoglycemic Drugs," *Journal of the American Medical Association* 232 (1975): 853–55; also see testimony of Thaddeus Prout, transcript of proceedings, FDA Panel, 83–84.

117. Testimony of Sidney M. Wolfe, transcript of proceedings, FDA Panel, 22.

118. Jeremiah Stamler, "The UGDP: One Year Later," Transcript from ADA Symposium, San Francisco, June 23, 1971, 191, FDA docket 75N-0062, vol. 3.

119. Kolata, "Controversy over Study of Diabetes Drug," 990.

120. Freis himself denied any such comparison, suggesting that, unlike the UGDP, the VA study answered the questions it was designed to answer and never sustained serious doubt as to the validity of its statistical methodology. Ibid., 990.

121. The Diabetes Control and Complications Trial, reported in 1993, is widely cited as a validation of the long-term benefit of diabetic management based on physiological parameters. However, as critics have pointed out, the DCCT was a trial of insulin in insulin-dependent diabetes, not of oral agents in adult-onset diabetes; since these constitute considerably different pathological and therapeutic processes, there has been extensive debate over how to generalize from one condition to the other. The DCCT did not resolve the UGDP controversy as much as it provided therapeutic rationalists with a more recent and less controversial reference point. Diabetes Control and Complications Trial Research Group, "The Effect of Intensive Treatment of Diabetes on the Development and Progression of Long-Term Complications in Insulin-Dependent Diabetes Mellitus," *New England Journal of Medicine* 329 (1993): 977–86.

CHAPTER 5: The Fall and Rise of a Risk Factor

Epigraph: This text appears (with apologies to Hamlet I:iii) in the preface of Louis Gilman and Alfred Gilman's *The Pharmacological Basis of Therapeutics,* 2nd ed. (New York: MacMillan, 1954).

1. The initial Framingham studies did not find that the increased risk of cardiovascular events associated with obesity and cigarette smoking were independently significant, though both were substantiated by later analyses. Thomas R. Dawber, Felix E. Moore, and George V. Mann, "Coronary Heart Disease in the Framingham Study," *American Journal of Public Health* 47 (1957): 3–24, quotation on 21.

2. See, e.g., Mariano J. Garcia, Patricia M. McNamara, Tavia Gordon, and William B. Kannell, "Morbidity and Mortality in Diabetics in the Framingham Population: Sixteen Year Follow-up Study," *Diabetes* 23, no. 2 (1974): 105–11.

3. Nancy Tomes, *The Gospel of Germs: Men, Women, and the Microbe in American Life* (Cambridge: Harvard University Press, 1997).

4. The idea of disease specificity predated the nineteenth-century advent of the germ theory. Edmund Pellegrino traces the concept back through the works of Paracelsus, Sydenham, Bichat, and Morgagni in "The Sociocultural Impact of 20th Century Therapeutics," in *The Therapeutic Revolution: Essays in the Social History of American Medicine*, ed. Morris Vogel and Charles Rosenberg (Philadelphia: University of Pennsylvania Press, 1979).

5. Greg Mitman, "When Pollen Became Poison: A Cultural Geography of Ragweed in America," in *The Moral Authority of Nature*, ed. Lorraine Daston and Fernando Vidal (Chicago: University of Chicago Press, 2003), 438–65; Carla Keirns, "Asthma Germs and Germicides: The Rise and Fall of Infectious Asthma, 1860–1925," paper presented at the May 2003 meetings of the American Association for the History of Medicine, Boston; Rima Apple, *Vitamania: Vitamins in American Culture* (New Brunswick, NJ: Rutgers University Press, 1996); Barbara Berney, "Round and Round It Goes: The Epidemiology of Childhood Lead Poisoning, 1950–1990," *Milbank Quarterly* 71 (1993): 3–39; David Rosner and Gerald Markowitz, *Deadly Dust: Silicosis and the Politics of Occupational Disease in Twentieth-Century America* (Princeton, NJ: Princeton University Press, 1991); Allan M. Kraut, *Goldberger's War: The Life and Work of a Public Health Crusader* (New York: Hill and Wang, 2003).

6. Essentially, to fulfill Koch's postulates for specific agent–specific disease causality, (1) the agent must be present in every case of the disease, (2) the agent must be isolated from the diseased host and grown in pure culture, (3) the specific disease must be reproduced when a pure culture of the agent is inoculated into a healthy susceptible host, and (4) the bacterium must be recoverable from the experimentally infected host.

7. Louis Katz, "Experimental Atherosclerosis," *Circulation*, January 1952, 9–32.

8. The study of the substance now known as cholesterol had preceded Anitschkow in the nineteenth-century physiological chemistry of Michel Eugène Chevreul and Marcellin Berthelot; its chemical formula was published in 1888, though the chemical structure of the compound was not elucidated until after Anitschkow's work. See J. D. Bernal, "Carbon Skeleton of the Sterols," *Chemistry and Industry* 51 (1932): 466; O. Rosenheim and H. King, "The Ring System of Sterols and Bile Acids," *Nature* 130 (1932): 315.

9. Willard Zinn and George Griffith, "Atherosclerosis: A Preventable Disease?" *Medical Clinics of North America* 36 (1952): 1001–12, quotation on 1003–4.

10. This initial definition of pathologically high cholesterol, or xanthomatosis, is explored in more detail in chapter 6. See David Davis, Beatrice Stern, and Gerson Lesnick, "The Lipid and Cholesterol Content of the Blood of Patients with Angina Pectoris and Atherosclerosis," *Annals of Internal Medicine* 8 (1935): 1436–74; C. Muller, "Angina Pectoris in Hereditary Xanthomatosis," *Archives of Internal Medicine* 64 (1939): 675–700; and E. P. Boas, A. D. Parets, and D. Aldersberg, "Hereditary Disturbance of Cholesterol Metabolism: A Factor in the Genesis of Atherosclerosis," *American Heart Journal* 35 (1948): 611–22.

11. Louis N. Katz and Jeremiah Stamler, *Experimental Atherosclerosis* (Springfield, IL: Charles C. Thomas, 1953), 9–10. Katz prefaced this quotation, "Until very recently, atherosclerosis was a step-child problem of medical research commanding very limited resources of personnel, equipment, plant and money [owing to] a fundamentally erroneous concept developed in the medical profession itself—the senescence 'theory' of the genesis of the arteriroscleroses. This 'theory' maintains that the arterioscleroses are

inevitable results of physiologic aging processes. This 'theory' likewise regards the specific entity atherosclerosis as such an unavoidable process of aging. The stagnating influence of this 'theory' upon medical research has been overwhelming. It engendered an atmosphere of helplessness and hopelessness that was for many years a serious brake on all investigations." A related discussion is found in Zinn and Griffith, "Atherosclerosis."

12. Robert E. Olson, "Atherosclerosis," in *Cecil-Loeb Textbook of Medicine*, 11th ed., ed. Paul B. Beeson and Walsh McDermott (Philadelphia: Saunders, 1963), 704–10, quotations on 705–6. A similar comparison between cholesterol and the tuberculosis bacillus was made in a 1950 article by William Dock: "We can produce the lesion, and modify its evolution, experimentally, just as we do with tuberculosis. We know its etiology, but, as in the case of most infections, we have only a vague idea as to its pathogenesis. Why one man develops meningitis and a dozen others remain carriers is as mysterious as why one man with hypercholesteremia is dead of a myocardial infarct at twenty, another hale, but spotted with xanthomata, at sixty." William Dock, "The Causes of Atherosclerosis," *Bulletin of the New York Academy of Medicine* 26 (1950): 184–90; quotation on 184.

13. Ancel Keys, Olaf Mickelsen, E. v. O. Miller, and Carleton B. Chapman, "The Relation in Man between Cholesterol Levels in the Diet and in the Blood," *Science* 112 (1950): 79–81; A. Keys, N. Kimura, A. Kusukawa, B. Bronte-Stewart, N. Larson, and M. H. Keys, "Lessons from Serum Cholesterol Studies in Japan, Hawaii, and Los Angeles," *Annals of Internal Medicine* 48 (1958): 83.

14. W. F. Enos Jr., J. C. Beyer, and R. H. Holmes, "Pathogenesis of Coronary Disease in American Soldiers Killed in Korea," *Journal of the American Medical Association* 158 (1955): 912.

15. David Kritchevsky, *Cholesterol* (New York: Wiley, 1958), 148.

16. Robert H. Fuhrman and Charles W. Robinson Jr. "Hypercholesterolemic Agents," *Medical Clinics of North America* 45, no. 4 (1961): 935–59, quotation on 935.

17. Popular literature on cholesterol and heart disease is visible in the 1950s, e.g., "Coronaries and Cholesterol," *Time*, December 1, 1952, 39; Blake Clark, "Is This the No. 1 Villain in Heart Disease?" *Reader's Digest*, December 1955, 130–36; S. M. Spenser, "Are You Eating Your Way to Arteriosclerosis?" *Saturday Evening Post*, December 1, 1956.

18. William E. Connor, "The Use of Drugs and Diets in the Control of Coronary Heart Disease," *Journal of the Iowa Medical Society* 56, no. 7 (1966): 657–60, quotation on 657.

19. For a review of early agents in cholesterol reduction, see David Kritchevsky, "The Use of Pharmacologic Agents in Atherosclerosis Therapy," *Annals of the New York Academy of Sciences* 149, no. 2 (1968): 1058–68, table on 1059.

20. See, e.g., Zinn and Griffith, "Atherosclerosis."

21. F. P. Palopoli, "Basic Research Leading to MER/29, Proceedings of the Conference on MER/29," *Progress in Cardiovascular Diseases* 2, supp. (1960): 489–91. W. Hollander and A. V. Chobanian, "Effects of Inhibitor of Cholesterol Biosynthesis, Triparanol (MER/29), in Subjects with and without Coronary Artery Disease," *Boston Medical Quarterly* 10 (1959): 1–8; W. Hollander, A. V. Chobanian, and Robert W. Wilkins, "The Effects of Triparanol (MER/29) in Subjects with and without Coronary Artery Disease," *Journal of the American Medical Association* 174 (1960): 87–88.

22. The aforementioned memorandum is cited from the work of investigational

journalist Sanford J. Ungar, "Get Away with What You Can," in *In the Name of Profit,* ed. Robert L. Helibroner, (Garden City, NY: Doubleday, 1971), 106–27, quotation on 109– 10. Ungar does not reveal the source of his documents but does reproduce the texts in full; it seems unlikely that he would go to the lengths of fabrication or text manipulation in such a case study.

23. Milton Moskowitz, "Bringing MER/29 to Market," *Drug and Cosmetic Industry* 89 (1961): 437.

24. Robert K. Plumb, "Cholesterol Cut in Tests of Drug: Preliminary Research with MER/29 on 86 Persons Reported to Biologists," *New York Times,* April 14, 1959, 47. Note that many of the physicians involved in this press release—John Moyer at Hahnemann in Philadelphia and Robert W. Wilkins at Boston University—were members of the same research groups that had reported early results on Diuril's use as an antihypertensive one year earlier (see chapter 1).

25. Moskowitz, "Bringing MER/29 to Market," 438.

26. As cited in Ungar, "Get Away with What You Can," 110.

27. Moskowitz, "Bringing MER/29 to Market," 551.

28. Ibid., 490–91.

29. Ibid., 437.

30. Early discussion of triparanol's side effects were generated when a competing team at Merck, testing samples of triparanol against its own unmarketed cholesterol-reducing compound, announced that test animals on triparanol developed cataracts and blindness. Fuhrman and Robinson, "Hypocholesterolemic Agents"; R. W. P. Achor, R. K. Winkelman, and M. O. Perry, "Cutaneous Side Effects from Use of Triparanol (MER29): Preliminary Data on Ichthyosis," *Proceedings of the Staff Meeting of the Mayo Clinic* 36 (1961): 217–18.

31. "Maker Withdraws Drug from Market," *New York Times,* April 17, 1962, 6.

32. For discussion on the role of thalidomide in the passage of the 1962 Kefauver-Harris amendments and the strengthening of FDA adverse-effects monitoring systems, see Arthur Daemmrich, *Pharmacopolitics: Drug Regulation in the United States and Germany* (Chapel Hill: University of North Carolina Press, 2004). The connection between thalidomide and MER/29 was apparent to contemporary commentators: "It is indeed unfortunate that it takes a "thalidomide tragedy" to vindicate [Senator Kefauver's] warnings and his call for additional safeguards. Just prior to this present disaster, a situation similar in type but perhaps less grave in import was uncovered in connection with another drug, MER/29—an anti-cholesterol agent." S. A. Rose, "Drug Control Advocated: Physician Supports Kefauver Bill as Offering Needed Safeguards," *New York Times,* August 20, 1962, 22.

33. Memorandum cited in Ungar, "Get Away with What You Can," 112.

34. The $80,000 fines were minuscule in comparison to Richardson-Merrell's $180 million revenue in that year alone. "Drug Maker Fined $80,000 in Failure to Tell of Danger," *New York Times,* June 5, 1964, 11.

35. "H. Robert Marschalk, 83, Ex-Drug Company Head" [obituary], *New York Times,* April 9, 1999, A21.

36. Robert K. Plumb, "Neomycin Found to Cut Blood Fat: Clue to Arteriosclerosis Is Seen in Effect of Drug on the Cholesterol Level," *New York Times,* April 21, 1959, 39.

37. Fuhrman and Robinson, "Hypocholesterolemic Agents"; AMA Council on

Drugs, "Supplemental Report on the Use of Heparin Sodium in Hyperlipidemia," *Bulletin of the Council on Drugs, A.M.A.*, August 28, 1957, 359–74.

38. R. Altschul, A. Hoffer, and J. D. Stephen, "Influence of Nicotinic Acid on Serum Cholesterol in Man," *Archives of Biochemistry and Biophysics* 54 (1955): 558; K. G. Berge, "Side Effects of Nicotinic Acid in Treatment of Hypercholesterolemia," *Geriatrics* 16 (1961): 416–22.

39. Fuhrman and Robinson, "Hypocholesterolemic Agents," 944.

40. "Clofibrate in Perspective," *New England Journal of Medicine* 279 (1968): 885–86; also see "U.S. to Use Drug on Cholesterol: Clofibrate, Backed by the F.D.A., on Sale May 1," *New York Times*, February 28, 1967, 23. Earlier research into fibric acid derivatives can be traced to J. M. Thorp and W. S. Waring, "Modification and Distribution of Lipids by Ethyl Chlorophenoxyisobutyrate," *Nature* 194 (1962): 948–49.

41. Initial reporting on the Coronary Drug Project was sparse; see "To Get Rid of Cholesterol" *New York Times*, March 5, 1967, 194. For an example of clofibrate's later uses, see "Cholesterol-Lowering Drug Ban Necessary," *F-D-C Reports*, May 30, 1977, 12.

42. Coronary Drug Project Research Group, "Clofibrate and Niacin in Coronary Heart Disease," *Journal of the American Medical Association* 231 (1975): 360. Also see Stan N. Finkelstein, Stephen B. Schectman, Edward J. Sondik, and Dana Gilbert, "Clinical Trials and Established Medical Practice: Two Examples," in *Biomedical Innovation*, ed. E. B. Roberts, R. I. Levy, S. N. Finkelstein, J. Moskowitz, and E. Sondik, (Cambridge: MIT Press, 1981), 200–215, esp. 205. A special issue of the journal *Controlled Clinical Trials* (vol. 4, 1983) was devoted entirely to the history, interpretation, and influence of the Coronary Drug Project. As backers of a particularly expensive negative study, its proponents performed many subsequent analyses of data in the 1970s in part to justify NIH (and particularly NHLBI) expenditures; see Robert I. Levy and Edward J. Sondik, "The Management of Biomedical Research: An Example for Heart, Lung, and Blood Diseases," in Roberts et al., *Biomedical Innovation*, 327–50.

43. Harold M. Schmeck, "2 Heart-Aid Drugs Discounted by Study," *New York Times*, January 23, 1975.

44. Robert Reinhold, "'Miracle' Drug Discredited; Health System Is Faulted," *New York Times*, December 9, 1980. Reinhold's article points out the role of Mary Lasker and Michael DeBakey in developing a public relations strategy "to generate publicity about the drug among doctors and patients alike" and in lobbying Congress to appropriate $40 million toward further clinical trials of cholesterol-lowering agents.

45. Connor, "Use of Drugs and Diets," 660.

46. For a more general exploration of "diseases of civilization" or luxus pathologies in American cultural history, see Allan Brandt, "Behavior, Disease, and Health in the Twentieth-Century United States: The Moral Valence of Individual Risk," 35–52; and Charles Rosenberg, "Banishing Risk: Continuity and Change in the Moral Management of Disease," 53–79, both in *Morality and Health*, ed. Allan M. Brandt and Paul Rozin (London: Routledge, 1997); also see Charles Rosenberg, "Pathologies of Progress: The Idea of Civilization as Risk," *Bulletin of the History of Medicine* 72, no. 4 (1998): 714–30. For accounts of the history of dietary interventions for cardiovascular risk, see Harry M. Marks, "You Gotta Have Heart," in his *The Progress of Experiment: Science and Therapeutic Reform in the United States, 1900–1990* (Cambridge: Cambridge University Press, 1997), 164–96; Karin Garrety, "Social Worlds, Actor-Networks, and Controversy:

The Case of Cholesterol, Dietary Fat, and Heart Disease," *Social Studies of Science* 27 (1997): 727–73.

47. See, e.g., A. Keys and K. T. Anderson, "The Relationship of the Diet to the Development of Atherosclerosis in Man," in *Symposium on Atherosclerosis* (Washington, DC: National Academy of Sciences, 1954), 181–87; Norman Joliffe, "Fats, Cholesterol, and Coronary Heart Disease: A Review of Recent Progress," *Circulation* 20 (1959): 109–27; A. Kagan, B. R. Harris, and W. Winkelstein Jr., "Epidemiological Studies of Coronary Heart Disease and Stroke in Japanese Men Living in Japan, Hawaii, and California: Demographic, Physical, Dietary, and Biochemical Characteristics," *Journal of Chronic Disease* 27 (1974): 345–64; W. B. Kannel, W. P. Castelli, and T. Gordon, "Cholesterol in the Prediction of Atherosclerotic Disease," *Annals of Internal Medicine* 90 (1979): 85–91; A. Keys, C. Aravanis, and H. Blackburn, *Seven Countries: A Multivariate Analysis of Death and Coronary Heart Disease* (Cambridge: Harvard University Press, 1980); popular accounts include "Personal Business: The 'Cholesterol Controversy' Which Has Blown Up in Recent Years Still Has Not Been Decided One Way or the Other," *Business Week,* January 30, 1960, 97; Leonard Engel, "Cholesterol: Guilty or Innocent?" *New York Times,* May 12, 1963, SM15.

48. See, e.g., Lester M. Morrison, "A Nutritional Program for Prolongation of Life in Coronary Atherosclerosis," *Journal of the American Medical Association* 159 (1955): 1425–28; H. B. Jones, J. W. Gofman, F. T. Lindgen, T. P. Lyon, D. M. Graham, B. Strisower, and A. V. Nichols, "Lipoproteins in Atherosclerosis," *American Journal of Medicine* 11 (1951): 358–79.

49. "New Light on Cholesterol Levels," *New York Times,* January 8, 1956.

50. I. H. Page, F. J. Stare, A. C. Corcoran, H. Pollack, and C. F. Wilkinson, "Atherosclerosis and the Fat Content of the Diet," *Circulation* 16 (1957): 163–78, quotation on 163.

51. "Fat and Oil Ads Disputed by U.S.," *New York Times,* December 11, 1959.

52. Peter Bart, "Advertising: Dairy Men Open Counterattack," *New York Times,* August 7, 1962, 36; Jane E. Brody, "Traditional American Breakfast Is Threatened by Heart Theories," *New York Times,* January 1, 1967, 55; for overviews of the economic interest of the dairy and beef industries and the politics of national nutrition policy, see Patricia Hausman, *Jack Sprat's Legacy: The Science and Politics of Fat and Cholesterol* (New York: Richard Marek, 1981); Marion Nestle, *Food Politics: How the Food Industry Influences Nutrition and Health* (Berkeley: University of California Press, 2001).

53. Central Committee for the Medical Community Program of the American Heart Association, "Dietary Fat and its Relation to Heart Attack and Strokes," *Circulation* 23 (1961): 389–91.

54. Jeremiah Stamler, *History of the National Diet Heart Study and the Deliberations of the Executive Committee on Diet and Heart Disease* (1968), cited in Marks, *Progress of Experiment,* 181.

55. National Heart and Lung Institute, Task Force on Arteriosclerosis, *Arteriosclerosis* (Washington, DC: National Institutes of Health, 1971), 2:65.

56. Jane E. Brody, "Health Aide Urges Federal Program for Conclusive Heart Disease Studies, *New York Times,* November 10, 1971, 22; "The Multiple Risk Factor Intervention Trial (MRFIT): A National Study of Primary Prevention of Coronary Heart Disease," *Journal of the American Medical Association* 235 (1976): 825–27.

57. Multiple Risk Factor Intervention Trial Research Group, "Multiple Risk Factor Intervention Trial: Risk Factor Changes and Mortality Results," *Journal of the American Medical Association* 248 (1982): 1465–77; Jane E. Brody, "Heart Disease Study Shows No Gain in Bid to Cut Risks," *New York Times*, September 17, 1982, A10; also see Thomas J. Moore, "The Cholesterol Myth," *Atlantic Monthly*, September 1989, 37.

58. The 1980 FNB Report immediately became the subject of an intense public debate over interest and science; several advocates of low-fat, low-cholesterol diets accused the FNB group of being inappropriately influenced by the beef and dairy industry. For a sample of articles in the *New York Times* in the weeks following the report's release, see Lawrence K. Altman, "Report about Cholesterol Draws Agreement and Dissent," May 28, 1980, D18; Jane E. Brody, "Dispute on Americans' Diets," May 29, 1980, D18; Brody, "When Scientists Disagree, Cholesterol Is in Fat City," June 1, 1980, E1; "A Confusing Diet of Fact," June 2, 1980, A18; Karen De Witt, "Scientists Clash on Academy's Cholesterol Advice," June 20, 1980, A15; Charles E. Rodgers Jr., "Cholesterol Diets: A New Low," June 29, 1980, WC18; Victor Herbert, "In Defense of the Cholesterol Report," July 13, 1980 (all in *New York Times*). For subsequent analysis of the FNB debate, see Hausman, *Jack Sprat's Legacy;* Garrety, "Social Worlds, Actor-Networks, and Controversy"; Nestle, *Food Politics.*

59. Regarding the LRC-CPPT's expense and public-relations handling by the NHLBI, see Garrety, "Social Worlds, Actor-Networks, and Controversy." Garrety presents a useful narrative framework of the diet-cholesterol controversy from 1950 to the mid-1980s; however, her analysis is more concerned with critiquing the empirical basis for actor-network theory than with exploring the broader questions of disease redefinition raised by her case study. The LRC-CPPT has also been the subject of a thorough media analysis of newspaper coverage of cholesterol from 1980 to 1985; see Jonathan R. Cole, "Dietary Cholesterol and Heart Disease: The Construction of a Medical 'Fact,'" in *Surveying Social Life*, ed. Hubert J. O'Gorman (Middletown, CT: Wesleyan University Press, 1988), 437–66.

60. Lipid Research Clinics Program (LRCP), "The Lipid Research Clinics Coronary Primary Prevention Trial Results: I. Reduction in Incidence of Coronary Heart Disease," *Journal of the American Medical Association* 251 (1984): 351–64, quotation on 351–52.

61. On the development of cholestyramine, see D. M. Tennent, H. Siegel, M. E. Zanetti, G-W. Kuron, W. H. Ott, and F. J. Wolf, "Plasma Cholesterol Lowering Action of Bile Acid Binding Polymers in Experimental Animals," *Journal of Lipid Research* 1 (1960): 469–73.

62. Albert Alberts, "Lovastatin Lunch," oral history conducted by Edward Shorter, February 1987, audiocassette, in "The Health Century" oral history collections NLM, OH 136, History of Medicine Division, National Library of Medicine, Bethesda, MD. Also see Edward Shorter, *The Health Century* (New York: Doubleday, 1987), 164.

63. LRCP, "Coronary Prevention Trial Results," 352.

64. Ibid. Compliance was closely monitored by study workers who rigorously counted the packets consumed every two months.

65. The full caption reads: "A million Americans have heart attacks every year. I was one. We are counting on you and the CPPT to help answer the cholesterol question. You can make the difference," "The Ball's in Your Court," images C00666 and C00667, Im-

ages from the History of Medicine Collection, History of Medicine Division, National Library of Medicine, Bethesda, MD.

66. The initial trial design had called for statistical proof at a probability of p < 0.01; that is to say, proof could be claimed if there was more than a 99% chance that the results could not be explainable by chance. Eventually, the investigators claimed proof at a 95% level (p < 0.05), which, though an acceptable value in itself, was in violation of study design.

67. As the initial publication read, "the LRC-CPPT was not designed to assess directly whether cholesterol lowering by diet prevents CHD. Nevertheless, its findings, taken in conjunction with the large volume of evidence relating diet, plasma cholesterol levels, and CHD, support the view that cholesterol lowering by diet also would be beneficial. The findings of the LRC-CPPT take on additional significance if it is acknowledged that it is unlikely that a conclusive study of dietary-induced cholesterol lowering for the prevention of CHD can be designed or implemented." LRCP, "Coronary Prevention Trial Results," 360.

68. Ibid.

69. Ibid.

70. Preparations for the Consensus Conference had been made in advance of the reporting of the LRC-CPPT results, in order to help coordinate the flow from evidence to consensus to action. "Consensus Conference Scheduled," *Nutrition Today,* September–October 1984, 34.

71. Consensus Conference, "Lowering Blood Cholesterol to Prevent Heart Disease," *Journal of the American Medical Association* 253 (1985): 2080–86.

72. Pediatric cholesterol reduction was directly addressed in the Consensus Conference; see "A Pediatrician Attends the National Institutes of Health Consensus Conference on Lowering Blood Cholesterol," *Pediatrics* 76, no. 1 (1985): 125–26.

73. Michael F. Oliver, "Consensus or Nonsensus Conferences on Coronary Heart Disease," *Lancet* (1985): 1087–89, quotation on 1087–88.

74. William R. Harlan and Geoffrey K. Stross, "An Educational View of a National Initiative to Lower Plasma Lipid Levels," *Journal of the American Medical Association* 253 (1985): 2087–90.

75. In translating a supposedly definitive clinical trial into justification for the creation of a prevention bureaucracy, NHLBI officials referred to the role of Edward Freis's VA study of antihypertensive medications in the founding of the NHBPEP. Claude Lenfant, in the press conference announcing the LRC-CPPT results, made the connection between the LRC-CPPT and the VA study explicit. "Science Press Briefing, Lipid Research Clinics Coronary Primary Prevention Trial Results, Lipid Metabolism-Atherogenesis Branch, National Heart, Lung, and Blood Institute, National Institutes of Health, Bethesda, Maryland," *Nutrition Today,* March–April 1984, 20–26.

76. Office of Prevention, Education, and Control, National Heart, Lung, and Blood Institute, National Institutes of Health (NHLBI-OPEC), *National Cholesterol Education Program, Planning Workshop for Professional and Patient Education, March 18–19, 1985, Summary Report* (Bethesda, MD: NHLBI-OPEC, 1985); NHLBI-OPEC, *National Cholesterol Education Program, Planning Workshop for Public Education, April 16–17, 1985, Summary Report* (Bethesda, MD, 1985); NHLBI-OPEC, *National Cholesterol Education Program and NHLBI Smoking Education Program, Planning Workshop for Worksite Programs, May 1985, Summary Report* (Bethesda, MD, 1985).

77. NHLBI-OPEC, *Workshop for Professional and Patient Education*, 4.

78. NHLBI-OPEC, *Workshop for Public Education*, 3.

79. NHLBI-OPEC, *Workshop for Professional and Patient Education*, 20.

80. NHLBI-OPEC, *Workshop for Public Education*, 11–12.

81. NHLBI-OPEC, *Workshop for Professional and Patient Education*, B3.

82. NHLBI-OPEC, *Workshop for Public Education*, 13.

83. NCEP, *Communications Strategy for Public Education*, December 1986 (Bethesda, MD: NCEP, 1986), 3.

84. Ibid.

85. Mike Gorman to Antonio Gotto, memorandum, c. 1985, box 2, folder 3, Mike Gorman Papers, MS C 462, National Library of Medicine, Bethesda, MD (hereafter cited as Gorman Papers).

86. Mike Gorman to Antonio Gotto, "Fact Sheet #1—New Frontiers in Lowering Cholesterol," c. 1985, box 2 folder 3, Gorman Papers.

87. Ibid.

88. All of these individuals are prominently listed on the "Citizens for Public Action on Cholesterol" letterhead that Gerald J. Wilson used to write to "Advisers, Sponsors, Friends of Citizens for Public Action on Cholesterol, and Citizens for the Treatment of High Blood Pressure," December 9, 1987, box 2, folder 3, Gorman Papers.

89. Ibid. Also see the first issue of *Cholesterol Update*, Fall 1987, box 2, folder 3, Gorman Papers.

90. See, e.g., Beth Schucker, J. T. Wittes, J. A. Cutler, K. Bailey, D. R. Mackintosh, D. J. Gordon, C. M. Haines, M. E. Mattson, R. S. Goor, and B. M. Rifkind, "Change in Physician Perspective on Cholesterol and Heart Disease"; and "Change in Public Perspective on Cholesterol and Heart Disease," *Journal of the American Medical Association* 258 (1987): 3521–26, 3527–31.

91. Irvine H. Page, *Nutrition Today*, September–October 1984, 28.

92. In this view Oliver was supported by many other prominent clinicians, including Edward H. Ahrens, "The Diet-Heart Question in 1985: Has It Really Been Settled?" *Lancet* (1985): 1085–87.

93. William C. Taylor, Theodore M. Pass, Donald S. Shepard, and Anthony L. Komaroff, "Cholesterol Reduction and Life Expectancy," *Annals of Internal Medicine* 106 (1987): 605–14.

94. Moore, "Cholesterol Myth"; House Subcommittee on Health and the Environment of the Committee on Energy and Commerce, *Cholesterol Education Program, Hearing before the Subcommittee on Health and the Environment of the Committee on Energy and Commerce*, serial no. 101-107, 101st Cong., 1st sess., December 7, 1989.

95. Gorman to Gotto, "Fact Sheet #1-New Frontiers in Lowering Cholesterol," Gorman Papers. As Gorman concluded, "We still don't have the ideal drug for lowering cholesterol and LDL cholesterol. What's needed is a safe, effective agent that will lower lipids by 30 to 40 percent."

96. The ATP guidelines were explicitly modeled after the JNC guidelines. The task force was specifically charged to create clear thresholds, or "cut points," for dietary and drug intervention and to generate a simple flow chart to identify major steps in screening, diagnosis, and treatment; see NHLBI-OPEC, *Workshop for Professional and Patient Education*, 7–8. See also NCEP, *Cholesterol: Current Concepts for Clinicians*, October 1988 (Bethesda, MD: NIH, 1988).

97. For an account of Mevacor's development as a direct justification for the high prices of prescription drugs, see P. Roy Vagelos, "Are Prescription Drug Prices High?" *Science* 252 (1991): 1080–84.

98. Ibid.; other chronologies of Mevacor's development include "Cholesterol Breakthrough," 5–13; Alberts, "Lovastatin Lunch."

99. For a review of the Japanese contribution to statin development, see Akira Endo, "Discovery and Development of the Statins," in *Statins: The HMG CoA Reductase Inhibitors in Perspective*, ed. Allan Gaw, Christopher J. Packard, and James Shepherd (Malden, MA: Martin Dunitz, 2000), 35–47.

100. Endo's initial discovery was reported in 1976: A. Endo, M. Kuroda, and Y. Tsujita, "ML-236A, ML-236B, and ML 236-C, New Inhibitors of Cholesterogenesis Produced by *Penicillium citrinum*," *Journal of Antibiotics (Japan)* 29 (1976): 1346–48; A. Endo, M. Kuroda, and K. Tanzawa, "Competitive Inhibition of 3-Hydroxy-3-methylglutaryl Coenzyme A Reductase by ML-236A and ML-236B, Fungal Metabolites Having Hypocholesteremic Activity," *FEBS Letter* 72 (1976): 323–26; animal tests were reported in dogs in 1979: Y. Tsujita, M. Kuroda, K. Tanzawa, N. Kitano, and A. Endo, "Hypolipidemic Effects in Dogs of ML-236B, a Competitive Inhibitor of 3-Hydroxy-3-methylglutaryl Coenzyme A Reductase," *Atherosclerosis* 32 (1979): 307–13; and the first clinical trials in humans were reported in A. Yamamoto, H. Suto, and A. Endo, "Therapeutic Effects of ML-236B in Primary Hypercholesterolemia," *Atherosclerosis* 35 (1980): 259–66.

101. H. Mabuchi, T. Sakai, Y. Sakai, A. Yoshimura, A. Watanabe, T. Wakasugi, J. Koizumi, and R. Takeda, "Reduction of Serum Cholesterol in Heterozygous Patients with Familial Hypercholesterolemia: Additive Effects of Compactin and Cholestyramine," *New England Journal of Medicine* 308 (1983): 609–13. The inherited condition of familial hypercholesterolemia—critical to both the FDA approval of Mevacor and the subsequent classification of high cholesterol as a pathological condition—is discussed in greater detail in chapter 6.

102. Alberts, "Lovastatin Lunch," tape 1, side 2.

103. For the United States and most of Europe, the patent was Merck's; see Vagelos, "Are Prescription Drug Prices High?" 24.

104. These rumors began in September 1980, when Sankyo halted clinical trials for compactin. Though never substantiated by any results or publications, they quickly cast a pall over the development of lovastatin. Sankyo expressly denied all rumors, claiming that trials on compactin were halted when the firm began to develop another, more promising statin. John A. Byrne, "The Miracle Company," *Business Week*, October 19, 1987, 88.

105. Alberts, as recorded in the "Lovastatin Lunch" audiotape, tape 1, side 2.

106. A large amount of this budget was spent on an early example of direct-to-consumer advertising in the form of a television commercial warning of the dangers of high cholesterol. Olivia Williams, Anne-Marie Jacks, Jim Davis, and Sabrina Martinez, "Merck(A) Mevacor," unpublished case study, University of Michigan Business School, September 2002, p. 4.

107. "Cholesterol Breakthrough."

108. Reference to Merck's initial contact is made by Mike Gorman and Gerald Wilson to Neal Packard, Merck Sharp & Dohme, October 16, 1981, box 8, folder 25, Gorman Papers.

109. As Fiskett wrote to Gorman shortly before the first one-hundred-thousand-dollar check was cut: "Frankly, we're delighted at the prospect, and consequently I see no reason for us to tell you what you already know how to do better than anyone else in Washington. Thus we are prepared to give you as much (or as little) assistance as you wish. What we view the National Committee as doing is set forth in our original proposal. As the proposal further stated, we have always viewed Merck's role to be primarily one of financial support." A. B. Fiskett, Merck Sharp & Dohme, to Mike Gorman, Citizens for the Treatment of High Blood Pressure, Inc., August 1, 1983, box 8, folder 25, Gorman Papers. Also see Fiskett to Gorman, January 25, 1983; Fiskett to Gorman, April 6, 1983; "National Glaucoma Advisory Committee: A Proposal to Mike Gorman and Mrs. Albert D. Lasker from Merck Sharp & Dohme," undated manuscript attached to letter of April 6, 1983; Gorman to Lasker, April 12, 1983, all in box 8, folder 25, Gorman Papers.

110. Gerald Wilson to the board of directors of the Citizens for the Treatment of High Blood Pressure, Legislative Action Network, Inc., December 4, 1987; typewritten notes entitled "Humphreys Conversation," December 17, 1987; both in box 5, folder 6, Gorman Papers. Also see Mike Gorman to Tim Ryder, American Security Bank, January 25, 1988, box 8, folder 25, Gorman Papers.

111. A copy of one small advertisement for *Cholesterol and You* is found in box 2, folder 3, Gorman Papers; also preserved are several responding letters, e.g., Leigh Anne Musser, University of Texas Health Science Center at Houston, to "Merck, Sharp, and Dohme, The Citizens for Public Action on Cholesterol," April 12, 1988, also in box 2, folder 3, Gorman Papers.

112. The articles included titles such as "Quick Quiz" and "Cholesterol and You" and appear to have been distributed with different content in different geographic areas; whether the articles were part of a market research project or the result of segmented marketing is unclear. At least eight marketing maps for ten different copy placements were generated and sent to Merck Sharp & Dohme for analysis and utilization in 1988; all these documents, under the title "Known Placements to Date of Your News Release," are in box 2, folder 4, Gorman Papers.

113. The full basis of the conflict is unclear, but it appears to have centered on a series of competing actions by Gorman and Gerald Wilson in which each attempted to force the other out of the organization through implications of financial impropriety.

114. Phillip J. Hilts, "FDA Approves Sale of More Effective Cholesterol-Lowering Drug," *Washington Post*, September 2, 1987, A3.

115. Williams et al., "Merck(A) Mevacor," 4.

116. Moore, "Cholesterol Myth," 54.

117. This is noted by Daniel Steinberg and Antonio Gotto in the retrospective account "Preventing Coronary Artery Disease by Lowering Cholesterol Levels: Fifty Years from Bench to Bedside," *Journal of the American Medical Association* 282 (1999): 2043–50.

118. J. A. Tobert, "New Developments in Lipid-Lowering Therapy: The Role of Inhibitors of Hydroxymethylglutaryl–Coenzyme A Reductase." *Circulation* 76, no. 3 (1987): 534–38, quotation on 538.

119. The key statin prevention trials of the 1990s include the 4S, WOSCOPS, and AFCAPS/TexCAPS trials, all of which are discussed in more detail in chapter 6. See, e.g., Scandinavian Simvastatin Survival Study Group, "Randomized Trial of Cholesterol

Lowering in 4444 Patients with Coronary Heart Disease: The Scandinavian Simvastatin Survival Study (4S)," *Lancet* 344, no. 8934 (1994): 1383–89; West of Scotland Coronary Prevention Study Group, "Prevention of Coronary Heart Disease with Pravastatin in Men with Hypercholesterolemia," *New England Journal of Medicine* 333 (1995): 1301–7; J. R. Downs, M. Clearfield, S. Weis, E. Whitney, D. R. Shapiro, P. A. Beere, A. Langendorfer, E. A. Stein, W. Kruyer, and A. M. Gotto Jr., "Primary Prevention of Acute Coronary Events with Lovastatin in Men and Women with Average Cholesterol Levels: Results of AFCAPS/TexCAPS: Air Force/Texas Coronary Atherosclerosis Prevention Study," *Journal of the American Medical Association* 279 (1998): 1615–22.

120. Michael F. Oliver, "Reducing Cholesterol Does Not Reduce Mortality," *Journal of the American College of Cardiologists* 12, no. 3 (1988): 14–17.

121. Michael F. Oliver, "Might Treatment of Hypercholesterolaemia Increase Non-Cardiac Mortality?" *Lancet* 337 (1991): 1529–31; Oliver, "Doubts about Preventing Coronary Heart Disease," *British Medical Journal* 304 (1992): 393–94.

122. See, e.g., Uffe Ravbskov, *The Cholesterol Myths: Exposing the Fallacy That Saturated Fat and Cholesterol Cause Heart Disease* (Washington, DC: New Trends, 2000).

123. Gaw, Packard, and Shepherd, *Statins*, vii.

CHAPTER 6: Know Your Number

Epigraph: Alfred A. Mannino, "Internal Controls," *Principles of Pharmaceutical Marketing,* 3rd ed., ed. Mickey C. Smith (Philadelphia: Lea and Febiger, 1983), 462.

1. Ira S. Nash, Lori Mosca, Roger Blumenthal, Michael H. Davidson, Sidney C. Smith Jr., and Richard C. Pasternak, "Contemporary Awareness and Understanding of Cholesterol as a Risk Factor: Results of an American Heart Association National Survey," *Archives of Internal Medicine* 163, no. 13 (2003): 1597–1600; R. M. Pieper, D. K. Arnett, P. G. McGovern, E. Shahar, H. Blackburn, and R. V. Luepker, "Trends in Cholesterol Knowledge and Screening and Hypercholesterolemia Awareness and Treatment, 1980–1992, The Minnesota Heart Study," *Archives of Internal Medicine* 157, no. 20 (1997): 2326–32; Earl S. Ford, Ali H. Mokdad, Wayne H. Giles, and George A. Mensah, "Serum Total Cholesterol Concentrations and Awareness, Treatment, and Control of Hypercholesterolemia among US Adults: Findings from the National Health and Nutrition Survey, 1999 to 2000," *Circulation* 107 (2003): 2185–89.

2. The idea of a drug having a biography or "life cycle" has been employed in a small number of pharmaceutical histories, e.g., Susan L. Speaker, "From Happiness Pills to National Nightmare: Changing Cultural Assessment of Minor Tranquilizers in America, 1955–1980," *Journal of the History of Medicine* 52 (1997): 338–77; Jordan Goodman and Vivien Walsh, *The Story of Taxol: Nature and Politics in the Pursuit of an Anti-Cancer Drug* (Cambridge: Cambridge University Press, 2001). The concept was also frequently invoked by post–World War II pharmaceutical marketers themselves, e.g., William E. Cox Jr. "Product Life Cycles as Marketing Models," *Journal of Business* 40 (1967): 376–81.

3. Scott Allen and Stephen Smith, "Statin Nation: Do We All Need to Lower Our 'Bad' Cholesterol?" *Boston Globe,* March 16, 2004; Ron Winslow, "Blood Feud: For Bristol Myers, Challenging Pfizer Was Big Mistake," *Wall Street Journal,* March 9, 2004, A1.

4. As a series of mergers in the 1980s and 1990s consolidated the industry into a

small number of large, publicly traded corporations, pressures mounted for pharmaceutical executives to concentrate their efforts on producing a small number of drugs with large-volume sales, as opposed to a large number of drugs with smaller markets. See B. Pecoul, P. Chirac, P. Trouiller, and J. Pinel, "Access to Essential Drugs in Poor Countries: A Lost Battle?" *Journal of the American Medical Association* 281 (1999): 361–67. On the relationship of the sulfa drugs and penicillin to previous moments in the history of the American pharmaceutical industry, see Harry F. Dowling, *Medicines for Man: The Development, Regulation, and Use of Prescription Drugs* (New York: Knopf, 1970); John Lesch, "Chemistry and Biomedicine in an Industrial Setting: The Invention of the Sulfa Drugs," in *Chemical Sciences in the Modern World*, ed. Seymour H. Mauskopf (Philadelphia: University of Pennsylvania Press, 1993), 158–215.

5. On Merck's decision to market Zocor as a competitor to Mevacor in the context of potential patent expiry and growing competition from other brands, see Olivia Williams, Anne-Marie Jacks, Jim Davis, and Sabrina Martinez, "Merck(A) Mevacor," unpublished case study, University of Michigan Business School, September 2002; Olivia Williams, Anne-Marie Jacks, Jim Davis, and Sabrina Martinez, "Merck(B) Zocor," unpublished case study, University of Michigan Business School, September 2002.

6. Merck had already sought maximal patent extensions for the prescription version of Mevacor. Rita Rubin, "Drugmakers Prescribe Move: Cholesterol Treatments Could Be First for Chronic Condition to Shift to OTC," *USA Today,* June 27, 2000, 1E; "Merck Seeks Over-the-Counter Status for Anticholesterol Drug," *Boston Globe,* April 7, 1999, E5.

7. By the late 1990s, the subject of Rx-to-OTC switching was gaining increasing attention within the industry as a means of extending brand life. Seventy pharmaceutical agents switched from Rx to OTC between 1975 and 2000, with a joint projected sales of $4.6 billion annually. "As Switch Market Grows, Everyone Benefits," *MMR,* March 8, 1999, 20.

8. "Nonprescription Mevacor(TM) FDA Advisory Committee Background Information," July 2000, NDA 21-213, Dockets Office, FDA, Rockville, MD.

9. Geoffrey Cowley, "Right Off the Shelf," *Newsweek,* July 10, 2000, 50.

10. "Drugmakers Prescribe Move: Cholesterol Treatments Could Be First for Chronic Condition to Shift to OTC," *USA Today,* June 27, 2000, 1D.

11. The psychosomatic literature on the health risks of hypercholesterolemic diagnosis focused on manifest symptoms that patients developed after diagnosis with hypercholesterolemia—such as dizziness, shortness of breath, and chest pain—with no discernible physiological basis except for the patients' identification as "sick" following hypercholesterolemic diagnosis. See, e.g., Allan S. Brett, "Psychologic Effects of the Diagnosis and Treatment of Hypercholesterolemia: Lessons from Case Studies," *American Journal of Medicine* 91 (1991): 642–47; R. C. Lefebvre, K. G. Hursey, and R. A. Carleton, "Labeling of Participants in High Blood Pressure Screening Programs: Implications for Blood Cholesterol Screenings," *Archives of Internal Medicine* 148 (1988): 1993–97; T. Tijmstra, "The Psychological and Social Implications of Serum Cholesterol Screening," *International Journal of Risk and Safety in Medicine* 1 (1990): 29–44.

12. See testimony of Jeffrey L. Anderson, in Department of Health and Human Services, FDA, "Public Hearing on FDA Regulation of Over-the-Counter Drug Products," June 29, 2000, FDA docket 00N-1256, 111–12:

There is a growing gap between primary preventive efforts and public concern about risk factors. The consumer already has moved to fill this gap, even if ill advised, through self-medication with so-called nutriceuticals . . . Though often relatively ineffective in cholesterol lowering and largely unsupported by randomized trials, these products form the fastest growing segment of the health product market, with $12 billion spent last year. Patients in my own practice regularly list self-selected health supplements in their medical histories. One of these, red yeast rice, contains Lovastatin in doses that approximate the proposed OTC dose and is available to the public and has generated a good deal of interest. We in the health care community should recognize this entrenched and growth public health movement towards self-medication for risk reduction and respond constructively.

13. Health and Human Services, *Regulation of Over-the-Counter Drug Products,* 7–8.

14. Robert Scott, then head of cardiovascular and metabolic products for Pfizer, Inc., as quoted in Rubin, "Drugmakers Prescribe Move." Data on Lipitor market share from IMS Health, "Developments in the Statin Market," www.ims-global.com/insight/news-story/news-story-000316a.htm, March 16, 2000.

15. Testimony of Edward Frohlich, in Health and Human Services, *Regulation of Over-the-Counter Drug Products,* 131–33.

16. Cross-examination, Robert Temple, ibid., 139–40.

17. For example, B. Waine Kong, CEO of the Association of Black Cardiologists, noted during the hearings that OTC cholesterol drugs would be more accessible to minority patients, a population in whom hypercholesterolemia was relatively undertreated. Victoria Stagg Elliott, "FDA Advisory Committee Vetoes OTC Status for Low-Dose Anti-Cholesterol Drugs," *AMNews,* August 7, 2000. Also see National Council of Negro Women to Jennie C. Butler, Dockets Management Branch, FDA, FDA docket OON-1256, Dockets Management Branch (HFA-305), FDA, Rockville, MD.

18. A parallel bid by Bristol-Myers Squibb to receive OTC approval for its own statin, Pravachol, was rejected the following day. "Panel Retains Status of Cholesterol Drugs," *New York Times,* July 18, 2000, F12; Elizabeth Mechcatie, "FDA Panel Rejects Making Low-Dose Statins Available OTC," *Family Practice News,* September 1, 2000.

19. The FDA Advisory Panel found that Merck demonstrated adequate safety and cholesterol-lowering efficacy of Mevacor OTC but failed to demonstrate that it would produce meaningful clinical end points in the target consumer population. Although the panel voted 11 to 1 against the submission, the vote on whether Merck had "presented adequate evidence that consumers *will be able* to use lovastatin 10mg safely in an OTC setting," was deeply split, 7 voting yes and 6 voting no. FDA, "Minutes of the Joint Meeting of Nonprescription Drugs Advisory Committee and the Endocrinological and Metabolic Drugs Advisory Committee, July 13 2000," NDA 21-213, Dockets Administration, FDA, Rockville, MD.

20. The changing textbook knowledge of cholesterol and atherosclerosis reflects changes in physician attitudes toward these topics—whether elevated cholesterol should be seen as a disease state and how it should be diagnosed and treated. As Ludwik Fleck observes, textbook knowledge always trails behind the leading edge of active scientific debate and journal reviews, but as such it is more likely to reflect the level of

knowledge of the average medical practitioner; it also serves as a consistent sampling device for measuring changing practice patterns, since textbook revisions occur on a regular basis, roughly every three to five years. The two textbooks used in my analysis, Russell Cecil's *Textbook of Medicine* (now known as the *Cecil-Loeb Textbook of Medicine*) and T. R. Harrison's *Principles of Internal Medicine* (now *Harrison's Principles of Internal Medicine*) took over the canonical role occupied by William Osler's *Principles and Practice of Medicine* around the mid-twentieth century and remain today as two central texts of American general medical practice.

21. Russell L. Cecil and Robert F. Loeb, eds., *Textbook of Medicine*, 9th ed. (Philadelphia: Saunders, 1955), 702–3; also see Kendall Emerson Jr., "Xanthomatosis and Lipidosis," *Principals of Internal Medicine*, 3rd ed., ed. T. R. Harrison (New York: McGraw-Hill, 1958), 742–45.

22. Howard A. Eder, "Primary Hyper- and Hypolipidemias," in *Cecil-Loeb Textbook of Medicine*, 11th ed., ed. Paul B. Beeson and Walsh McDermott (Philadelphia: Saunders, 1963), 1334 (emphasis added).

23. Ibid., 1333.

24. Ibid., 1332–35. A contemporary medico-philosophical framework for equating Gaussian conceptions of normality with definition of healthy physiology was provided by Elias Amador, who preferred to distinguish three sorts of normality in medicine: qualitative normality, quantitative normality, and molecular normality. "Nowadays," Amador argued, "a value compatible with health is called 'normal,' a term that has led to confusion because of its biological and statistical uses. The two usages, 'normal' indicating health and 'normal' indicating Gaussian are not synonymous." Amador, "Health and Normality," *Journal of the American Medical Association* 232 (1975): 953–54, quotation on 954; also see William E. Stempsey, *Disease and Diagnosis: Value-Dependent Realism* (Boston: Kluwer Academic, 1999).

25. On the history of statistics, see Theodore M. Porter, *The Rise of Statistical Thinking: 1820–990* (Princeton, NJ: Princeton University Press, 1986); Ian Hacking, *The Taming of Chance* (Cambridge: Cambridge University Press, 1990). On the labeling of statistical outliers as medically deviant, see J. Rosser Matthews, *Quantification and the Quest for Medical Certainty* (Princeton, NJ: Princeton University Press, 1995); William G. Rothstein, *Public Health and the Risk Factor: A History of an Uneven Medical Revolution* (Rochester, NY: University of Rochester Press, 2003), 36–49.

26. Daniel Steinberg and Antonio M. Gotto, "Preventing Coronary Artery Disease by Lowering Cholesterol Levels: Fifty Years from Bench to Bedside," *Journal of the American Medical Association* 282 (1999): 2043–50.

27. For a discussion of the renal threshold in diabetes, see Claude Bernard, *Leçons sur le diabète et la glycogenèse animale* (1877), as translated in Georges Canguilhem, *The Normal and the Pathological*, trans. Carolyn R. Fawcett (1943; New York: Zone Books, 1991), 70.

28. The periodical literature of the 1950s and 1960s was saturated with material on the epidemic of heart disease and attempts to control it, e.g. "Last Ten Years: Giant Steps against Heart Disease," *Today's Health*, July 1959, 30; J. Stuart, "Any Man Can Have a Heart Attack," *Today's Health*, November 1961, 12–13; L. Kavaler, "Will There Soon Be a Drug That Might Ultimately Prolong Your Husband's Life?" *Good Housekeeping*, October 1969, 112. For a more general exploration of "diseases of civilization" or luxus pathologies in American cultural history, see Allan Brandt, "Behavior, Disease, and

Health in the Twentieth-Century United States: The Moral Valence of Individual Risk," 35–52; and Charles Rosenberg, "Banishing Risk: Continuity and Change in the Moral Management of Disease," 53–79, both in *Morality and Health*, ed. Allan M. Brandt and Paul Rozin (London: Routledge, 1997); also see Rosenberg, "Pathologies of Progress: The Idea of Civilization as Risk," *Bulletin of the History of Medicine* 72, no. 4 (1998): 714–30.

29. A. Keys and K. T. Anderson, "The Relationship of the Diet to the Development of Atherosclerosis in Man," in *Symposium on Atherosclerosis* (Washington, DC: National Academy of Sciences, 1954), 181–87; Norman Joliffe, "Fats, Cholesterol, and Coronary Heart Disease: A Review of Recent Progress," *Circulation* 20 (1959): 109–27; A. Kagan, B. R. Harris, and W. Winkelstein Jr., "Epidemiological Studies of Coronary Heart Disease and Stroke in Japanese Men Living in Japan, Hawaii, and California: Demographic, Physical, Dietary, and Biochemical Characteristics," *Journal of Chronic Disease* 27 (1974): 345–64. This is only the most recent chapter in a long history of medical primitivism, in which practices and objects of "preindustrialized" life are invested with an a priori sense of healthfulness and presented as balms to the harried denizens of the modern age.

30. Consensus Conference, "Lowering Blood Cholesterol to Prevent Heart Disease," *Journal of the American Medical Association* 253 (1985): 2080–86, quotation on 2083.

31. Ibid.

32. For example, the 1970 entry in *Harrison's* listed xanthomatosis as a "morphological term" in contrast to the preferred molecular taxonomy of "primary disturbances in lipid metabolism." See Donald S. Fredrickson, "Disorders of Lipid Metabolism and Xanthomatosis," in *Harrison's Principles of Internal Medicine*, 6th ed., ed. Maxwell M. Wintrobe, George W. Thorn, Raymond D. Adams, Ivan L. Bennett, Eugene Braunwald, Kurt J. Isselbacher, and Robert G. Petersdorf (New York: McGraw-Hill, 1970), 630.

33. Ibid., 633: "A prompt fall to normal cholesterol levels in a diet restricted in cholesterol and saturated fats and high in polyunsaturated fats is both a good diagnostic test and adequate therapy. It has not yet been proved that hypolipidemic agents are useful or necessary in treatment of acquired type II."

34. In this system there is no longer any acquired form, only genetics that is well understood (single-gene) and genetics that is murky and poorly understood (multifactorial). Michael S. Brown and Joseph L. Goldstein, "The Hyperlipoproteinemias and Other Disorders of Lipid Metabolism," in *Harrison's Principles of Internal Medicine*, 9th ed., ed. Kurt J. Isselbacher, Raymond D. Adams, Eugene Braunwald, Robert G. Petersdorf, and Jean D. Wilson (New York: McGraw-Hill, 1980), 507–18, quotation on 507.

35. Clifford Geertz has described the value of such "catch-all" or "wastebasket" categories in holding together thought systems (such as taxonomies) that might otherwise be threatened by the anomalous; he makes specific reference to Evans-Pritchard's ethnography of knowledge among the Azande. Geertz describes the Zande concept of witchcraft as "a kind of dummy variable in the system of common-sense thought. Rather than transcending that thought, it reinforces it by adding to it an all-purpose idea which acts to reassure the Zande that their fund of commonplaces is, momentary appearances to the contrary notwithstanding, dependable and accurate." From "Common Sense as a Cultural System," in Geertz, *Local Cultures: Further Essays in Interpretive Anthropology* (New York: Basic Books, 1983); 73–93, quotation on 79.

36. Brown and Goldstein, "Hyperlipoproteinemias and Other Disorders," 516.

37. Marketing decisions regarding pharmaceutical products in development began long before launch. For example in early 1985, Jonathan Tobert, the head of the Mevacor product team inside of Merck, was disturbed to learn that the pharmaceutical staff had created Mevacor as a yellow tablet. Protesting that "yellow is the color of butter" and therefore had no place in a cholesterol-reducing product, Tobert mandated that the pharmacologist redesign Mevacor as a "soothing" light blue tablet instead. John A. Byrne, "The Miracle Company," *Business Week*, October 19, 1987, 88.

38. "Cholesterol Breakthrough: Mevacor Caps a Decades-Long Research Effort," *Merck World* 8, no. 5 (1987): 4–13, quotation on 13.

39. The 1987 entry in *Harrison's Principles of Internal Medicine* mentions Mevacor only in conjunction with FH. Michael S. Brown and Joseph L. Goldstein, "The Hyperlipoproteinemias and Other Disorders of Lipid Metabolism," in *Harrison's Principles of Internal Medicine*, 11th ed., ed. Eugene Braunwald, Kurt J. Isselbacher, Robert J. Petersdorf, Jean D. Wilson, Joseph B. Martin, and Anthony S. Fauci (New York: McGraw-Hill, 1987), 1658.

40. "Cholesterol Breakthrough," 13.

41. The early Mevacor clinical trials focused on patients with heterozygous FH; see, e.g., D. W. Bilheimer, S. M. Grundy, M. S. Brown, and J. L. Goldstein, "Mevinolin and Colestipol Stimulate Receptor-Mediated Clearance of Low Density Lipoprotein from Plasma in Familial Hypercholesterolemia Heterozygotes," *Proceedings of the National Academy of Sciences of the United States of America* 80, no. 13 (1983): 4124–28; S. M. Grundy and D. W. Bilheimer, "Inhibition of 3-Hydroxy-3-methylglutaryl-CoA Reductase by Mevinolin in Familial Hypercholesterolemia Heterozygotes: Effects on Cholesterol Balance," *Proceedings of the National Academy of Sciences of the United States of America* 81 (1984): 2538–42; J. M. Hoeg, M. B. Maher, L. A. Zech, K. R. Bailey, R. E. Gregg, K. J. Lackner, S. S. Fojo, M. A. Anchors, M. Bojanovski, and D. L. Sprecher, "Effectiveness of Mevinolin on Plasma Lipoprotein Concentrations in Type II Hyperlipoproteinemia," *American Journal of Cardiology* 57 (1986): 933–39; R. J. Havel, D. B. Hunninghake, D. R. Illingworth, R. S. Lees, E. A. Stein, J. A. Tobert, S. R. Bacon, J. A. Bolognese, P. H. Frost, and G. E. Lankin, "Lovastatin (Mevinolin) in the Treatment of Heterozygous Familial Hypercholesterolemia: A Multicenter Study," *Annals of Internal Medicine* 107, no. 5 (1987): 609–15. A few case reports by the same investigators also focused on homozygotes; see C. East, S. M. Grundy, and D. W. Bilheimer, "Normal Cholesterol Levels with Lovastatin (Mevinolin) Therapy in a Child with Homozygous Familial Hypercholesterolemia Following Liver Transplantation," *Journal of the American Medical Association* 256 (1986): 2843–48.

42. "Cholesterol Breakthrough," 12.

43. Alfred Alberts, "Lovastatin Lunch," oral history conducted by Edward Shorter, February 1987, tape 1, side 1, in "The Health Century" oral history collections NLM, OH 136, History of Medicine Division, National Library of Medicine, Bethesda, MD.

44. Merck's investment in non-FH clinical trials began only shortly after favorable results began to appear supporting Mevacor in FH, through the efforts of a group called the Lovastatin Study Group. See Lovastatin Study Group II, "Therapeutic Response to Lovastatin (Mevinolin) in Nonfamilial Hypercholesterolemia: A Multicenter Study," *Journal of the American Medical Association* 256 (1986): 2829–34; Lovastatin Study

Group III, "A Multicenter Comparison of Lovastatin and Cholestyramine Therapy for Severe Primary Hypercholesterolemia," *Journal of the American Medical Association* 260 (1988): 359–66.

45. See, e.g., Philip J. Hilts, "FDA Approves Sale of More Effective Cholesterol-Lowering Drug," *Washington Post*, September 2, 1987, A3: "[Mevacor] will be used to treat high cholesterol levels in the 400,000 Americans who have an inherited condition of high cholesterol that is unaffected by changes in diet and exercise. The drug also may be prescribed for others with 'severe' cholesterol problems that are not helped by at least three months of adjusted diet and exercise levels."

46. Gotto's extensive institutional affiliations are discussed in Timothy J. Moore, "The Cholesterol Myth," *Atlantic Monthly*, September 1989, 54.

47. Hilts, "FDA Approves Cholesterol-Lowering Drug."

48. "Cholesterol Breakthrough," 13.

49. Additional goals for the EXCEL trial included extended safety monitoring for cataracts and myopathy and comparison of the effectiveness of variable dose-regimens. Reagan H. Bradford, C. L. Shear, A. N. Chremos, C. Dujovne, M. Downton, F. A. Franklin, A. L. Gould, J. Higgins, and D. P. Hurley, "Expanded Clinical Evaluation of Lovastatin (EXCEL) Study: Design and Patient Characteristics of a Double-Blind, Placebo Controlled Study in Patients with Moderate Hypercholesterolemia," *American Journal of Cardiology* 66 (1990): 44B–55B; R. H. Bradford, C. L. Shear, A. N. Chremos, C. Dujovne, M. Downton, F. A. Franklin, A. L. Gould, J. Higgins, and D. P. Hurley, "Expanded Clinical Evaluation of Lovastatin (EXCEL) Study Results. I. Efficacy in Modifying Plasma Lipoproteins and Adverse Event Profile in 8245 Patients with Moderate Hypercholesterolemia," *Archives of Internal Medicine* 151, no. 1 (1991): 43–49; R. H. Bradford, C. L. Shear, A. N. Chremos, C. Dujovne, M. Downton, F. A. Franklin, A. L. Gould, J. Higgins, and D. P. Hurley, "Expanded Clinical Evaluation of Lovastatin (EXCEL) Study Results: III. Efficacy in Modifying Lipoproteins and Implications for Managing Patients with Moderate Hypercholesterolemia," *American Journal of Medicine* 91, no. 1B (1991): 18S–24S.

50. Michael F. Oliver, "Consensus or Nonsensus Conferences on Coronary Heart Disease," *Lancet* (1985): 1087–89; Edward H. Ahrens, "The Diet-Heart Question in 1985: Has It Really Been Settled?" *Lancet* (1985): 1085–87.

51. Moreover, the gap between the pathological level of 240 and the ideal level of 200, coupled with the fact that most practicable diets, even when effective, typically lowered blood cholesterol only by 10% or less, helped to guarantee a wider role for Mevacor than might be apparent from a superficial reading of the NCEP flowchart. This argument was made by product manager Jonathan Tobert as early as 1987 in support of broader Mevacor usage; see J. A. Tobert, "New Developments in Lipid-Lowering Therapy: The Role of Inhibitors of Hydroxymethylglutaryl–Coenzyme A Reductase," *Circulation* 76, no. 3 (1987): 534–38.

52. NCEP, *Second Report of the Expert Panel on Detection, Evaluation, and Treatment of High Blood Cholesterol in Adults* (Bethesda, MD: NHLBI, 1993), III-3–III-10.

53. "Bristol-Myers Gets Approval on Drug," *New York Times*, November 2, 1991, 39.

54. Bradford et al. "EXCEL Study: Design and Patient Characteristics," 45B.

55. Shortly after the publication of EXCEL, Clinical Research International, Inc., was acquired by a larger company named ClinTrials Research, Inc., which proceeded to

acquire other smaller outfits and by 1998 shifted its central offices to Glasgow to take advantage of the favorable high-tech labor market overseas. See Kenny Kemp, "Microcosm: A Glasgow Company Serving the Industry," *Scotsman*, February 10, 1998; "CEO and President Named at FHI," press release, Family Health International, www.fhi.org/en/AboutFHI/Media/Releases/prArchived/CEO_Pres_FHI_1-26-98.htm, January 16, 1998.

56. Adriana Petryna, "The Human Subjects Research Industry," in *Global Pharmaceuticals: Ethics, Markets, Practices,* ed. Adriana Petryna, Andrew Lakoff, and Arthur Kleinman (Durham, NC: Duke University Press, 2006). For comments on the early history of the CRO industry, see J. E. Beach, "Clinical Trials Integrity: A CRO Perspective," *Accountability in Research* 8, no. 33 (2001): 245–60; also see R. K. H. Wyse and R. G. Hughes, *Pharmaceutical Contract Research in the 1990s* (London: Technomark Consulting Services, 1993). The emergence of a more comprehensive client-centered approach is evident in David Fin, "Temporary Help Stretches to Long Term: Contract Research Organizations, Outsourcing Lets Consolidating Companies Cut Costs and Eliminate Excess Capacity," *Financial Times,* July 15, 1999. The slogan of Quintiles Transnational, the world's largest CRO during the mid-1990s, was "Follow the Molecule," emphasizing the firm's full capacity for providing services to guide a client's drug through any stage of the development process.

57. Fin, "Temporary Help Stretches to Long Term," 5. Only five years earlier the number of CROs was only one thousand worldwide, and the total market size was $3 billion. Clive Cookson, "Survey of Pharmaceuticals," *Financial Times,* March 23, 1994.

58. K. A. Getz, *AMCs Rekindling Clinical Research Partnerships with Industry* (Boston: Centerwatch, 1999), as cited in Thomas Bodenheimer, "Uneasy Alliance: Clinical Investigators and the Pharmaceutical Industry," *New England Journal of Medicine* 342 (2000): 1539–44.

59. Frank Hurley, as quoted in Kathleen Day, "Test-Driving Pharmaceuticals: Biometric Research Finds Success in Saving Drug Manufacturers Time and Money," *Washington Post,* March 21, 1994, F5; also see Cookson, "Survey of Pharmaceuticals."

60. The Lovastatin-Pravastatin Study Group was later criticized for comparing unequal dosages of Mevacor and Pravachol, pitting a 20 mg dose of lovastatin against a 10 mg dose of pravastatin. Merck also conducted an analogous fleet of studies comparing its new product Zocor to Pravachol with equally favorable results for Merck. See Lovastatin Pravastatin Study Group, "A Multicenter Comparative Trial of Lovastatin and Pravastatin in the Treatment of Hypercholesterolemia," *American Journal of Cardiology* 71 (1993): 810–15; also see M. R. Weir, M. I. Berger, C. L. Liss, and N. C. Santanello, "Comparison of the Effects on Quality of Life and of the Efficacy and Tolerability of Lovastatin versus Pravastatin: The Quality of Life Multicenter Group," *American Journal of Cardiology* 77 (1996): 475–79; see also "Merck Claims Efficacy Advantage for Statins," *Pharma Marketletter,* June 1, 1992.

61. The analysis of competitive trial data is from Cathy Kelley, Mark Helfand, Chester Good, and Michael Ganz, *Drug Class Review: Hydroxymethylglutaryl-Coenzyme-A Reductase Inhibitors (Statins)* (Washington, DC: Veterans Health Administration, Pharmacy Benefit Management, Strategic Healthcare Group, 2002), 3–10.

62. The delineation of secondary prevention indications for cholesterol-lowering drugs became a broad theme for discussion among cardiologists in the early 1990s; see,

e.g., Jonathan S. Silberberg and David A, Henry, "The Benefits of Reducing Cholesterol Levels: The Need to Distinguish Primary from Secondary Prevention," *Medical Journal of Australia* 155 (1991): 665–74.

63. On the language of signal and noise in clinical trial organization, see the ethnographic work of Andrew Lakoff, "The Mousetrap: Managing the Placebo Effect in Antidepressant Trials," *Molecular Interventions* 2 (2002): 72–76.

64. Between the ATP-I in 1987 and the ATP-III in 2001, the cutoff lines dividing "ideal" from "borderline" and "borderline" from "high" cholesterol levels remained constant at 200 mg/dL and 240 mg/dL, respectively.

65. NCEP, *Second Report of the Expert Panel on Detection, Evaluation, and Treatment of High Blood Cholesterol in Adults (Adult Treatment Panel II): Executive Summary* (Bethesda, MD: NHLBI, 1993), 14–20.

66. Both Merck and BMS began their secondary prevention trials in the late 1980s, well before the NCEP guidelines were revised.

67. Zocor recorded sales of $700 million in 1992; in that year Mevacor was already capturing over $1 billion in revenue. Williams et al., "Merck(B) Zocor," 2.

68. For the design of the 4S trial, see Scandinavian Simvastatin Survival Study Group (SSSS), "Design and Baseline Results of the Scandinavian Simvastatin Survival Study of Patients with Stable Angina and/or Previous Myocardial Infarction," *American Journal of Cardiology* 71, no. 5 (1993): 393–400; SSSS, "Randomized Trial of Cholesterol Lowering in 4444 Patients with Coronary Heart Disease: The Scandinavian Simvastatin Survival Study (4S)," *Lancet* 344, no. 8934 (1994): 1383–89. For a review of the study's significance, see S. Guptha, "Profiling a Landmark Clinical Trial: Scandinavian Simvastatin Survival Study," *Current Opinion in Lipidology* 6, no. 5 (1995): 251–53.

69. Clive Cookson, "Drugs Improve Heart Patient's Prospects," *Financial Times,* November 18, 1994, 4; Gina Kolata, "Cholesterol Drugs Found to Save Lives," *New York Times,* November 17, 1994, A1.

70. "FDA Approves *Zocor* Label Change," *Pharma Marketletter,* July 10, 1995; Williams et al., "Merck(B) Zocor."

71. Frank M. Sacks, Marc A. Pfeffer, Lemuel A. Moye, Jean L. Rouleau, John D. Rutherford, Thomas G. Cole, Lisa Brown, J. Wayne Warnica, J. Malcolm, O. Arnold, Chuan-Chuan Wun, Barry R. Davis, and Eugene Braunwald, "The Effects of Pravastatin on Coronary Events after Myocardial Infarction in Patients with Average Cholesterol Levels," *New England Journal of Medicine* 335 (1996): 1001–9. Press coverage in the *New York Times* was not front-page; see "Wider Use Seen for Treatment of Cholesterol," *New York Times,* March 27, 1996, A17.

72. Estimated cost of trial from "Wider Use Seen for Treatment of Cholesterol," A17.

73. NCEP, *Cholesterol Lowering in the Patient with Coronary Heart Disease* (Bethesda, MD: NHLBI, 1997), 4.

74. "Statin Trial Results Imminent," *Pharmaceutical Business News,* September 27, 1995.

75. Ron Winslow and Elyse Tanouye, "Bristol-Myers Enters Battle of Heart Drugs," *Wall Street Journal,* November 16, 1995, B1; J. Shepherd, S. M. Cobbe, I. Ford, C. G. Isles, A. R. Lorimer, P. W. MacFarlane, J. H. Killop, and C. J. Packard, "Prevention of Coronary Heart Disease with Pravastatin in Men with Hypercholesterolemia," *New England Journal of Medicine* 333 (1995): 1301–7.

76. Jane E. Brody, "Benefit to Healthy Men Is Seen from Cholesterol-Cutting Drug:

Study Finds Reduced Incidence of Heart Attack," *New York Times*, November 16, 1995, A1.

77. The WOSCOPS investigators used a higher threshold level than NCEP, 252 mg/dL instead of 240, and 155 instead of 130 for LDL cholesterol. Shepherd et al., "Prevention of Coronary Heart Disease."

78. As quoted in Brody, "Benefit Seen from Cholesterol-Cutting Drug."

79. Torje Pedersen, "Lowering Cholesterol with Drugs and Diet," *New England Journal of Medicine* 333 (1995): 1350–51.

80. "F.D.A. Allows Drug as a Heart Medicine," *New York Times*, July 9 1996; "Bristol-Myers Squibb Co.: FDA approves Ad Claims on Drug, First Heart Attack," *Wall Street Journal*, July 9, 1996, B6.

81. Promotion of Mevacor and Pravachol utilized direct-to-consumer (DTC) media a decade before the 1997 decision. See Milton Liebman, "DTC's Role in the Statin Bonanza," *Medical Marketing and Media* 36, no. 11 (2001): 86–90.

82. "Pravachol Helps Prevent First Heart Attacks," two-page advertisement, *New York Times*, first printed September 8, 1996, 33–34.

83. Thomas A. Pearson, "Lipid-Lowering Therapy in Low-Risk Patients," *Journal of the American Medical Association* 279 (1998): 1659–61.

84. J. R. Downs, P. A. Beere, E. Whitney, M. Clearfield, S. Weis, J. Rochen, E. A. Stein, D. R. Shapiro, A. Langendorfer, and A. M. Gotto Jr., "Design and Rationale of the Air Force/Texas Coronary Atherosclerosis Prevention Study (AFCAPS/TexCAPS)," *American Journal of Cardiology* 80, no. 3 (1997): 287–93; J. R. Downs, M. Clearfield, S. Weis, E. Whitney, D. R. Shapiro, P. A. Beere, A. Langendorfer, E. A. Stein, W. Kruyer, and A. M. Gotto Jr., "Primary Prevention of Acute Coronary Events with Lovastatin in Men and Women with Average Cholesterol Levels: Results of AFCAPS/TexCAPS. Air Force/Texas Coronary Atherosclerosis Prevention Study," *Journal of the American Medical Association* 279 (1998): 1615–22,

85. "Mevacor Study May Extend Lipid-Lowering Therapy," *Pharma Marketletter*, November 13, 1997.

86. "Mevacor Approved for Primary Prevention," *Pharma Marketletter*, March 18, 1999.

87. NCEP, *Third Report of the National Cholesterol Education Program (NCEP) Expert Panel on Detection, Evaluation, and Treatment of High Blood Cholesterol in Adults (Adult Treatment Panel III)* (Bethesda, MD: NHLBI, 2001).

88. Thomas M. Burton and Chris Adams, "New U.S. Guidelines Would Triple Use of Cholesterol-Lowering Drugs," *Wall Street Journal*, May 16, 2001, B1. The difficulty of locating a reviewer with sufficient expert knowledge in a medical field who has not received funding from a pharmaceutical company with interests in the field is discussed by Marcia Angell, "The Pharmaceutical Industry—To Whom Is It Accountable? *New England Journal of Medicine* 342 (2000): 1902–4.

89. Anderson testimony, Merck Research Laboratories, in Health and Human Services, *Regulation of Over-the-Counter Drug Products*, 108–11.

90. Rubin, "Drugmakers Prescribe Move"; Elliott, "FDA Vetoes OTC Status."

91. Testimony of Jerome D. Cohen, St. Louis University, in Health and Human Services, *Regulation of Over-the-Counter Drug Products*, 84–85.

92. Anderson testimony, Merck Research Laboratories, ibid., 107; population estimates come from available survey data: "Serum Total Cholesterol (mg/dL) Levels for

Persons 20 years of Age or Older, United States, 1988–94," in NCEP, *Third Report on High Blood Cholesterol*, III-A-1.

93. For example, the NHLBI held a conference in 1990 with the title "Low Blood Cholesterol: Mortality Associations"; see Stephen B. Hulley, Judith M. B. Walsh, and Thomas B. Newman, "Health Policy on Blood Cholesterol: Time to Change Directions," *Circulation* (1992): 1026–29; also George Davey Smith and Juha Pekkanen, "Should There Be a Moratorium on the Use of Cholesterol Lowering Drugs?" *British Medical Journal* 304 (1992): 431–34. Arguments of the risk of low cholesterol and possible harm via cholesterol-lowering were directly countered by proponents of the major statin prevention trials; e.g., Peter H. Jones, "Low Serum Cholesterol Increases the Risk of Non-cardiovascular Events: An Antagonist Viewpoint," *Cardiovascular Drugs and Therapy* 8 (1994): 871–74. By the end of the 1990s, "hypocholesterolemia" contained only the briefest entry in medical textbooks, and the point at which cholesterol levels were low enough to be a possible threat to the organism were defined as less than 100 mg/dL, a number so far below the range of the general population as to offer no practical lower boundary to the diagnosis of treatable cholesterol.

94. Allen and Smith, "Statin Nation"; Winslow, "Blood Feud," A1.

95. Toronto Working Group on Cholesterol Policy, "Asymptomatic Hypercholesterolemia: A Clinical Policy Review," *Journal of Clinical Epidemiology* 43, no. 10 (1990): 1021–22.

96. See Alan M. Garber, "Where to Draw the Line against Cholesterol," *Annals of Internal Medicine* 111 (1989): 625–27; also see A. M. Garber, B. Littenberg, H. C. Sox Jr., M. E. Gluck, J. L. Wagner, and B. M. Duffy, *Costs and Effectiveness of Cholesterol Screening in the Elderly*, U.S. Congress, Office of Technology Assessment (Washington, DC: U.S. Government Printing Office, 1989).

97. Toronto Working Group on Cholesterol Policy, "Asymptomatic Hypercholesterolemia," 1093.

98. A. Williams, "Screening for Risk of CHD: Is It a Wise Use of Resources?" in *Screening for Risk of Coronary Heart Disease*, ed. M. Oliver, M. Ashley Miller, and D. Wood (London: Wiley, 1986), 97–106; G. Oster and A. M. Epstein, "Cost-Effectiveness of Antihyperlipidemic Therapy in the Prevention of Coronary Heart Disease: The Case of Cholestyramine," *Journal of the American Medical Association* 258 (1987): 2381–87. Also see William Taylor, "Screening for High Blood Cholesterol and Other Lipid Abnormalities," in *Guide to Clinical Preventive Services: Report of the U. S. Preventive Services Task Force*, 2nd ed. (Baltimore, MD: Williams and Wilkins, 1989). Although the general literature on cost-effectiveness seems to begin in the late 1970s, e.g., M. C. Weinstein and W. B. Stason, "Foundations of Cost-Effectiveness Analysis for Health and Medical Practices," *New England Journal of Medicine* 296 (1977): 716–21, its use by the mid-1980s was still limited to a relatively small section of the medical literature, e.g., P. Doubilet, M. C. Weinstein, and B. J. McNeil, "Use and Misuse of the Term 'Cost-Effective' in Medicine," *New England Journal of Medicine* 314 (1986): 253–56; G. W. Torrance, "Measurement of Health States Utilities for Economic Appraisal: Review," *Journal of Health Economics* 5 (1986): 1–30; D. U. Himmelstein and S. Woolhandler, "Free Care, Cholestyramine, and Health Policy," *New England Journal of Medicine* 311 (1984): 1511–14; M. C. Weinstein and W. B. Stason, "Cost-Effectiveness of Interventions to Prevent or Treat Coronary Heart Disease," *Annual Review of Public Health* 6 (1985): 41–63.

99. Toronto Working Group on Cholesterol Policy, "Asymptomatic Hypercholesterolemia," 1100.

100. E.g., W. B. Stason, "Costs and Benefits of Risk Factor Reduction for Coronary Heart Disease: Insights from Screening and Treatment of Serum Cholesterol," *American Heart Journal* 119 (1990): 718–24; M. D. Kelley, "Hypercholesterolemia: The Cost of Treatment in Perspective," *Southern Medical Journal* 83, no. 12 (1990): 1421–25; L. Goldman, M. C. Weinstein, P. A. Goldman, and L. W. Williams, "Cost-Effectiveness of HMG-CoA Reductase Inhibition for Primary and Secondary Prevention of Coronary Heart Disease," *Journal of the American Medical Association* 265 (1991): 1145–51; Silberberg and Henry, "Benefits of Reducing Cholesterol Levels"; Ivar S. Kristiansen, Anne E. Eggen, and Dag S. Thelle, "Cost Effectiveness of Incremental Programmes for Lowering Serum Cholesterol Concentration: Is Individual Intervention Worth While?" *British Medical Journal* 302 (1991): 1119–22.

101. NCEP, *Second Report of the Expert Panel on Detection, Evaluation, and Treatment of High Blood Cholesterol in Adults (Adult Treatment Panel II): Final Report* (Bethesda, MD: NHLBI, 1993), IV-13.

102. *World Development Report 1993, Investing in Health: World Development Indicators* (New York: Oxford University Press, 1993), see particularly chapter 3, "The Roles of the Government and the Market in Health"; also see Paul Dolan, "Output Measures and Valuation in Health," in *Economic Evaluation in Health Care: Merging Theory with Practice,* ed. Michael Drummond and Alistair McGuire (New York: Oxford University Press, 2001).

103. NCEP, *Second Report on High Blood Cholesterol: Final Report,* IV-15; for the origins of these thresholds in the field of health economics, see M. C. Weinstein, "From Cost-Effectiveness Ratios to Resource Allocation: Where to Draw the Line?" in *Valuing Health Care: Costs, Benefits, and Effectiveness of Pharmaceuticals and Other Medical Technologies,* ed. F. A. Sloan (New York: Cambridge University Press, 1995).

104. The NCEP's evaluation on this level had been published a year earlier as L. Goldman, D. J. Gordon, B. M. Rifkind, S. B. Hulley, A. S. Detsky, D. W. Goodman, B. Kinosian, and M. C. Weinstein, "Cost and Health Implications of Cholesterol Lowering," *Circulation* 85, no. 5 (1992): 1960–68. Note that among the authors are both NCEP advocates (Rifkind and Goodman) and those whose earlier writings were essential in bringing the cost-effectiveness critique to bear upon the NCEP (Goldman, Kinosian, and Weinstein).

105. J. Lyle Bootman, Raymond J. Townsend, and William F. McGhan, *Principles of Pharmacoeconomics* (Cincinnati: Harvey Whitney, 2001).

106. J. W. Hay, E. H. Wittels, and A. M. Gotto Jr., "An Economic Evaluation of Lovastatin for Cholesterol Lowering and Coronary Artery Disease Reduction," *American Journal of Cardiology* 67, no. 9 (1991): 789–96, quotation on 789.

107. See, e.g., D. Thompson and G. Oster, "Cost-Effectiveness of Drug Therapy for Hypercholesterolaemia: A Review of the Literature," *Pharmacoeconomics* 2, no. 1 (1992): 34–42; the authors were now attached to a company called Policy Analysis, Inc., based in Brookline, Massachusetts, and this favorable review emphasized the relative cost-effectiveness of Mevacor and Zocor over cholestyramine.

108. "Cholesterol Pill Linked to Lower Hospital Bills," *New York Times,* March 27, 1995; for examples of cost-effectiveness research incorporated into secondary and pri-

mary statin prevention trials, see R. B. Goldberg, M. J. Mellies, F. M. Sacks, L. A. Moye, B. V. Howard, W. J. Howard, B. R. Davis, T. G. Cole, M. A. Pfeffer, and E. Braunwald, "Cardiovascular Events and Their Reduction with Pravastatin in Diabetic and Glucose-Intolerant Myocardial Infarction Survivors with Average Cholesterol Levels: Subgroup Analyses in the Cholesterol and Recurrent Events (CARE) Trial," *Circulation* 98 (1998): 2513–19; J. Caro, W. Klittich, A. McGuire, I. Ford, J. Norrie, D. Pettitt, J. McMurray, and J. Shepherd, "The West of Scotland Coronary Prevention Study: Economic Benefit Analysis of Primary Prevention with Pravastatin," *BMJ* 315 (1997): 1577–84; A. M. Gotto, S. J. Boccuzzi, J. R. Cook, C. M. Alexander, J. B. Roehm, G. S. Meyer, M. Clearfield, S. Weis, and E. Whitney, "Effect of Lovastatin on Cardiovascular Resource Utilization and Costs in the Air Force/Texas Coronary Atherosclerosis Prevention Study (AFCAPS/TexCAPS)," *American Journal of Cardiology* 86, no. 11 (2000): 1176–81.

109. See, e.g., David Atkins and Carolyn DiGuiseppi, "Screening for High Blood Cholesterol and Other Lipid Abnormalities," in *Guide to Clinical Preventive Services: Report of the U.S. Preventive Services Task Force*, 2nd ed., ed. Carolyn DiGuiseppi, David Atkins, and Steven H. Woolf (Baltimore, MD: Williams and Wilkins, 1996), 15–38; L. A. Prosser, A. A. Stinnett, P. A. Goldman, L. W. Williams, M. G. Hunink, L. Goldman, and M. C. Weinstein, "Cost-Effectiveness of Cholesterol-Lowering Therapies according to Selected Patient Characteristics," *Annals of Internal Medicine* 132 (2000): 769–79. The latter study was funded by a grant from the Agency for Healthcare Research and Quality.

110. Prosser et al., "Cost-Effectiveness of Cholesterol-Lowering Therapies." As the study's authors admitted, their own critique threatened to dissolve into a relativism of metrics: "We chose to present the societal perspective. Using alternative perspectives, such as that of the individual patient or a managed care organization, would probably result in different cost-effectiveness ratios, along with different cost-effectiveness thresholds for each perspective" (779).

111. See, e.g., M. Johannesson, "Economic Evaluation of Lipid Lowering—A Feasibility Test of the Contingent Valuation Approach," *Health Policy* 20, no. 3 (1992): 309–20.

112. J. M. Gaspoz, J. W. Kennedy, E. J. Orav, and L. Goldman, "Cost-Effectiveness of Prescription Recommendations for Cholesterol-Lowering Drugs: A Survey of a Representative Sample of American Cardiologists," *Journal of the American College of Cardiology* 27, no. 5 (1996): 1232–37.

113. M. Mitka, "Expanding Statin Use to Help More At-Risk Patients Is Causing Financial Heartburn," *Journal of the American Medical Association* 290 (2003): 2243–45.

114. Mickey C. Smith, *Small Comfort: A History of the Minor Tranquilizers* (New York: Praeger, 1985).

115. In the case of statins and cholesterol, early critics delineated a danger of "medicalization" that was independent of economical costs and adverse effects; see, e.g., Adam Linton, "Cholesterol: Consensus and Controversy," *Journal of Clinical Epidemiology* 43, no. 10 (1990): 1021: "The cost implications were considerable, but quality of care and the impact on patients were also major issues, given the problem of inaccurate testing, drug side-effects, and 'medicalization' of a substantial segment of the population."

116. E.g., Clifton Meador, "The Last Well Person," *New England Journal of Medicine* 330 (1994): 440–41; Victoria Stagg Elliott, "Are We All Sick? Doctors Debate the 'Medicalization' of Life," *AMNews*, September 20, 2004. On the broader shift from views of

the body as inherently healthy to seeing it as inherently ill, see Joseph Dumit's forthcoming book, *Drugs for Life: Managing Identity with Facts and Pharmaceuticals.*

117. The concept of "pharmacological Calvinism" was initially defined by Gerald Klerman as the following: "If a drug makes you feel good, it's either somehow morally wrong, or you're going to pay for it with dependence, liver damage, chromosomal change, or some other form of secular theological retribution." Although Klerman's analysis is limited to the pharmacology of hedonism, the same point can be extended to the newer pharmacology of enhancement. See Gerald Klerman, "Drugs and Social Values," *International Journal of the Addictions* 5 (1970): 313–19; Smith, *Small Comfort.*

118. For historical and philosophical accounts of the enhancement problem, see David Rothman and Sheila Rothman, *The Pursuit of Perfection: The Promise and Perils of Medical Enhancement* (New York: Random House, 2004); Carl Elliott, *Better than Well: American Medicine Meets the American Dream* (New York: Norton, 2003).

119. See, e.g., David Noonan, "You Want Statins with That?" *Newsweek*, July 14, 2004. Similar concerns that pharmaceutical developments might lead to a laxity of regimen can be found in the history of diabetes; see, e.g., Chris Feudtner, "The Want of Control: Ideas, Innovations, and Ideals in the Modern Management of Diabetes Mellitus," *Bulletin of the History of Medicine* 69 (1995): 66–90.

120. See, e.g., Sylvia Carter, "New York Strip House," *New York Newsday,* November 23, 2001: "Service and wine are both exemplary. Drink red, though, to cancel out the bad effects of fat. Bring along some Lipitor, too." Also see John S. Long, "Friday!" *Cleveland Plain Dealer,* October 1, 2004, 20: "It's time to take your Lipitor and head to downtown Cleveland to the Town Fryer, a new restaurant at 3859 Superior Ave. dedicated to fried foods."

121. Al Branch Jr., "J&J/Merck Seek OTC Status for *Mevacor*," *Pharmaceutical Executive,* November 2002, 22.

CONCLUSION: The Therapeutic Transition

Epigraph: Committee on the Care of the Diabetic, "Settling the UGDP Controversy?" *Journal of the American Medical Association* 232 (1975): 816.

1. Roger Smith, "Editor's Choice: The Most Important *BMJ* for 50 Years?" *British Medical Journal* 326, no. 7404 (2003): 0–f.

2. Nonetheless, several incredulous readers wrote to the editor inquiring if the piece was a send-up, a spoof, or a belated April Fool's joke; e.g., David A. Brodie, "Is the Paper a Spoof?" rapid response, bmj.com, posted July 3, 2003.

3. On the planned commercial production of the Polypill, see "Competing Interests: NW and MI Have Filed a Patent Application on the Formula of a Combined Pill to Simultaneously Reduce Four Cardiovascular Risk Factors, as well as a Trademark Application for the Name Polypill," *British Medical Journal* 327 (2003): 810.

4. N. J. Wald and M. R. Law, "A Strategy to Reduce Cardiovascular Disease by More than 80%," *British Medical Journal* 326 (2003): 1419–20, quotation on 1419.

5. Ibid.

6. As reported by Richard Smith, "Editor's Choice: Polypill May Be Available in Two Years," *British Medical Journal* 327 (2003): 0–g.

7. See, e.g., Das Sabapathy, "Polypill—Why Not? Polypharmacy Is Practiced!" rapid response, bmj.com, posted October 4, 2003.

8. See, e.g., A. Mark Clarfield, "Polypill-Pr, a new Formulation for the Mature Male," rapid response, bmj.com, posted January 30, 2004; Jeremy G. Jones, "Re: Polypill—Adding Osteoporosis Treatment Restores Gender Equality," rapid response, bmj.com, posted January 30, 2004; Clarfield made reference to J. D. McConnell, C. D. Roehrborn, O. M. Bautista, et al., "The Long-Term Effect of Doxazosin, Finasteride, and Combination Therapy on the Clinical Progression of Benign Prostatic Hyperplasia," *New England Journal of Medicine* 349 (2003): 2387–98. Other respondents suggested the desirability of specialized Polypills for Asians, Blacks, diabetics, and postmenopausal women.

9. As cited by Smith, "Polypill May Be Available in Two Years."

10. As the authors noted, individual risk factors, "though etiologically important, are poor predictors of future cardiovascular disease events." Wald and Law, "Strategy to Reduce Cardiovascular Disease," 1421.

11. Although Syndrome X had existed in the medical literature since 1991, widespread popularization of the diagnosis appears to have begun with two books published in the first months of 2000: Jack Challem, Burton Berkson, and Melissa Diane Smith, *Syndrome X: The Complete Nutritional Program to Prevent and Reverse Insulin Resistance* (New York: Wiley, 2000); Gerald Reaven, Terry Kristen Strom, and Barry Fox, *Syndrome X, the Silent Killer: The New Heart Disease Risk* (New York: Simon and Schuster, 2000).

12. Jane E. Brody, "Syndrome X and Its Dubious Distinction," *New York Times*, October 10, 2000, F8.

13. Reaven, Strom, and Fox, *Syndrome X, the Silent Killer*, 72–93, 132–42.

14. The rapid expansion of genetic screening and pharmaco-genomics, along with increased attention to racial, ethnic, and sex differences in risk stratification and pharmacological fate, would indicate otherwise, suggesting instead a progressively heightening individualization of risk and, indeed, a proliferation of risk factors and subspecies of risk factors, all to be treated differently. See, e.g., K. Kajinami, N. Takekoshi, M. E. Brousseau, and E. J. Schaefer, "Pharmacogenetics of HMG-CoA Reductase Inhibitors: Exploring the Potential for Genotype-Based Individualization of Coronary Heart Disease Management," *Atherosclerosis* 177, no. 2 (2004): 219–34; W. E. Evans and H. McLeod, "Pharmacogenomics: Drug Disposition, Drug Targets, and Side Effects," *New England Journal of Medicine* 348 (2003): 538–49.

15. The present study is necessarily limited in scope, and other stories could have been written about Diuril, Orinase, and Mevacor, focusing more on the science of their discovery, the experience of consumption, or several other dimensions. A project linking other agents for the same diseases would have exposed a different set of thematic subjects; had I chosen the antidiabetic Rezulin (troglitazone) or the statin Baycol (cerivastatin), both of which were removed from the market in scandal after dangerous side effects were revealed, a much darker portrayal of the pharmaceutical industry would have emerged. One could similarly map out entirely different relationships between drugs and diseases by adding more drugs to the sample or by looking at qualitatively different classes of drugs, such the popular symptom-relief products Tylenol, Tagamet, and Claritin.

16. This sort of heavy-handed determinism is found in many critical accounts of pharmaceutical marketing in relation to disease categories, e.g. Thomas Szasz's *Pharmacracy: Medicine and Politics in America* (London: Praeger, 2001).

17. The role of pharmaceutical innovations in the reductions of infectious disease mortality has been a highly contentious point in the history of medicine after Thomas McKeown and John and Sonia McKinlay demonstrated that the introduction of antibiotics came late in a preexisting pattern of declining infectious disease mortality. Nonetheless, within the scope of the twentieth century, antibiotics and vaccines did help materially to "defang" many formerly feared infectious disease categories, such as pneumonia, tuberculosis, polio, and smallpox. John B. McKinlay and Sonja A. McKinlay, "The Questionable Contribution of Medical Measures to the Decline of Mortality in the United States in the 20th Century," *Milbank Memorial Fund Quarterly: Health and Society* 55 (1977): 405–28; Thomas McKeown, *The Role of Medicine: Dream, Mirage, or Nemesis?* (Princeton, NJ: Princeton University Press, 1979).

18. For two recent philosophical and historical approaches to the problem of enhancement and therapeutic legitimacy, see Carl Elliott, *Better Than Well: American Medicine Meets the American Dream* (New York: Norton, 2003); David Rothman and Sheila Rothman, *The Pursuit of Perfection: The Promise and Perils of Medical Enhancement* (New York: Random House, 2004).

19. Hypocholesterolemia is currently defined as cholesterol levels less than 100 mg/dL, a number so far below the range of the general population as to offer no practical lower boundary to the diagnosis of "treatable" cholesterol. Nonetheless, as noted in chapter 6, a literature on low cholesterol flourished before the major statin trials; the NHLBI, for example, held a conference in 1990 entitled "Low Blood Cholesterol: Mortality Associations."

20. See, e.g., S. D. Nesbitt and S. Julius, "Prehypertension: A Possible Target for Antihypertensive Medication," *Current Hypertension Reports* 2, no. 4 (2000): 356–61.

21. See, e.g., M. A. Krousel-Wood, P. Muntner, and P. K. Whelton, "Primary Prevention of Essential Hypertension," *Medical Clinics of North America* 88, no. 1 (2004): 223–38.

22. John E. McKeen, "Contributions of the Pharmaceutical Industry to Medical Science," unpublished transcript of a talk given before the Research and Development Section of the Pharmaceutical Manufacturers Association, The Greenbrier, White Sulphur Springs, WV, November 4, 1959, in the Kremers Reference Files, C 36 (e) I: PMA, American Institute for the History of Pharmacy, Madison, WI.

23. See, e.g., Harry Marks, *The Progress of Experiment: Science and Therapeutic Reform in the United States, 1900–1990* (Cambridge: Cambridge University Press, 1997).

24. Ron Winslow, "Blood Feud: For Bristol Myers, Challenging Pfizer Was Big Mistake," *Wall Street Journal*, March 9, 2004, A1.

25. Recent scandals and ethical breaches on the part of the industry are amply documented in the recently published literature of pharmaceutical industry critiques, e.g., Marcia Angell, *The Truth about the Drug Companies: How They Deceive Us and What to Do about It* (New York: Random House, 2004); Jerome Avorn, *Powerful Medicines: The Benefits, Risks, and Costs of Prescription Drugs* (New York: Knopf, 2004); and John Abramson, *Overdo$ed America: The Broken Promise of American Medicine* (New York: HarperCollins, 2004).

26. See, e.g., E. Zucato, D. Calamari, M. Natangelo, and R. Fanelli, "Presence of Therapeutic Drugs in the Environment," *Lancet* 355, no. 9217 (2000): 1789–90.

27. Data from IMS Health, *2001 World Review*, www.imshealth.com.

28. Much of this argument was set out in the context of mild hypertension by Lach-

Ian Forrow, Steven A. Wartman, and Dan W. Brock, "Science, Ethics, and the Making of Clinical Decisions," *Journal of the American Medical Association* 259 (1988): 3161–67.

29. World Health Organization Expert Committee on Malaria, *WHO Expert Committee on Malaria, 20th Report*, WHO Technical Report Series 892 (Geneva: WHO, 1999).

30. "African Trypanosomiasis," in *WHO Report on Global Surveillance of Epidemic-Prone Infectious Diseases*, by World Health Organization (Geneva: WHO, 2000), 95–100.

31. Although malarone (in the treatment of malaria) and eflornithine (in the treatment of trypanosomiasis) represent two promising new developments in these fields, it should be noted that the development of both agents has not been prompted from within the pharmaceutical industry itself but through concerted partnership efforts by international public health bodies.

32. On market failure and the essential drugs movement, see B. Pecoul, P. Chirac, P. Trouiller, and J. Pinel, "Access to Essential Drugs in Poor Countries: A Lost Battle?" *Journal of the American Medical Association* 281 (1999): 361–67.

Index

gifts, pharmaceutical, 39–43. *See also* marketing, pharmaceutical; sales representatives, pharmaceutical
glaucoma, 182
Glucophage (metformin), 113–14
glucose tolerance test (GTT), 101, 107–9, 120
glycemia, 85
glycosuria, 85, 87, 103–4, 106
Goldring, William, 72
Goldstein, Joseph L., 175–76, 180–81, 185, 198, 227
Gorman, Mike, 75–77, 175–77, 181–83
Gotto, Antonio, 175–76, 183–84, 201, 215
Grundy, Scott, 176, 180–81, 184–85, 210
guanethedine (Ismelin), 48

Hahnemann Symposium on Hypertensive Disease, 66–67
heart disease: as disease of civilization, 10–11, 196–97; as epidemic, 12, 13, 221–25; cholesterol and, 131–57, 164–67, 168–77; diabetes and, 2–3, 115–23; hypertension and, 21, 71. *See also* cholesterol; chronic disease; diabetes; epidemiology, hypertension; population health; prevention
HIV/AIDS, 16, 200, 226–28
HMG-CoA reductase, 178–79, 181, 185
HMG-CoA reductase inhibitors. *See* statins
Hoechst AG, 89
Hoffman, Carl, 178
Hollander, W., 27, 31
house organs, 32–33, 253nn47–48
Huff, Jesse, 178
Hull, Richard, 38
Hydralazine, 61, 64
HydroDiuril, 71
Hygroton (chlorthalidone), 71
hypercholesterolemia. *See* cholesterol: elevated levels as disease
hyperglycemia. *See* diabetes
hypertension: asymptomatic, 53–54, 62–63; autonomic-nervous hypothesis, 60; benign, 62–63; changing basis of mortality, 21, 71; as compensatory mechanism, 54; degenerative model, 67, 71; diagnosis, 55; early history, 54–55; electrotherapy, 60; essential, 54, 57–58; etiology, 48, 57–58, 62; ganglionic blockers, 23, 61; Hahnemann Symposium on Hypertensive Disease, 66–67; home blood pressure monitors, 192–94; hydralazine, 23, 64; hypertensive emergency, 62; insurance industry and, 55–56; malignant, 60, 62; as market, 27–

28; mild, 53–54, 63; moderate, 53–54, 58–59, 63, 73–76; mosaic theory of etiology, 58–59; nitrates, 23; numerical diagnosis, 67–69; Platt-Pickering controversy, 59; primary vs. secondary, 54, 57–58, 62–63, 259n21; pyrotherapy, 60; quantification and measurement, 55; radiotherapy, 60; *Rauwolfia* compounds, 23; redefinition, 53–54; reserpine, 64; retinal grades of severity, 55–56; rice diet, 61; risks of widespread therapy, 72; severe, 63; sphygmomanometry, 55–56; surgical sympathectomy, 23, 60–61, 260n30; symptomatic, 57; as syndrome, 57; therapeutic nihilism, 54, 63, 68–69; therapeutics, 59–62; threshold for treatment, 67–68; threshold of normotension vs. hypertension, 58–59; VA Cooperative Study, 73–74; *Veratrum* alkaloids, 23. *See also* antihypertensive agents; clinical guidelines; chronic disease; Diuril; population health; prevention

ichthyosis, 161
I. G. Farben, 89
illness and over-the-counter medications, 191–93
incommensurability, 54, 138
infectious disease mortality, 10. *See also* epidemiological transition; population health
insulin: and changing definition of diabetes, 16, 86; discovery, 86; as miracle drug, 88; tyranny of the needle, 94
insurance industry, 55–56
Ismelin (guanethedine), 48

Jackson, W. P. U., 106–7
Janbon, M. J., 89
Janeway, Theodore, 57
Johnson, Anita, 140–41
Joint Commission on Chronic Disease, 12, 13
Joint National Committee (JNC) Guidelines, 77–78, 205, 265n94
Jordan, Beulah, 162
Joslin, Elliott P., 83, 86, 88, 98, 103–5, 114, 119
Joslin Clinic, 90–92, 129
journal advertising. *See* advertising, pharmaceutical

Kansas Medical Society, 125
Katz, Louis, 154–55
Kefauver, Estes, Senate Hearings, 32, 44–47
Kefauver-Harris Amendments, 130, 161, 277n55
Keith, Norman, 57
Kempner, William, 61–62